Committee on Remediation of PCB-Contaminated Sediments

Board on Environmental Studies and Toxicology

Division on Life and Earth Studies

National Research Council

NATIONAL ACADEMY PRESS
Washington, D.C.

NATIONAL ACADEMY PRESS 2101 Constitution Ave., N.W. Washington, D.C. 20418

NOTICE: The project that is the subject of this report was approved by the Governing Board of the National Research Council, whose members are drawn from the councils of the National Academy of Sciences, the National Academy of Engineering, and the Institute of Medicine. The members of the committee responsible for the report were chosen for their special competences and with regard for appropriate balance.

This project was supported by Grant No. R 827175-01 between the National Academy of Sciences and the U.S. Environmental Protection Agency. Any opinions, findings, conclusions, or recommendations expressed in this publication are those of the author(s) and do not necessarily reflect the view of the organizations or agencies that provided support for this project.

International Standard Book Number 0-309-07321-9

Library of Congress Control Number: 2001089191

Additional copies of this report are available from:

National Academy Press
2101 Constitution Ave., NW
Box 285
Washington, DC 20055

800-624-6242
202-334-3313 (in the Washington metropolitan area)
http://www.nap.edu

Printed in the United States of America

THE NATIONAL ACADEMIES

National Academy of Sciences
National Academy of Engineering
Institute of Medicine
National Research Council

The **National Academy of Sciences** is a private, nonprofit, self-perpetuating society of distinguished scholars engaged in scientific and engineering research, dedicated to the furtherance of science and technology and to their use for the general welfare. Upon the authority of the charter granted to it by the Congress in 1863, the Academy has a mandate that requires it to advise the federal government on scientific and technical matters. Dr. Bruce M. Alberts is president of the National Academy of Sciences.

The **National Academy of Engineering** was established in 1964, under the charter of the National Academy of Sciences, as a parallel organization of outstanding engineers. It is autonomous in its administration and in the selection of its members, sharing with the National Academy of Sciences the responsibility for advising the federal government. The National Academy of Engineering also sponsors engineering programs aimed at meeting national needs, encourages education and research, and recognizes the superior achievements of engineers. Dr. William A. Wulf is president of the National Academy of Engineering.

The **Institute of Medicine** was established in 1970 by the National Academy of Sciences to secure the services of eminent members of appropriate professions in the examination of policy matters pertaining to the health of the public. The Institute acts under the responsibility given to the National Academy of Sciences by its congressional charter to be an adviser to the federal government and, upon its own initiative, to identify issues of medical care, research, and education. Dr. Kenneth I. Shine is president of the Institute of Medicine.

The **National Research Council** was organized by the National Academy of Sciences in 1916 to associate the broad community of science and technology with the Academy's purposes of furthering knowledge and advising the federal government. Functioning in accordance with general policies determined by the Academy, the Council has become the principal operating agency of both the National Academy of Sciences and the National Academy of Engineering in providing services to the government, the public, and the scientific and engineering communities. The Council is administered jointly by both Academies and the Institute of Medicine. Dr. Bruce M. Alberts and Dr. William A. Wulf are chairman and vice chairman, respectively, of the National Research Council.

COMMITTEE ON REMEDIATION OF PCB-CONTAMINATED SEDIMENTS

Members

JOHN W. FARRINGTON *(Chair)*, Woods Hole Oceanographic Institution, Woods Hole, MA
RAYMOND C. LOEHR *(Vice-Chair)*, University of Texas at Austin, Austin, TX
ELIZABETH L. ANDERSON, Sciences International, Inc., Alexandria, VA
W. FRANK BOHLEN, University of Connecticut, Groton, CT
YORAM COHEN, University of California at L.A., Los Angeles, CA
KEVIN J. FARLEY, Manhattan College, Riverdale, NY
JOHN P. GIESY, JR., Michigan State University, East Lansing, MI
DIANE S. HENSHEL, Indiana University, Bloomington, IN
STEPHEN U. LESTER, Center for Health, Environment and Justice, Falls Church, VA
KONRAD J. LIEGEL, Preston Gates & Ellis, LLP, Seattle, WA
PERRY L. MCCARTY, Stanford University, Stanford, CA
JOHN L. O'DONOGHUE, Eastman Kodak Company, Rochester, NY
JAMES J. OPALUCH, University of Rhode Island, Kingston, RI
DANNY D. REIBLE, Louisiana State University, Baton Rouge, LA

Staff

ROBERTA M. WEDGE, Project Director
EILEEN N. ABT, Staff Officer
MICHELLE C. CATLIN, Staff Officer
RUTH E. CROSSGROVE, Editor
MIRSADA KARALIC-LONCAREVIC, Information Specialist
LUCY V. FUSCO, Project Assistant
PAMELA FRIEDMAN, Project Assistant
JENNIFER E. SAUNDERS, Project Assistant
BRYAN P. SHIPLEY, Project Assistant

Sponsor

U.S. ENVIRONMENTAL PROTECTION AGENCY

BOARD ON ENVIRONMENTAL STUDIES AND TOXICOLOGY[1]

---

[1]This study was planned, overseen, and supported by the Board on Environmental Studies and Toxicology.

OTHER REPORTS OF THE BOARD ON ENVIRONMENTAL STUDIES AND TOXICOLOGY

Toxicological Effects of Methylmercury (2000)
Strengthening Science at the U.S. Environmental Protection Agency: Research-Management and Peer-Review Practices (2000)
Scientific Frontiers in Developmental Toxicology and Risk Assessment (2000)
Modeling Mobile-Source Emissions (2000)
Toxicological Risks of Selected Flame-Retardant Chemicals (2000)
Copper in Drinking Water (2000)
Ecological Indicators for the Nation (2000)
Waste Incineration and Public Health (1999)
Hormonally Active Agents in the Environment (1999)
Research Priorities for Airborne Particulate Matter: I. Immediate Priorities and a Long-Range Research Portfolio (1998); II. Evaluating Research Progress and Updating the Portfolio (1999)
Ozone-Forming Potential of Reformulated Gasoline (1999)
Risk-Based Waste Classification in California (1999)
Arsenic in Drinking Water (1999)
Brucellosis in the Greater Yellowstone Area (1998)
The National Research Council's Committee on Toxicology: The First 50 Years (1997)
Toxicologic Assessment of the Army's Zinc Cadmium Sulfide Dispersion Tests (1997)
Carcinogens and Anticarcinogens in the Human Diet (1996)
Upstream: Salmon and Society in the Pacific Northwest (1996)
Science and the Endangered Species Act (1995)
Wetlands: Characteristics and Boundaries (1995)
Biologic Markers (5 reports, 1989-1995)
Review of EPA's Environmental Monitoring and Assessment Program (3 reports, 1994-1995)
Science and Judgment in Risk Assessment (1994)
Ranking Hazardous Waste Sites for Remedial Action (1994)
Pesticides in the Diets of Infants and Children (1993)
Issues in Risk Assessment (1993)
Setting Priorities for Land Conservation (1993)
Protecting Visibility in National Parks and Wilderness Areas (1993)
Dolphins and the Tuna Industry (1992)
Hazardous Materials on the Public Lands (1992)
Science and the National Parks (1992)
Animals as Sentinels of Environmental Health Hazards (1991)

Assessment of the U.S. Outer Continental Shelf Environmental Studies Program, Volumes I-IV (1991-1993)
Human Exposure Assessment for Airborne Pollutants (1991)
Monitoring Human Tissues for Toxic Substances (1991)
Rethinking the Ozone Problem in Urban and Regional Air Pollution (1991)
Decline of the Sea Turtles (1990)

Copies of these reports may be ordered from
the National Academy Press
(800) 624-6242
(202) 334-3313
www.nap.edu

Preface

The management of PCB-contaminated sediments has been the subject of much scientific enquiry, technical innovation, regulatory confusion, public debate, and litigation. At contaminated sites around the country, management plans have been proposed, implemented, or completed, with varying levels of satisfaction from the involved parties. PCBs have been associated with a variety of risks to humans and ecosystems. To address these myriad risks, Congress and the U.S. Environmental Protection Agency asked the National Research Council (NRC) to develop a consistent and clear process for assessing the risks from PCB-contaminated sediments and the risks that might be posed by the various technologies that may be used to manage them.

In this report, the Committee on the Remediation of PCB-Contaminated Sediments proposes a framework and guidance for assessing the risks of managing PCB-contaminated sediment. Because of the national attention to this issue and the controversy surrounding management decisions that are pending at several contaminated sites, the committee released the Executive Summary of the report on December 29, 2000, in advance of the full report. The NRC and the committee were aware that many persons were anticipating that this report would recommend a preferred remedial technology for PCB-contaminated sites. However, the committee concluded that such a recommendation could not be made in light of the need to consider site-specific conditions that would affect the efficacy, cost, and risks for each potential remedial technology. The full report provides the detailed documentation and site-specific examples to support the conclusions and recommendations given in the Executive Summary.

The committee (whose biographical sketches are given in Appendix A) gratefully acknowledges the many individuals who made presentations to the committee at its public meetings (a list of these individuals is in Appendix B) and provided background information (a complete list of these materials is in Appendix C). The committee also wishes to thank the following EPA staff: Dennis Timberlake, Office of Research and Development; Timothy Oppelt, Office of Research and Development; Larry Reed, Office of Emergency and Remedial Response; Tudor Davies, Office of Water; and Larry Zaragoza, Office of Emergency and Remedial Response.

The Executive Summary and full report have been reviewed in draft form by individuals chosen for their diverse perspectives and technical expertise, in accordance with procedures approved by the NRC's Report Review Committee. The purpose of this independent review is to provide candid and critical comments that will assist the institution in making its published report as sound as possible and to ensure that the report meets institutional standards for objectivity, evidence, and responsiveness to the study charge. The review comments and draft manuscript remain confidential to protect the integrity of the deliberative process. We wish to thank the following individuals for their review of this report: Steven D. Aust, Biochemistry Center, Utah State University, Logan, Utah; G. Allen Burton, Jr., Institute for Environmental Science, Wright State University, Dayton, Ohio; David E. Daniel, University of Illinois at Urbana-Champaign, Urbana, Illinois; David A. Dzombak, Carnegie Mellon University, Pittsburgh, Pennsylvania; Judith T. Kildow, Hancock Institute for Marine Studies, University of Southern California, Los Angeles, California; Allan B. Okey, University of Toronto, Toronto, Ontario, Canada; Gilbert S. Omenn, University of Michigan, Ann Arbor, Michigan; Hilary Sigman, Rutgers University, New Brunswick, New Jersey; David E. Williams, Oregon State University, Corvallis, Oregon.

Although the reviewers listed above have provided many constructive comments and suggestions, they were not asked to endorse the conclusions or recommendations, nor did they see the final draft of the report before its release. The review of this report was overseen by Edwin H. Clark II, Washington, DC, and Ronald W. Estabrook, University of Texas Southwestern Medical Center. Appointed by the National Research Council, they were responsible for making certain that an independent examination of this report was carried out in accordance with institutional procedures and that all review comments were carefully considered. Responsibility for the final content of this report rests entirely with the authoring committee and the institution.

The committee is grateful for the assistance of the NRC staff in preparing the report. Staff members who contributed to this effort are Roberta Wedge, program director for risk analysis; Eileen Abt, staff officer; Michelle Catlin, staff officer; Ruth E. Crossgrove, editor; Lucy Fusco, project assistant; Pamela Friedman, project assistant; Jennifer Saunders, project assistant; Bryan

Shipley, project assistant; and Mirsada Karalic-Loncarevic, information specialist.

Finally, I would like to thank all members of the committee for their dedicated efforts throughout the development of this report.

John W. Farrington, Chair
Committee on the Remediation
of PCB-contaminated Sediments

Contents

A Risk-Management Strategy for PCB-Contaminated Sediments

Executive Summary

BACKGROUND

Polychlorinated biphenyls (PCBs) are synthetic organic chemicals that are widespread environmental contaminants found in air, water, sediments, and soils around the globe. PCBs are not simple compounds, but are complex mixtures of individual chlorobiphenyls that contain 1 to 10 chlorine atoms. They were manufactured in the United States from 1929 to 1977. Their low reactivity and high chemical stability made them useful in a number of industrial applications, particularly in electrical transformers and capacitors. These same qualities make many individual chlorobiphenyls slow to degrade upon their release to the environment relative to most other organic chemicals. PCBs bind strongly to organic particles in the water column, atmosphere, sediments, and soil. The deposition of particle-bound PCBs from the atmosphere and the sedimentation of them from water are largely responsible for their accumulation in sediments and soils.

As PCBs move through the environment, the absolute and relative concentrations of individual chlorobiphenyls change over time and from one environmental medium to another because of physical and chemical processes and selective bioaccumulation and metabolism by living organisms. These processes result in mixtures that are substantially different from the original mixtures that were released to the environment. The identification, quantification, and risk assessments are complicated by these changes in the composition of the PCB mixtures.

Numerous bodies of water in the United States contain PCB-contaminated sediments that pose current and potential future risks. PCBs in sediments can enter the aquatic food chain, thus contaminating aquatic organisms, including fish, and ultimately placing humans and wildlife at risk of adverse health effects from consumption of these organisms. Acknowledging the human health risks posed by exposure to PCBs at many contaminated sites, some state health and environmental agencies have issued fish and wildlife consumption advisories to caution sport fishers and hunters and their families against eating the fish or wildlife from these sites. The risks of PCB-contaminated sediments, however, extend beyond direct health effects to humans and wildlife. For example, the establishment of fish and wildlife advisories might result in economic hardship for people who rely on the consumption of fish and in erosion of culture for native communities that have a fishing tradition. The presence of contaminated sediments might curtail the recreational use of the body of water for swimming or fishing or lead to restrictions on maintenance dredging, thereby potentially affecting water-borne transportation.

In recent years, substantial progress has been made in the scientific understanding of the dynamics of PCBs in the environment and the effects of PCBs on humans and ecosystems. However, important issues remain regarding the overall risks of PCB-contaminated sediments and the management strategies best suited to reduce them.

Effective management of PCB-contaminated sediments is often challenging. Many PCB-contaminated sediment sites are large, measured in acres or miles—or in tons of sediment. The sheer volume and mass of PCB-contaminated sediments at these sites makes the application of any remediation option a difficult task. The implementation of a comprehensive risk-management strategy is even more complex. Management of these sites is further complicated by the fact that many of the sediments also contain other chemicals of concern, including polycyclic aromatic hydrocarbons, metals, and pesticides. The time required to design and implement a management strategy and to evaluate its effectiveness might reasonably range from years to decades. Thus far, management strategies have been evaluated fully at only a few contaminated sites. Some but not all of these contaminated sites have been designated as Superfund sites under the Comprehensive Environmental Response, Compensation, and Liability Act (CERCLA) of 1980.

THE COMMITTEE'S TASK

In an effort to address these complexities and to understand the risks associated with the management of PCB-contaminated sediments, the U.S. Congress directed EPA to "enter into an arrangement with the National

Academy of Sciences to conduct a review which evaluates the availability, effectiveness, costs, and effects of technologies for the remediation of sediments contaminated with polychlorinated biphenyls, including dredging and disposal." In response to this congressional request, the National Research Council (NRC) convened the Committee on Remediation of PCB-Contaminated Sediments, which prepared this report. The committee was charged to address the following tasks:

- Select, refine, and apply a scientific, risk-based framework for assessing the remediation alternatives for exposure of humans and other living organisms to PCBs in contaminated sediments.
- Evaluate the likelihood that the specified remediation technologies will achieve their remedial objectives, by considering different site-specific conditions such as water and sediment dynamics.
- For a few selected sites and using the framework, estimate human and ecological risks associated with each of the specified remediation approaches for contaminated sediments containing PCBs in light of the availability, costs, and effectiveness of the various approaches.
- Where applicable, recommend areas for future research.

The Committee's Approach

During its deliberations, the NRC committee held three public sessions (Washington, DC; Green Bay, Wisconsin; and Albany, New York) to gather information from a broad audience with interest in the remediation of PCB-contaminated sediments. Two of these meetings were held in areas with known PCB contamination (i.e., the Fox River in Wisconsin and the upper Hudson River in New York) so that the committee could hear from affected parties about their understanding of the risks posed by the sediments and of possible management options. Numerous affected parties attended the meetings and/or submitted written materials to the committee. The committee considered these materials in the preparation of this report.

In the sections below, the committee presents its conclusions regarding the need for a framework to evaluate the overall risks associated with the management of PCB-contaminated sediments. The committee identifies an appropriate framework and, in the report, uses selected actual sites to illustrate key aspects of the framework. The committee highlights its general conclusions based on its recognition of the uniqueness of each contaminated site, and makes recommendations for further scientific and engineering research.

Furthermore, the committee provides a general assessment of the human health and ecological impacts associated with management approaches that may be used at contaminated sites.

After considerable deliberation, the committee does not believe that it is possible to state unequivocally whether dredging, capping, monitored natural attenuation, or any particular remediation option is applicable in general to PCB-contaminated sediment sites. Because each PCB-contaminated site is unique, the selection of remediation options and a risk-management strategy must be based on site-specific factors and risks. Therefore, the committee finds that, without detailed knowledge of a particular site, it is inappropriate to make generalizations concerning whether an option will be effective.

The committee is aware that many readers expect this report to recommend remediation options that are most suitable for reducing the risks associated with PCB-contaminated sediments or on the options that would be most applicable to specific sites. However, the committee strongly believes that making such recommendations is not appropriate, because selection of remediation options must be based on numerous site-specific factors that require evaluation by all affected parties, including local communities and federal and state regulatory agencies. In the committee's view, the adequacy of the site-specific decisions depends upon the extent to which they are consistent with the risk-management process that the committee recommends.

MAJOR CONCLUSIONS AND RECOMMENDATIONS

The committee's major conclusions and recommendations concerning the risks posed by PCB-contaminated sediments and the options that may be used to manage them are given below. The following sections explain, amplify, and provide support for these conclusions and recommendations. Additional detailed information related to these conclusions and recommendations are provided in the chapters of the report.

1. The committee's review of recent scientific information supports the conclusion that exposure to PCBs may result in chronic effects (e.g., cancer, immunological, developmental, reproductive, neurological) in humans and/or wildlife. Therefore, the committee considers that the presence of PCBs in sediments may pose long-term public health and ecosystem risks.

2. The paramount consideration for PCB-contaminated sediment sites should be the management of overall risks to humans and the environment rather than the selection of a remediation technology (e.g., dredging, capping or natural attenuation).

3. Risk management of PCB contaminated sediment sites should comprehensively evaluate the broad range of risks posed by PCB contaminated sediments and associated remedial actions. These risks should include societal, cultural, and economic impacts as well as human health and ecological risks.

4. Risk management of PCB-contaminated-sediment sites should include early, active, and continuous involvement of all affected parties and communities as partners. Although the need for involvement of the affected communities has often been recognized, it has not been implemented on a consistent basis.

5. All decisions regarding the management of PCB-contaminated sediments should be made within a risk-based framework. The framework developed by the Presidential/Congressional Commission on Risk Assessment and Risk Management provides a good foundation that should be used to assess the broad range of risks associated with PCB-contaminated sediments and the various management options for a site.

6. Risk assessments and risk-management decisions should be conducted on a site-specific basis and should incorporate all available scientific information.

7. Identification and adequate control of sources of PCB releases to sediments should be an essential early step in site risk management.

8. There should be no presumption of a preferred or default risk-management option that is applicable to all PCB-contaminated-sediment sites. A combination of technical and non-technical options is likely to be necessary at any given site.

9. Current management options can reduce risks but cannot completely eliminate PCBs and PCB exposure from contaminated sediment sites. Because all options will leave some residual PCBs, the short- and long-term risks that they pose should be considered when evaluating management strategies.

10. Long-term monitoring and evaluation of PCB-contaminated sediment sites should be conducted to evaluate the effectiveness of the management approach and to ensure adequate, continuous protection of humans and the environment.

11. Further research is recommended in several areas of investigation. These research areas concern:

- A better assessment of human health and ecological risks associated with mixtures of individual chlorobiphenyls present in specific environmental compartments.
- The impact of co-contaminants (e.g., polycyclic aromatic

hydrocarbons and metals) on PCB risk assessments and risk-management strategies.

- Processes governing the fate of PCBs in sediments, including erosion, suspension, transport of fine cohesive sediments, pore water diffusion, biodegradation, and bioavailability.
- Improvement of ex situ and in situ technologies associated with removal or containment of PCB-contaminated sediments, treatment of PCB-contaminated material, and disposal of such sediments.
- Pilot scale testing of innovative technologies, such as biodegradation and in situ active treatment caps, to assess their effectiveness and applicability to various sites.
- The impact of continuing PCBs releases and global environmental cycling on site-specific risk assessments.

DISCUSSION

1. The committee's review of recent scientific information supports the conclusion that exposure to PCBs might result in chronic effects (e.g., cancer, immunological, developmental, reproductive, neurological) in humans and/or wildlife. Therefore, the committee considers that the presence of PCBs in sediments may pose long-term public health and ecosystem risks.

The toxicity of PCBs is complicated because PCBs are mixtures and not individual chemicals. The toxicity of different PCB mixtures varies because the dose-effect relationships differ for individual chlorobiphenyls. The more chlorinated PCBs are less likely to be metabolized in humans and wildlife and, therefore, bioaccumulate to a greater extent. The less chlorinated PCBs are more water soluble and have shorter half-lives in the body because of more rapid metabolism and excretion. The greater metabolism and more rapid excretion of the less chlorinated PCBs does not necessarily indicate less concern for toxicity, because some metabolites of these PCBs may also be toxic. Consequently, the health and ecological risks associated with PCB mixtures can vary as the chemical composition changes as a function of space, time, and trophic level. Organisms at the top of the food chain, including humans, tend to accumulate PCBs in their tissues, placing them at risk for adverse health effects.

Toxicological studies have implicated PCBs in a variety of adverse effects, including increased risk of cancer in workers and developmental and neurological effects in infants. Recent toxicological studies have associated the less chlorinated PCBs with immunotoxic, neurotoxic, and endocrine effects.

Wildlife exposed to PCBs have also exhibited adverse effects ranging from subtle biochemical changes to population-level impacts. These effects include the induction of certain enzymes, liver damage, depletion of important compounds such as vitamin A, embryo lethality, birth defects, and neurobehavioral deficits.

2. The paramount consideration for PCB-contaminated sediment sites should be the management of overall risks to humans and the environment. The selection of a remediation option or technology (e.g., dredging, capping or natural attenuation) should be made within a risk-management context.

It is the conclusion of the committee that decision-making often focuses too quickly on defining appropriate remediation technologies. All remediation technologies have advantages and disadvantages when applied at a particular site, and it is critical to the risk management that these be identified individually and as completely as possible for each site. For example, managing risks from contaminated sediment in the aqueous environment might result in the creation of additional risks in both aquatic and terrestrial environments. These additional risks might occur either in the same communities and ecosystems affected by the in situ sediments or in other communities or ecosystems affected by the transport, treatment, or disposal of contaminated dredged material. The evaluation of sediment management and remediation options should take into account all costs and potential changes in risks over time for the entire sequence of activities and technologies that constitute each management option. Removal of contaminated materials can adversely impact existing ecosystems and can remobilize contaminants, resulting in additional risks to humans and the environment. Thus, management decisions at a contaminated site should be based on the relative risks of each alternative management action.

3. Risk management of PCB-contaminated-sediment sites should comprehensively evaluate the broad range of risks posed by PCB-contaminated sediments and associated remedial actions. These risks should include societal, cultural, and economic impacts as well as human health and ecological risks.

The committee found that the risks from PCB-contaminated sediments extend beyond traditional human health and ecological risk assessments as practiced by EPA and other regulatory agencies. The committee emphasizes that societal, cultural, and economic risks should also be considered when developing and implementing a risk-management strategy for the contaminated sediments. These risks are discussed in Chapters 5, 6, and 7 of the report. For example, restrictions on commercial and recreational fishing can impact local communities, as occurred in New Bedford Harbor where PCB-contaminated sediments resulted in economic losses to the commercial lobster fishery. Cultural impacts can result when subsistence use of a resource is lost, affecting such traditions as sharing among the community or passing on indigenous knowledge to younger generations, as occurred among the Mohawk Community of Akwesasne on the St. Lawrence River. Marine transportation can be affected by restrictions on dredging due to the need to handle contaminated sediments. Use of a framework that will allow consideration of this broader definition of risks is essential for successful risk management.

In general, the committee found that regulatory agencies do not give sufficient attention to such risks as ecological effects, impacts on the local economy, or effects on cultural traditions. Furthermore, little consideration appears to be given to the risks to affected parties or ecosystems located near disposal sites in the case where the removal of contaminated sediments is chosen as the remediation option.

4. Risk management of PCB-contaminated-sediment sites should include early, active and continuous involvement of all affected parties and communities as partners. Although the need for involvement of the affected communities has often been recognized, it has not been implemented on a consistent basis.

Affected parties include government regulators at all levels, community groups and individuals, elected officials, environmental organizations, trade associations, and industry. Because an understanding of the risks posed by PCB-contaminated sediments extends to community values and concerns beyond traditional scientific and technical considerations, the involvement of the affected parties, including the local communities and others who might be affected by the contamination and potential remediation activity, is integral to a successful management process. These affected parties, particularly community groups, should be treated as partners in all stages of the risk-management process and have access to the resources necessary to allow their participation in this process. It is important that such involvement be started early and be continuous, active, and transparent.

5. All decisions regarding the management of PCB-contaminated sediments should be made within a risk-based framework. The framework developed by the Presidential/ Congressional Commission on Risk Assessment and Risk Management provides a good foundation that should be used to assess the broad range of risks associated with PCB-contaminated sediments and the various management options for a site.

Much of the dissension that occurs among parties at a given site often appears to focus on the selection of a remediation technology to remove and/or treat the PCB-contaminated sediments. This argument often occurs before the risks at the site have been clearly identified and before the need for their management is established. At some sites, there might be a desire to reduce a specific risk even if such a reduction would mean that the risk is transferred from one area to another, or if mitigation of one risk might result in a greater risk elsewhere. For a site, it is important to consider "overall" or "net" risk in addition to specific risks. A comprehensive approach is needed to address all the risks—societal, cultural, economic, ecological, and human health—of a PCB-contaminated site, as well as the changes in risk that occur with various management approaches. A risk-based framework helps risk managers—whether they are governmental officials, private businesses, or individual members of the public—make good risk-management decisions.

The committee considered a number of frameworks for risk assessment and risk management that had been developed by various organizations, including those proposed in the 1983 NRC "red book," *Risk Assessment in the Federal Government: Managing the Process*, EPA's 1991 *Risk Assessment Guidelines for Superfund* (RAGS), and EPA's 1999 *Guidelines for Ecological Risk Assessment*. Although several of these frameworks are useful for conducting standard health and ecological risk assessments, the committee sought a framework that is inclusive of the broader range of risks that are associated with PCB-contaminated sediments. In addition, the committee sought a framework that would be applicable both to newly identified sites and to sites where the management process is already in progress.

The committee selected the framework developed by the Presidential/Congressional Commission on Risk Assessment and Risk Management, *Framework for Environmental Health Risk Management* (1997) (see Figure ES-1), as appropriate for managing the risks posed by PCB-contaminated sediments, potential remediation options, and risks that remain when the remediation is complete. This framework provides a systematic approach to risk management and includes the following stages:

- Involve the affected parties early and actively in the process.
- Define the problem.
- Set risk-management goals.
- Assess risks.
- Evaluate remediation options.
- Select a risk-management strategy.
- Implement the risk-management strategy.
- Evaluate the success of the risk-management strategy.

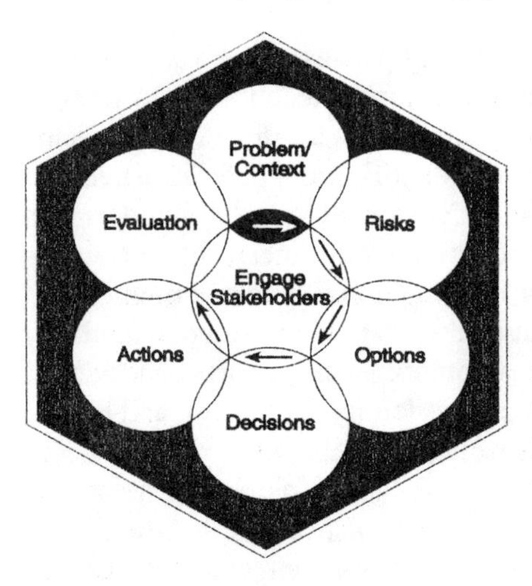

FIGURE ES-1 Framework for environmental health risk management.

The major advantages of this risk-management framework are that (1) it can be applied to any PCB-contaminated-sediment site that might have both new and ongoing remediation; (2) it is iterative, allowing any stage in the framework to be revisited as new information about the site, its environs, remediation technologies, environmental dynamics, or health effects of PCBs becomes available; (3) it can be used to address risks ranging from human health to economic impacts at a site; and (4) it involves all affected parties in all stages of the management process.

6. Risk assessments and risk-management decisions should be conducted on a site-specific basis and should incorporate all available scientific information.

Comparative assessments of overall short-term and long-term risks from various risk-management options are site-specific and depend upon thorough, integrated assessments of human health, ecological, social, cultural, and economic risks. In addition, the broad range of risks at a site—before, during, and after application of a risk-management option—should be assessed so that the overall risk reduction from application of the option is clear. Some examples of these broad-ranging risks include economic impacts, such as changes in the use of a waterway for recreational or commercial purposes, or changes in cultural norms, such as loss of fishing to a culture where fishing is at its core.

Current studies on the toxicity and fate of PCBs in the environment should be used to inform risk assessments at contaminated sites. In recent years, there has been important progress in the scientific understanding of the human health and ecological effects of PCBs and their environmental dynamics. Risk assessments based on the original PCB mixture that entered the environment are not sufficient determinants of either the persistence and toxicity of the weathered PCB mixture present in the sediment or the risks to humans and the ecosystem posed by the weathered mixture. Risk characterizations—and sampling and monitoring to support them—should be performed on the basis of specific congeners and the total mixture of congeners that exist at each site, rather than on the basis of "total PCBs" (all PCB congeners) or Aroclor (commercial PCB mixtures). This method will allow for an accounting of the differences in the physicochemical, biochemical, and toxicological behavior of the different congeners in the risk calculations.

Many PCB-contaminated sites contain elevated concentrations of other chemicals of concern such as DDT, polycyclic aromatic hydrocarbons, dioxins, furans, and metals. However, the knowledge base for addressing multiple chemical risks is severely limited. As new information becomes available on PCB interactions with other chemicals of concern, it should be factored into ongoing risk assessments. The presence of contaminants other than PCBs at a site can affect the degree of risk reduction achievable by a given risk-management strategy.

Traditional human health and ecological risk assessments are based on the analysis of the hazards and the potential for exposures to the chemical in the environment. For this purpose, exposure models can be used to describe all relevant PCB-exposure pathways from the contaminated sediments through the aquatic food web and to specific organisms. These models should factor in exposures to sensitive populations. With regard to human health, these populations include but are not limited to the elderly, pregnant women, infants, children, and culturally or economically unique populations. For ecosystems, sensitive populations and threatened and endangered wildlife and their habitats should be considered.

PCB mass balance and bioaccumulation models to project future PCB exposure levels have been developed for a number of sites. These models have most often been applied to evaluate natural attenuation scenarios, but in some cases they have also been used to examine the efficacy of other remediation options. The model formulations are reasonably well-developed, but even at sites where extensive data collection has been performed, model calibration and application require a certain degree of professional judgment. The scientific basis for model parameter specification, model calibration procedures, and model assumptions (e.g., of future loading conditions) should be carefully reviewed. Where possible, models should be calibrated and applied on a congener-specific basis to provide a more rigorous calibration and a more representative description of PCB behavior. All these models and their results, which have inherent uncertainty, should be peer reviewed.

The ultimate use of mass balance and bioaccumulation models needs to be tied to risk-management goals. This is a key point since the reduction of PCB mass in sediments is not necessarily equivalent to reduction in exposure or risk. Exposure to, and thus risks from PCBs, is mainly a function of the biological availability of PCBs in the surface sediments, and not the total mass of PCBs in the sediments, particularly PCBs in sediments below the biologically active zone. Intrusive remediation technologies such as dredging, and the mixing of buried PCBs into the biologically active sediment layer have the potential to disperse buried PCBs and thereby, increase risk in the short term; however, the slow leakage of PCBs from deeper sediments to overlying surface sediments by diffusive processes may serve as a longer-term source.

Contaminated sites might also have contributions from the global redistribution of PCBs. Therefore, such continual global contributions, as well as continuing sources at or near the site, should be considered in the overall risk assessment and in the selection of the management strategy.

7. Identification and adequate control of sources of PCB releases to sediments should be an essential early step in site risk management.

Source identification and control should be the first goal of any risk-management strategy. In some cases, it might be necessary to reassess the risk-management goals and the potential effectiveness of any prescribed remediation technology if it appears that there are continuing sources that cannot be identified or curtailed at a site. If a significant external source of PCBs is not identified or is allowed to persist, then efforts to reduce contaminant levels through other management options are not likely to be successful; for example, this occurred on the Hudson River in 1991, when a previously unidentified source of PCBs was found at an abandoned paper mill (see Chapter 7). Full development of an accurate, verifiable, material-balance-based mathematical model of the site remains one way to identify as-yet-unidentified sources. Lack of source control might make sediment remediation efforts to reduce site-specific risks unsuccessful. In other cases, a continuing source, if not too significant, might simply limit the reduction that is achievable in contaminant levels.

8. There should be no presumption of a preferred or default risk-management option that is applicable to all PCB-contaminated-sediment sites. A combination of technical and nontechnical options is likely to be necessary at any given site.

The development of a successful risk-management strategy at a site requires a combination of technical and nontechnical options. Technical options include source control, dredging, capping, and bioremediation; nontechnical options include natural attenuation and institutional controls (e.g., fish consumption advisories or covenants). A risk-management strategy may include some combination of the following options; each of which is described below:

- Institutional controls.
- Source control (discussed previously).
- Natural attenuation.
- In situ treatments, which include
  - — Capping.
  - — Biological degradation.
- Multicomponent removal and ex situ treatments, which include
  - — Dredging technologies.
  - — Treatment technologies for dredged materials before disposal.
  - — Ex situ treatment and disposal technologies.
  - — Technologies for management of residual contaminants.

Institutional controls are "interim controls" implemented to control exposure to contaminants and reduce risk to humans and the environment until risks can be reduced to acceptable levels by other remediation options. There are four general categories of institutional controls: government controls; proprietary controls; enforcement tools with institutional-control components; and informational devices.

Natural attenuation processes will be a part of any remediation strategy, because some residual PCBs are expected to remain at a site despite efforts to remove all contamination. Natural attenuation processes consist of sedimentation and/or biodegradation. These processes are most effective in areas that are hydrodynamically stable and where deposition of clean sediments is occurring, resulting in the burial of the contaminated sediments. Biodegradation might occur either anaerobically or aerobically depending on the composition of the PCB mixture and nature of the sediments.

In situ treatment options include capping and enhanced biological degradation. The use of capping is limited to sites where adequate placement and maintenance of the cap is feasible. For example, in situ containment by thick-layer capping and armoring can be an effective means of reducing risks where the cap can be maintained because of (1) a hydrodynamically stable environment, (2) adequate design of protective structures, and (3) adequate monitoring and maintenance of the containment system. See Chapter 7 for a description of in situ capping in Hamilton Harbor, Lake Ontario. Other innovative in situ treatments, such as enhanced biological degradation and active treatment caps, are still in the experimental stage and are not yet practical options for remediating PCB-contaminated sediments.

Ex situ remediation technologies, such as dredging and dry excavation, might have limited applicability due to their high cost, difficulty in controlling contaminants during removal, and lack of disposal options for post-treatment residuals. However, ex situ remediation technologies may be effective for exposed and accessible "hot spots" that pose significant risks. Removal options such as dredging and dry excavation require pre-treatment (dewatering and volume equalization) and appropriate treatment and disposal options for the excavated sediments (landfilling, treatment, incineration, or placement in a confined disposal facility) and for any separated liquids. Dredging at sites such as Manistique Harbor, Michigan, and the Grasse River at Massena, New York, is discussed in Chapter 7. The committee concluded that there have been substantial improvements in the ability of removal technologies to target and process specific sediment zones. However, there have been few improvements in methods to contain contaminants during removal and subsequent treatment and disposal. None of the ex situ options is completely effective in eliminating risks. Therefore, these residual risks must be considered when comparing in situ versus ex situ management options.

The optimal risk-management strategy to be chosen for a particular site depends upon site-specific factors and conditions, such as sediment depth, currents, ecosystems, extent of contamination, and cocontaminants, as well as local social, legal, cultural, and economic considerations. The effectiveness of any strategy is dependent on those site-specific conditions and cannot be predicted without a full understanding of the hydrogeological setting and the risk-reduction potentials of the management options appropriate for that site. Selection of the risk-management strategy will depend upon which risks need to be addressed.

9. Current management options can reduce risks but cannot completely eliminate PCBs and PCB exposure from contaminated sediment sites. Because all options will leave some residual PCBs, the short- and long-term risks that they pose should be considered when evaluating management strategies.

Because of the dimensions of many PCB-contaminated-sediment sites (some covering many miles), complete removal of all PCBs from a site is neither feasible nor practical. Even after the application of a remediation technology, some level of residual contamination will remain. The efficacy and adequacy of any option to manage residual contamination depends on site-specific factors, such as water currents, type of sediment, and topography of the river bed.

There are uncertainties inherent in the assessment and application of any remediation technology. These uncertainties include predictions of failure of the technology (e.g., stability of the cap), estimates of the level of residual contamination, and the financial costs expected at a particular site. Specific areas of uncertainty include (1) the long-term stability of sediment and sediment caps and the types of failure that might occur if caps are destabilized; (2) assessment of residual PCB mass and concentration levels resulting from the inability to capture or target all contaminated sediments; (3) assessment of the bioavailability of PCBs in the surface sediments; and (4) estimates of the financial costs for a remediation strategy due to inadequate site characterization.

Decision-makers selecting a risk-management strategy for a site should be sensitive to how the affected parties are informed about, perceive, and accept not only the short-term and long-term risks from PCBs, but also those risks resulting from the implementation of any remediation technologies. For example, a community might consider the risk of a critical habitat loss during remediation to be a priority, particularly if there are threatened or endangered species present.

10. Long-term monitoring and evaluation of PCB-contaminated-sediment sites should be conducted to evaluate the effectiveness of the management approach and to ensure adequate, continuous protection of humans and the environment.

Long-term evaluation at a site is crucial to determining the success of the chosen management strategy. Monitoring information is available from only a few sites where a risk-management strategy has been implemented and fewer still where it has been completed—for example, Massena, New York; New Bedford, Massachusetts; Duwamish Waterway, Washington, and Manistique Harbor, Michigan. Long-term monitoring results are sparse, in part because most actual management efforts were conducted within the past 5 years, and only a few were conducted as long as 10 years ago. There are significant disincentives to conducting long-term monitoring, including costs and a need for closure. Nevertheless, such monitoring is critical to evaluating the effectiveness of any management strategy, both at that site and at other similar sites where the management options might be applicable.

Presently available monitoring information has been gathered mainly during implementation to (1) measure ambient exposures to PCBs to protect human health; (2) monitor PCB releases to water and PCB concentrations in either wild-caught or caged fish and other aquatic organisms in an effort to minimize ecological risks; and (3) assess bioavailable PCBs in the surface sediments. In addition to monitoring during implementation, adequate long-term monitoring is needed to ensure that the protection of human health and the environment has occurred.

The collection of baseline data for new sites before risk management is undertaken is essential. For ongoing sites where additional remediation is likely, the collection of data during the implementation of the current management strategy may form the basis for future management decisions. Adequate data for pre-remediation baseline assessment are often lacking at sites currently undergoing remediation, making evaluation of the effectiveness of the risk-management strategy difficult.

Short-term and long-term assessments of the efficacy of the risk-management strategy require carefully planned and adequately funded monitoring. Information gathered from assessments of completed and ongoing management projects should be used in the risk assessments, and within the risk-management framework, to inform decisions about remediation options and management strategies for other sites. The information to be gathered should not be restricted to that identified in the remedial investigation/feasibility study guidelines or in the guidelines for conducting human health or ecological risk assessments. Rather, data-gathering efforts should be directed to determine the successful management of all types of risk, including societal, cultural, and economic risks. Therefore, the types of information that might need to be gathered could include, but not necessarily be limited to, data such as number of fish caught by sport fishers, loss of revenues to marinas, and restrictions on navigation.

Each site should have a communication mechanism by which the affected parties can have rapid and easy access to monitoring data and a clear understanding of the implications of the data. Various mechanisms may be used to provide this access; interactive websites and a central repository for the data such as a public library may be used. These mechanisms need to be coupled with an agreed upon mechanism for involvement of all parties in the management process if the monitoring data indicate significant deviations from the expected results.

1

Introduction

Management of polychlorinated biphenyl (PCB)-contaminated sediments is a difficult challenge that has been evolving for decades. In this report, the National Research Council's Committee on the Remediation of PCB-Contaminated Sediments reviews the nature of the challenge; provides an overview of current knowledge about the inputs, fates, and effects of PCBs; recommends a risk-based framework for assessing remediation technologies and risk-management strategies; elaborates on this framework as it applies to PCB-contaminated sediments specifically; and provides recommendations for research that should enhance the nation's ability to manage PCB-contaminated sediments effectively.

This chapter briefly reviews why PCBs in sediments are of environmental concern, states the tasks addressed by the committee, sets forth the committee's activities and deliberative process in developing the report, and explains the organization of the report.

Overview of PCBs in Sediments

PCBs are a contaminant of the nation's waterways as a result of both deliberate and inadvertent releases to the environment, primarily during the 1950s and 1960s when their production and use were greatest. Although PCBs are distributed globally to waters, soils, and sediments by a combination of atmospheric and aquatic transport processes, they are found typically in the greatest concentrations in marine and freshwater sediments near the sources of their environmental release. Most PCBs do not degrade easily in the environment and may persist for years in sediments. Furthermore, PCBs are bioaccumulated by aquatic and terrestrial organisms and thus can enter the food web. Humans and wildlife that consume contaminated organisms, such as fish, can accumulate PCBs in their tissues. Such accumulations are of concern, because they may lead to body burdens of PCBs that could have adverse health effects in humans and wildlife. PCBs can affect not only individual organisms but ultimately whole ecosystems.

The presence of PCBs in sediments might trigger several regulatory actions, such as designation of part of a water body as a hazardous waste site, possible inclusion on the U.S. Environmental Protection Agency's (EPA's) National Priorities List (NPL), issuance of a fish consumption advisory for some or all fish species in a water body, and restrictions on permits for navigational dredging of the sediments, among others. Of the 1,230 sites on the NPL, 535 contain PCBs, of which 122 sites have contaminated sediments. As of 1998, there were 679 fish consumption advisories for PCBs in the United States (J. Bigler, EPA, personal commun., October 12, 2000). Many NPL sites are quite large and have "hot spots" where PCB concentrations are considerably higher than those in other parts of the site. Such sites include the upper Hudson River, Commencement Bay in Washington, and New Bedford Harbor in Massachusetts.

Legitimate concerns over the potential health, environmental, economic, and social impacts of the various strategies for managing PCB-contaminated sediments have led to extensive debate among industry, regulators, communities, and other interested or affected parties on the best course of action for dealing with these sediments. The debate often focuses on certain questions: What are the immediate and the long-term (decades) human health and ecological risks associated with PCB concentrations found in various parts of a specific ecosystem? If PCB concentrations must be reduced, how much reduction is needed, and what is the best approach to achieve the reduction? What are the immediate and the long-term human health and ecological risks associated with the various processes to reduce PCB concentrations both locally and for a wider geographic area? What are the economic and social costs of various options?

THE COMMITTEE'S TASK

The challenges of managing PCB-contaminated sediments have drawn the attention of the U.S. Congress. In 1998, the U.S. House of Representatives

Report 105-107 to accompany H.R. 2158, Department of Veterans Affairs and Housing and Urban Development, and Independent Agencies Appropriations Bill, 1998, called for the EPA to enter into an agreement with the National Academy of Sciences to conduct a study that evaluates the risks, availability, effectiveness, costs, and effects of technologies for the management of sediments contaminated with PCBs, including dredging and disposal.

In response to this request, the National Research Council (NRC) convened the Committee on Remediation of PCB-Contaminated Sediments. The committee was given the following charge:

> This NRC study will provide a scientific risk-based framework for evaluating different approaches for remediating PCB-contaminated, submerged sediments in terms of the efficacy and human and ecological risks associated with each approach. In developing a scientific framework, the committee will evaluate data from specific sites with PCB contamination such as the Hudson River, New Bedford Harbor, etc. Remediation approaches to be assessed and compared include natural recovery, source control, dredging, capping, and contaminated-sediment disposal.

This report responds to the EPA request and to the scientific and technical issues associated with the congressional concerns.

To accomplish its charge, the committee undertook the following tasks:

- Select, refine, and apply a risk-based framework for assessing the remediation alternatives for exposure of humans and other biota to PCBs in contaminated sediments.
- Evaluate the likelihood that the specified remediation technologies will achieve their remedial objectives, considering different site-specific conditions, and water and sediment dynamics.
- For a few selected sites and using the framework, estimate human and ecological risks associated with each of the specified remediation approaches for contaminated sediments containing PCBs in light of the availability, costs, and effectiveness of the various approaches.
- Where applicable, recommend areas for future research.

The committee determined that differences in environmental dynamics and toxicity of various PCB congeners were to be taken into consideration with respect to the above tasks. In addition, the committee considered PCBs in combination with other contaminants that might be present in sediment.

THE COMMITTEE'S APPROACH

To accomplish its task, the committee convened a series of seven meetings between June 1999 and December 2000. At the initial meeting, the committee decided to visit several PCB-contaminated sites to hear the perspectives and experiences of federal, state, and local government officials; industry representatives; environmental and community groups; and local citizens and to view first-hand the terrain and waterways in areas affected by PCB-contaminated sediments.

At three of the six committee meetings (Washington, DC; Green Bay, Wisconsin; and Albany, New York), public sessions were held where committee members heard from regulators, industry groups, environmental organizations, local citizens, researchers, and legislators, about the risks posed by the sediments and possible management options. At two of the meetings (Green Bay and Albany), committee members toured nearby PCB-contaminated sites, i.e., the Fox River and Hudson River, respectively. (Appendix B contains a list of speakers at the public meetings.)

At the public sessions, the committee devoted substantial time to listening to a variety of interested organizations and individuals. The committee emphasized, however, that its purpose in visiting the contaminated sites was not to solve or recommend any solutions for any specific site, but to obtain information that would be valuable for addressing the management of PCB-contaminated sites nationwide. The committee also reviewed written materials submitted by EPA; state and local government agencies; affected industries, such as the Fox River Group and General Electric Corporation; environmental groups, such as the Scenic Hudson and Sierra Club; numerous business and community groups; American Indian organizations; and many concerned and knowledgeable individuals. A complete list of written materials provided to the committee is given in Appendix C.

On the basis of the information provided to the committee and its own expertise, the committee determined that, although there are many concerns that are applicable to PCBs in general, management strategies to address the risks posed by PCB-contaminated sediments and any selection of management options must be developed on a site-by-site basis. The committee also found that a framework for assessing risks associated with the remediation technologies and developing a risk-management strategy was imperative. Such a framework should be flexible, inclusive of all affected parties, comprehensive, transparent, and applicable to a variety of sites and situations. It should provide a consistent approach that would be of use to all parties in the risk-management process. The committee adopted and refined the framework set forth by the Presidential/Congressional Commission on Risk Assessment and

Risk Management (PCCRARM 1997) for application to water bodies with PCB-contaminated sediments.

After considerable deliberation, the committee concludes that it is impossible to state unequivocally whether dredging, capping, monitored natural attenuation, or any particular risk-management option is applicable in general to PCB-contaminated sediment sites. Because each PCB-contaminated site is unique, the selection of management options and a risk-management strategy must be based on site-specific factors and risks. Therefore, the committee finds that it is inappropriate to make generalizations about the effectiveness of an option for a particular site without detailed knowledge of that site.

The committee is aware that many readers expect this report to recommend risk-management options that are most suitable for reducing the risks associated with PCB-contaminated sediments or that would be most applicable to specific sites. However, the committee strongly believes that making such recommendations is not appropriate, because selection of management options must be based on numerous site-specific factors that require evaluation by all affected parties, including local communities and federal and state regulatory agencies. In the committee's view, the adequacy of the site-specific decisions depends upon the extent to which they are consistent with the risk-management process that the committee recommends.

REPORT ORGANIZATION

Chapter 2 provides a brief review of the nature of PCBs, their inputs, fates, and effects in the environment, striking a balance between providing an exhaustive literature review and highlighting significant aspects of the environmental chemistry and ecological and human health effects of PCBs. Further information on the toxicity of PCBs to humans and wildlife may be found in Appendix G. Chapter 3 explains the risk-based framework recommended by the committee and provides a brief review of other frameworks considered by the committee. Chapter 4 addresses the importance of community involvement in dealing with PCB-contaminated sediments, drawing on concerns raised at public meetings held by the committee. Chapter 5 presents the first stage of the risk-based framework, defining the problem of PCB-contaminated sediments and setting risk-management goals. This chapter provides an approach for identifying the affected parties, defining the problems associated with the contaminated sediments, determining the extent of the contamination and its sources, and setting management goals. Chapter 6 addresses the issue of characterizing existing and potential risks at PCB-contaminated sediment sites and provides an overview of the environmental risk-assessment process, including the use and limitations of scientific information in this process. Chapter 7 describes the options available to achieve risk-management goals at a contaminated site and analyzes their benefits, effectiveness, costs, and feasibility. Chapter 8 discusses the issues that must be considered when choosing among the various risk-management options and making a decision that is most likely to achieve the management goals established in the first stage of the framework. Chapter 9 discusses the factors that must be considered when implementing the risk-management strategy. Chapter 10 addresses the need for evaluation in assessing the efficacy of risk-management projects and reviews recent evaluations that have been conducted on selected contaminated sites.

REFERENCE

PCCRARM (Presidential/Congressional Commission on Risk Assessment and Risk Management). 1997. Framework for Environmental Health Risk Management: Final Report. Washington, DC: The Commission.

2

PCBs in the Environment

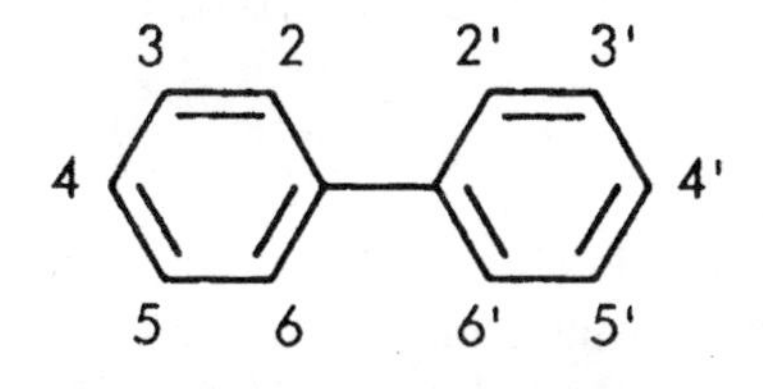

This chapter provides an overview of PCBs in the environment as a background to understanding their history of use, sources of input to the environment, distribution in the environment, and their human health and ecological effects. Because PCBs are such complex chemicals, knowledge of their chemical and physical properties is needed to understand their transport, fate, and toxicity. Considerable new information has become available in the past two decades and new information from field and laboratory studies is published regularly.

DEFINING PCBs

Polychlorinated biphenyls, or PCBs, as they are commonly called, have been used industrially since 1929 (Jensen 1972), and are entirely of anthropogenic origin. The backbone of the chemical structure is a biphenyl, consisting of two hexagonal "rings" of carbon atoms connected by carbon-carbon bonds. The specific manner by which the carbon atoms share electrons forming the hexagonal rings leads to the biphenyl being an "aromatic" compound. Polychlorinated biphenyls have between 1 and 10 chlorine atoms substituting for hydrogen atoms on the biphenyl rings (Figure 2-1). The various number and positions of the chlorine atoms on the biphenyl molecule result in up to 209 possible chemical structures designated as congeners in the scientific literature. PCBs are subdivided into groups based on the degree of chlorination or number of chlorine atoms per biphenyl molecule (e.g., trichlorobi-

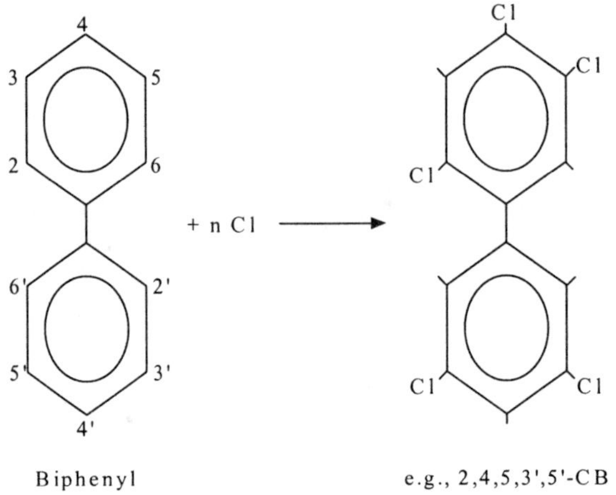

FIGURE 2-1 Synthesis of PCBs (e.g., 2,3′,4,5,5′-pentachlorobiphenyl) by direct chlorination of biphenyl.

phenyls (three chlorines) and tetrachlorobiphenyls (four chlorines)). The PCBs within a series of structures of specific chlorine content are known as homologues (i.e., the mono-, di-, tri-, tetra-, penta-, hexa-, hepta-, octa-, nona-, and decachlorobiphenyl homologues). Within a homologue group (e.g., the trichlorobiphenyls), the individual chlorobiphenyl molecules are isomers of each other, meaning that they each have the same number of chlorine atoms, but these chlorine atoms are arranged at different positions on the biphenyl rings. Examples of chemical structures of PCBs are provided in Figures 2-2 and 2-3. A complete list of congeners is in Appendix H. Since the chlorine atom is part of the group of elements known as halogens (others are fluorine, bromine, and iodine), polychlorinated biphenyls are part of a larger group of chemicals known as halogenated aromatic compounds.

Industrial PCBs were complex mixtures composed of up to 50 or 60 congeners (or individual chlorobiphenyls). The composition of the PCB mixture was governed by the reaction conditions and the reaction properties by which they were manufactured. These conditions and properties favor the production of specific congeners; thus, there are different relative proportions of congeners within given industrial mixtures. These mixtures exist as liquids to viscous solids. Between 1930 and 1977, when their industrial manufacture was banned in the United States, these mixtures were produced almost exclusively by Monsanto under the commercial name of Aroclors. Each Aroclor has a code number (e.g., Aroclor 1242, Aroclor 1248, and Aroclor 1254), the

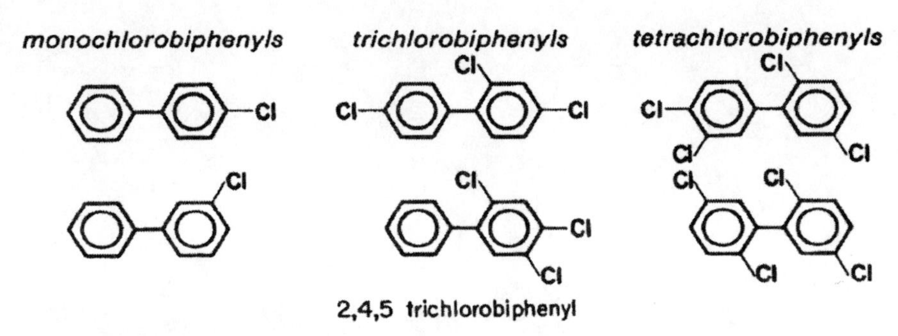

dichlorobiphenyls

Cl Cl

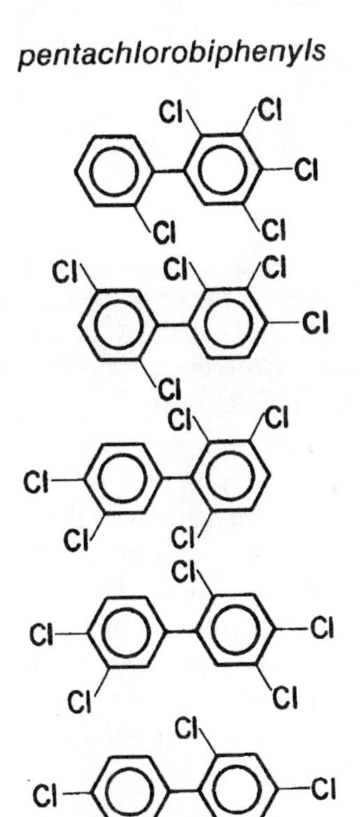

hexachlorobiphenyls

Cl

hept-[7], octa-[8], nona-[9] chlorobiphenyls and the single decachlorobiphenyl are not as common in use or in the environment

FIGURE 2-2 Examples of PCB homologues.

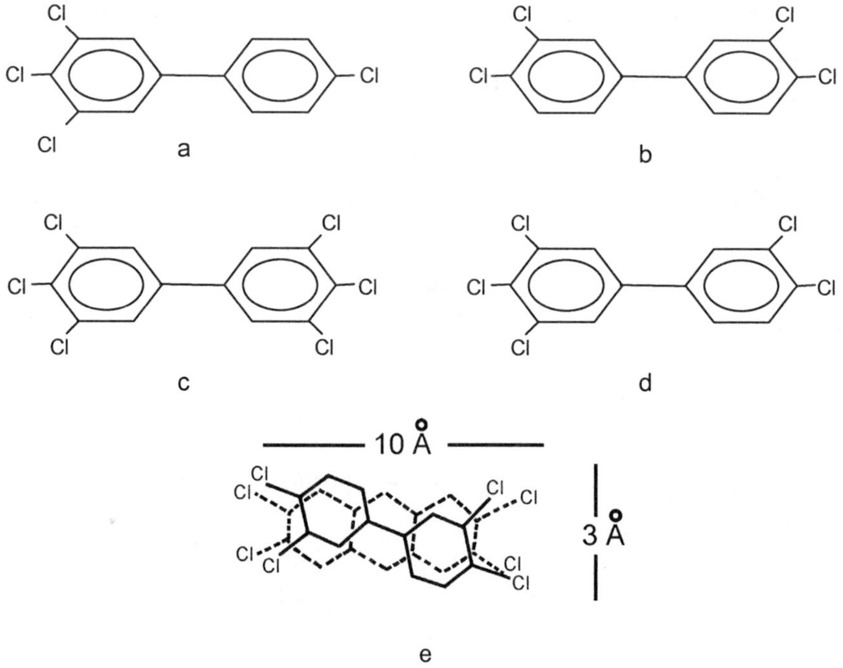

FIGURE 2-3 Coplanar PCBs with no ortho-chlorines: (a) 3,4,4′,5-tetra-chlorinated biphenyl; (b) 3,3′,4,4′-tetra-chlorinated biphenyl; (c) 3,3′,4,4′,5,5′-hexa-chlorinated biphenyl; (d) 3,3′,4,4′,5-penta-chlorinated biphenyl; (e) comparison of the shape and size of a coplanar chlorinated biphenyl to 2,3,7,8-TCDD.

last two numbers of which generally, but not always, refer to the percent by weight of chlorine in the mixture. For example, Aroclor 1254 is 54% chlorine by weight. Manufacturers of PCBs in other countries used different commercial names for PCBs—for example, Kanechlor (Japan), Santotherm (Japan), Phenolclor (France), Pyralene (France), Fenclor (Italy), Soval (Soviet Union), and Clophens (Germany). It is important to note that use of Aroclor as a trade name was not restricted to PCBs but was used for other polyhalogenated aromatic mixtures as well.

History of PCBs

PCBs, manufactured in the United States from 1929 to 1977, were used widely as insulating fluids in electrical equipment such as transformers and capacitors, as well as in hydraulic systems, surface coatings, and flame retardants. Their chemical properties, such as nonflammability, chemical and thermal stability, dielectric properties, and miscibility with organic compounds, were responsible for many of their industrial applications. Their primary domestic uses in the United States as of 1970 are summarized in Table 2-1.

Between 1929 and 1977, approximately 700,000 tons of PCBs were manufactured in the United States; 625,000 tons were used domestically and 75,000 tons were exported. Use of PCBs peaked in 1970 at 42,500 tons annually. The U.S. Environmental Protection Agency (EPA) estimates that over half of the PCBs sold in the United States were disposed of before enactment of federal regulations in 1976 (EPA 1999b).

Although PCBs are no longer commercially manufactured and their disposal from existing industrial equipment is heavily regulated, there are several potential sources for continuing environmental releases. These sources include (1) continued use and disposal of PCB-containing products such as transformers, capacitors, and other electrical equipment that were manufactured before 1977, (2) combustion of PCB-containing materials, (3) recycling of PCB-contaminated products, such as carbonless copy paper, and (4) releases of PCBs from waste storage and disposal. Old consumer goods and household waste might also contain PCBs and their use and disposal are unregulated.

The EPA database for registered electrical transformers (EPA 2000a) shows that, as of 1998, the 18,714 transformers listed contained a total of about 54,000 tons of PCBs, and as of 1988, 141,000 tons of PCBs remained in service in electrical equipment. Due to the long service life of this equipment, considerable amounts of PCBs are likely to remain in use for many

TABLE 2-1 Domestic Uses of PCBs

Category	Type of Product	New Total Use
Closed electrical systems	Transformer, capacitors, other (minor) electrical insulating and cooling applications	61% before 1971; 100% after 1971
Nominally closed systems	Hydraulic fluids, heat transfer fluids, lubricants	13% before 1971; 0% after 1971
Open-end applications	Plasticizers, surface coatings, ink and dye carriers, adhesives, pesticide extenders, carbonless copy paper, dyes	26% before 1971; 0% after 1971

Source: NRC (1979).

years. Spills of PCBs during handling or transport are an additional source of contamination. Between 1989 and 2001, there were 2,611 such spills (USCG 2001). Spills of greater than 1 pound of PCBs are reported to the EPA National Response Center.

The National Toxics Inventory, an inventory conducted every 3 years by EPA under the Clean Air Act Amendments of 1990, reports atmospheric releases of hazardous air pollutants, including PCBs, from mobile and stationary sources. Point source air emissions of PCBs were 136 pounds per year from 127 maximum achievable control technology sources, which included utility boilers, industrial boilers, waste incineration, sewerage sludge incineration, portland cement manufacturing, municipal landfill, and other biological incineration (EPA 2000b). Unquantified emissions include accidental fires or uncontrolled combustion sources. The Toxic Release Inventory for 1998 reported that 3,747,166 pounds of PCBs were released (to air, surface water, land, and underground injection) from all industries in the United States (EPA 2000c). Facilities with effluent discharges are required to report PCB releases for permit compliance purposes under the Clean Water Act. On an annual basis, most of these releases are quite small.

Under the Toxic Substances Control Act (TSCA) of 1976, all uses of PCBs are banned with certain exceptions. These exceptions include totally enclosed activities, such as certain electrical equipment—an authorized use, or exempted use under a special petition. These uses or activities are allowed because EPA has determined that they provide no unreasonable risk to human health or the environment. In general, materials containing PCBs at less than 50 parts per million (ppm) are considered "non-PCB" items by EPA and are not regulated under TSCA. Exemptions under TSCA for manufacturing, processing and distributing PCBs in commerce have been provided for their use in microscopy oils and research and development activities.

TSCA also regulates the inadvertent generation of PCBs. EPA has estimated that more than 200 chemical processes can inadvertently generate PCBs. Products that might be a new source of PCBs include chlorinated solvents, paints, printing inks, agricultural chemicals, plastic materials, and detergent bars. The annual PCB concentration in wastes from these manufacturing processes or imported in the United States must average no more than 25 ppm, with a maximum level of 50 ppm; detergent bars must contain less than 5 ppm of PCBs. Releases of inadvertently produced PCBs from manufacturing operations must contain less than 10 ppm for air releases and less than 100 ppb for water discharges (EPA 1999b).

Recently, a number of unrecognized uses of PCBs have been identified under the category of nonliquid PCBs (EPA 1999a). Currently in use, these solid materials were manufactured with PCBs as an intermediary reactant, and include insulation (wool felt, foam rubber, and fiberglass), sound-dampening materials, paints, water-proofing materials, coatings for water pipes and storage tanks, and other materials, many of which have been found in federal buildings or military equipment, such as naval vessels. Although these solid materials might not present a current health risk from PCB exposure, they might become a significant source of PCB exposure as their utility expires.

Another continuing source of PCBs are recycling activities that keep PCBs in circulation for many years. Materials that might contain PCBs include automobile and truck parts (e.g., nonmetallic parts such as glass and plastic), military equipment (e.g., ship parts), plastics, asphalt-roofing materials, and paper.

In most other countries, PCB production is also banned. However, PCBs are reported to be manufactured in Russia and might also be manufactured in North Korea (Carpenter 1998). If that is the case, PCBs might be entering the environment both in those countries and in other countries that buy their PCB-containing products. Although these sources of PCBs are likely to be relatively small, they are a new source of PCBs in the environment. Unfortunately, estimates of continuing worldwide production of PCBs are not available. Such information would improve our understanding of the global balance of PCBs in the environment and the potential long-term impact of site management efforts.

These continuing and new sources of PCBs to the global environment are important to consider as various physical, chemical, and biological processes transport PCBs regionally and globally. This issue is discussed in more detail in the next section.

DISTRIBUTION AND DYNAMICS OF PCBS IN THE ENVIRONMENT

The chemical properties of PCBs, such as stability and low reactivity, made them ideal for many industrial uses. PCBs are slow to biodegrade in the environment in comparison with many other organic chemicals and are generally persistent in all media. PCBs have relatively low water solubility and low vapor pressures (Erickson 1997) that allow them to partition between water and the atmosphere. Once released into the environment, PCBs tend to partition to the more organic components of the environment. For that reason, PCBs adsorb to organic matter in soils and sediments. As a result, PCBs can be found in almost every compartment in the environment (Tanabe 1988).

PCBs adhere to the surfaces of organic particles in the water column, resulting in their eventual deposition and accumulation in sediments. The highest concentrations of PCBs are typically found in fine-grained, organically rich sediments. Horizontal and vertical variations in PCB concentrations in sediments are common and are dependent on the history of PCB inputs to the ecosystem and on the temporal and spatial deposition patterns of fine- and coarse-grained sediments.

At sites without new inputs of PCBs, the greatest concentrations tend to be found below the surficial sediments, where contaminated sediments are buried by less-contaminated sediments. The distribution of PCBs in sediments is affected by such factors as continuing use and disposal of PCBs; leaching from disposal sites; resuspension by turbulence; redeposition (hydrodynamic forces); chemical changes; and physical and biological mixing of the sediment. The different physical and chemical properties of the individual congeners determine their behavior during those various dynamic processes. As a result, identifying the specific environmental characteristics of PCB-contaminated sediments is challenging. Sediment characterization typically involves a combination of sampling techniques that include direct measurement of PCBs by high-resolution analytical methods and direct and indirect measurement of sediment properties.

PCBs are considered to exist in three phases in the sediment and overlying water: freely dissolved, associated with dissolved organic carbon (DOC),[1] and sorbed to particles.[2] PCBs sorbed to particles are subject to settling, resuspension, and burial. Particles suspended in the water column are affected by hydrodynamic conditions. PCBs that are freely dissolved or associated with DOC can cross the sediment-water interface and move between the deeper sediments (below the bioturbation or bioactive surface sediment) and the surface sediment. This movement is largely a function of diffusion between the sediment pore water, and the overlying water column. It is dependent on the detailed hydrodynamic structure at the water-sediment interface and can be greatly enhanced by bioturbation caused by organisms living in the sediments. Freely dissolved PCBs in the water column are also subject to volatilization across the air-water interface. Such loss can be substantial, especially in systems that provide substantial time for the water-air interactions.

Transformations of PCBs can also occur in aquatic systems by microbial degradation (in aerobic water columns and surficial sediments), reductive dechlorination (in anaerobic sediments), and metabolism via organisms that

---

[1]The term, "associated with DOC" is used, because the exact mechanism of interaction of PCBs with DOC is not well-defined. DOC can include colloidal materials that are mostly organic matter.

[2]The term "sorbed" suggests that a combination of adsorptive and absorptive processes are involved, depending upon the types of particles.

take up the PCBs. Metabolism by microorganisms (Mavoungou et al. 1991) and animals (McFarland and Clarke 1989) can cause relative proportions of some congeners to increase while others decrease (Boon and Eijgenraam 1988; Borlakoglu and Walker 1989).

Because the susceptibility of PCBs to degradation and bioaccumulation is congener specific, the composition of PCB congener mixtures that occur in the environment differs substantially from that of the original industrial mixtures released into the environment (Zell and Ballschmiter 1980; Giesy and Kannan 1998; Newman et al. 1998). In addition to environmental transformation products of PCBs, other chemicals, such as polyaromatic hydrocarbons (PAHs), polychlorinated dibenzofurans (PCDFs), polychlorinated dibenzo-*p*-dioxins (PCDDs), pesticides, and metals, might be present in contaminated sediments.

Generally, the less-chlorinated congeners are more water soluble, more volatile, and more likely to biodegrade. Therefore, lower concentrations of these congeners are found in sediments compared with the original concentrations of Aroclors that entered the environment. Higher-chlorinated PCBs are often more resistant to degradation and volatilization and sorb more strongly to particulate matter. Some of these more-chlorinated PCBs tend to bioaccumulate to greater concentrations in tissues of animals than do lower-molecular-weight PCBs. The more-chlorinated PCBs can also biomagnify in food webs (see Box 2-1), and other higher-molecular-weight congeners have specific structures that make them susceptible to metabolism by enzymes once these congeners are taken up by such species as fish, crustacea, birds, and mammals.

The low vapor pressure of PCBs, coupled with air, water, and sediment transport processes, means that they are readily transported from local or regional sites of contamination to remote areas (Risebrough et al. 1968; NRC 1979; Atlas and Giam 1981; Subramanian et al. 1983). PCBs can enter a global biogeochemical cycle that transports them far from their initial source of input. This global biogeochemical cycling of PCBs is the result of volatilization losses from tropical and subtropical waters to the atmosphere. These atmospheric PCBs move from warmer regions to polar regions, especially in the northern hemisphere, where they are deposited to soil and water surfaces (Muir et al. 2000). Table 2-2 presents some atmospheric concentrations of PCBs from various regions of the world, illustrating the scale and variability of their global distribution.

POTENTIAL EXPOSURE PATHWAYS

Humans and wildlife can be exposed to PCBs either directly from contact with contaminated air, sediments, or water or indirectly through the diet.

BOX 2-1 Definitions

Bioaccumulation—The net accumulation of PCBs by an organism as a result of uptake from all routes of exposure (i.e., water, sediment, food, or air) (Suter 1993).

Bioconcentration—The net accumulation of PCBs directly from water by aquatic organisms (Suter 1993).

Food Web Transfer—The movement of PCBs in the tissue of prey to the tissue of the predator, repeated one or more times in the food web, where the predator of the first transfer is the prey in the next step (Van Leeuwen and Hermens 1995).

Biomagnification—The tendency of PCBs to accumulate to higher concentrations at higher levels in the food web through dietary accumulation (Suter 1993).

Bioavailability—The ratio of the amount of PCBs taken into the organism and thus available to internal tissues, compared to the amount of PCBs ingested into the gut, inhaled into the lungs, or in direct contact with the skin (Suter 1993).

When considering exposure pathways, it is imperative to assess the biologically available fraction of PCBs. In sediments, PCBs can be buried below the biologically active zone and, therefore, are less available for uptake by aquatic organisms. The biologically active zone is the top layer of sediments, typically 5-10 centimeters (cm) deep. This layer is continuously reworked by sediment-dwelling organisms and remains in contact with the overlying water. PCBs that are strongly sorbed to organic sediment particles in the biologically active zone tend to have reduced bioavailability to organisms that ingest or are exposed to these sediments (EPA 1994).

Consumption of PCB-contaminated foods is the most significant route of exposure to PCBs for the general human population (Newhook 1988; Birmingham et al. 1989; Fitzgerald et al. 1996). This exposure occurs as a result of bioaccumulation of PCBs through the food chain. For example, PCBs can enter the aquatic food web via uptake by benthic invertebrates that are in close contact with the contaminated sediments. These invertebrates are eaten by other aquatic organisms, such as fish, and thus the PCBs migrate up the food

TABLE 2-2 Global Atmospheric PCB Concentrations in Ambient Outdoor Air

Location	Concentration (ng/m^3)	Reference
Antarctic coast	0.06-0.2	Tanabe et al. 1983
Canadian Arctic (81° N)	0.1-0.3	Bidleman et al. 1990
Remote	0.02-0.5	Eisenreich et al. 1983
Great Lakes	0.1-5	Eisenreich et al. 1983
Rural	0.1-2	Eisenreich et al. 1983
Urban	0.5-30	Eisenreich et al. 1983
Lake Superior, U.S. (peak in spring)	0.2	Hornbuckle et al. 1994
Lake Superior, U.S. (low point in fall)	0.065	Hornbuckle et al. 1994
Various U.S. locations	0.02-36	NRC 1979
Marine air	0.05-2.0	NRC 1979
Atlantic Ocean	0.05	WHO 1976
Gulf of Mexico	0.2-0.9	Giam et al. 1976
North Pacific Ocean	0.54	Atlas and Giam 1981
North Atlantic Ocean	1.84	Tanabe at al. 1982
West Pacific Ocean	0.06-1.2	Tanabe et al. 1982
Bermuda	0.4	Panshin and Hites 1994a
Bloomington, IN	0.7-2.5	Panshin and Hites 1994b
North Atlantic Ocean	<0.05-1.6	Bidleman et al. 1976, 1981
Lake Baikal, Siberia	0.009-0.023	Iwata et al. 1995
Several oceans and seas	0.004-0.6	Iwata et al. 1993
Arctic	0.002-0.013 (other studies also reviewed)	Bidleman et al. 1990
Tokyo, Japan	20	Kimbrough 1980
Matsuyama, Japan	2-5	Kimbrough 1980
Sweden	<0.8-3.9	WHO 1976
Germany	5-10	Benthe et al. 1992
United States	5	WHO 1976
Landfills, U.S.	2-18	MacLeod 1979
Electrical substations, U.S.	1-47	MacLeod 1979
Transformer manufacturer, U.S.	17-5,900	MacLeod 1979
Spill site, U.S.	10-10,800	MacLeod 1979

Source: Adapted from Erickson (1997).

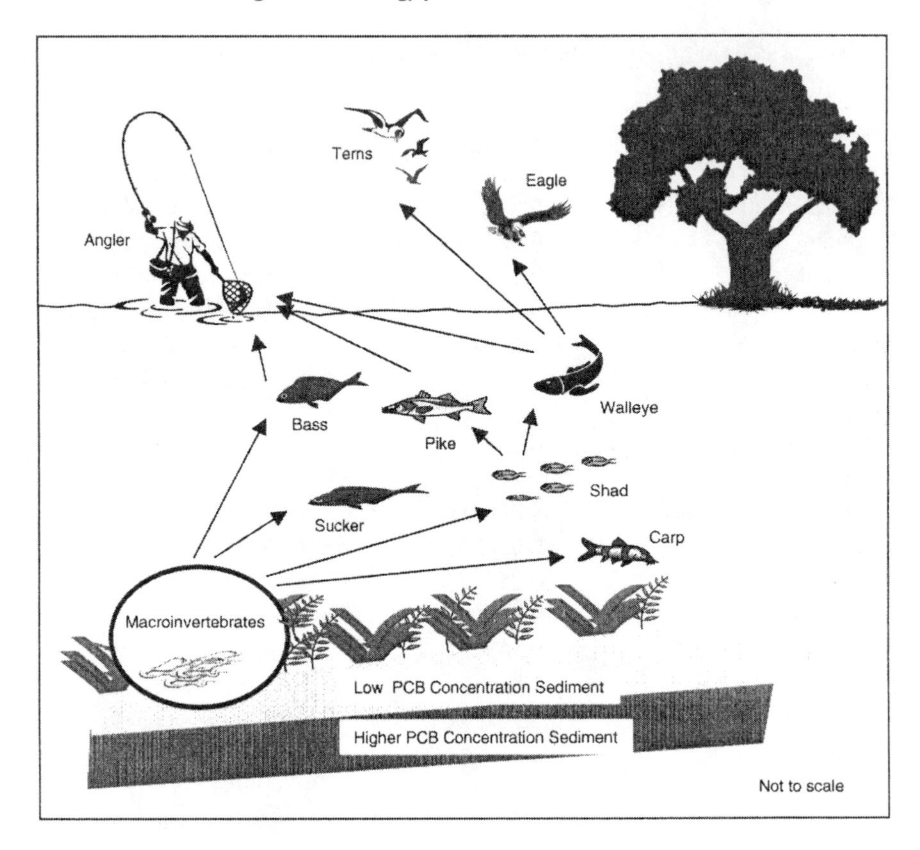

FIGURE 2-4 Transfer of PCBs in the food chain.

chain (Figure 2-4). Fish can accumulate PCBs by direct absorption through the gills and by eating contaminated sediments, insects, and smaller fish. Evans et al. (1991) showed that PCBs in Lake Michigan bioconcentrate by a factor of about 13 from plankton to fish.

Studies have shown that there is a significant correlation between the amount of fish consumed and the organochlorine body burden in humans (Fitzgerald et al. 1996; Schantz et al. 1996). Some populations, such as the Inuits of northern Canada, whose diet consists largely of fish and marine mammals (e.g., whales, seals, and polar bears), might be exposed to high concentrations of PCBs (Dewailly et al. 1993); serum lipid concentrations of PCBs at 4.1 milligrams per kilogram (mg/kg) of body weight have been reported (Ayotte et al. 1997).

Although dermal contact and absorption through the skin are possible exposure routes for PCBs, such exposures are typically limited to the occupational environment. At some sites, however, such as along parts of the Housatonic River in Massachusetts, PCB concentrations in riverbank surfaces and sediments represent a potential route of exposure for populations in the vicinity. At these sites, PCB concentrations have been high enough to warrant concerns and public-health warnings against any dermal contact with the sediments (NRC 1999b).

TOXICITY OF PCBS

Mode of Action

The biological activity of PCBs is congener specific, and, therefore, different mixtures of PCBs will have different biological and toxicological activity. Many of the effects of PCBs are mediated through interaction with the arylhydrocarbon receptor (AhR). 2,3,7,8-Tetrachlorodibenzo-*p*-dioxin (TCDD) is the prototypical ligand for the AhR, and the effects mediated through the AhR are described as "dioxin-like." Like TCDD, non-ortho-substituted and, in some cases, mono-ortho-substituted congeners that are substituted in the 3, 4, or 5 lateral positions (3, 4, 5 or 3', 4', and 5' positions) can exist in a planar conformation (i.e., the coplanar PCBs; Figure 2-3) and bind to the AhR. The potency with which individual PCB congeners bind to the AhR is correlated with their ability to elicit dioxin-like effects. The potency with which individual PCB congeners elicit dioxin-like effects (compared with the potency of TCDD itself) gives rise to the concept of TCDD-toxic equivalents, or toxic equivalence factors (TEF). These factors provide a means of pooling and comparing different mixtures of PCB congeners. One of the main endpoints on which TEFs are based is the induction of CYP1A1 (sometimes described as induction of arylhydrocarbon hydroxylase or microsomal cytochrome P-450 enzymes, or 3-methylcholanthrene-like induction). This process involves binding of the dioxin-like PCB molecule with the AhR in cytosol, association of the bound complex with a nuclear translocation factor, translocation of this ternary complex to the nucleus, and binding of the complex to a specific DNA sequence, the dioxin-responsive element (DRE). Binding at the DRE elicits transcription of the gene corresponding to CYP1A1, resulting ultimately in enhanced synthesis of the CYP1A1 enzyme. This endpoint, however, is not necessarily an adverse health effect. The congeners that exhibit the highest TEF values tend to be the planar, most highly substituted forms, with lateral chlorine substitution. Noncoplanar congeners and congeners with low levels of chlorination are rated at very low TEF values; yet they have been associated with immunological and neurobehavioral endpoints (see review by Fischer et al. 1998). The toxicity of the noncoplanar PCBs is not mediated by the AhR, and their toxicity is not accounted for in the TEF approach. Therefore, although TEFs are useful in PCB risk assessments, they should not be relied on without considering other risks.

In addition to the effects of planar PCBs, studies have shown that noncoplanar PCBs elicit neurotoxic effects in exposed animals and in cell cultures (Kodavanti et al. 1993; Rice 1995; Rice and Hayward 1998). Recently, experiments have been conducted to investigate the mechanism or mechanisms that underlie the neurotoxic effects of the noncoplanar PCBs; some of those effects are not mediated by the Ah receptor (Kodavanti and Tilson 1997) but rather by the signal transduction pathways. PCBs have been shown to affect tyrosine kinase, protein kinase C, and phospholipase A2. Intracellular calcium homeostasis is also affected by noncoplanar PCBs (see reviews by Kodavanti and Tilson 1997; Tilson et al. 1998). Some PCB congeners also appear to have estrogenic and antiestrogenic effects, possibly mediated by interactions with one or more steroid receptors. PCBs affect the metabolism of thyroid hormones through the induction of enzymes involved in thyroid hormone metabolism. PCBs also affect the immune system. Effects on the immune system seem to occur through both AhR- and non-AhR-mediated mechanisms. PCBs might also increase oxidative stress, which might contribute directly to carcinogenesis (Amaro et al. 1996; Oakley et al. 1996). This mechanism would explain PCB-induced cancer with no direct involvement of any receptor and is consistent with observations that PCBs are generally negative in conventional mutagenicity bioassays and have not been shown to form DNA adducts. PCBs can be metabolized and that metabolism can affect their toxicity, sometimes increasing toxic potential. For example, hydroxylated PCBs that interact with the estrogen receptor and exert estrogenic or antiestrogenic effects can be formed (Connor et al. 1997). Evidence also exists for the metabolic activation of PCBs through arene epoxides to yield covalent protein, RNA and DNA adducts (reviewed in Safe 1994). A more detailed discussion of the mechanisms of PCB toxicity is presented in Appendix G.

Potential Human Health Effects of PCBs

The human health effects of PCBs were first formally documented in the Yusho and Yucheng incidents. The Yusho incident occurred in Japan in 1968 and involved more than 1,600 individuals ingesting rice oil contaminated with PCBs. Reported health effects of those exposed included acneform dermatitis, hyperpigmentation of the skin, aches and pain, peripheral nerve damage, and severe headaches. The children born to affected mothers showed similar effects, in addition to decreased birth weight and impaired intellectual development. Studies of the human health ramifications of this incident continue and studies of the incident have been documented in numerous reports (Kimbrough 1987; Erickson 1997; Safe 1994). A similar incident occurred in 1979 in Taiwan, where about 2,000 people consumed rice oil contaminated with PCBs (Erikson 1997). It was called Yucheng, which means "oil disease" in Chinese. However, in both the Yusho and Yucheng incidents, exposures to not only PCBs but also polychlorinated dibenzofurans, terphenyls, and quaterphenyls were considered to play an important role in the observed toxicity (Erikson 1997).

Most of the data on human health effects from exposures to PCBs are based on occupational exposures or consumption of contaminated fish. These studies have correlated relatively high levels of exposure to PCBs with potential subclinical health effects (WHO 1992; EPA 1999a; NRC 1999a; ATSDR 2000). Studies reported that increased serum PCB levels were statistically associated with neurobehavioral and developmental deficits in children exposed in utero and disruption of reproductive function and systemic health effects in adults (self-reported liver disease and diabetes and effects on thyroid and immune system function). In the National Research Council report *Hormonally Active Agents in the Environment* (NRC 1999a), PCBs were some of the agents for which the research literature was critically evaluated for evidence of hormonal effects attributable to PCBs and other agents. The NRC (1999a) reported evidence that human prenatal exposure to PCBs was associated with lower birth weight and shorter gestation (Fein et al. 1984; Taylor et al. 1984, 1989; Jacobson et al. 1990) and neurological deficits and delays in neuromuscular development (Gladen et al. 1988; Rogan and Gladen 1991; Jacobson et al. 1992; Jacobson and Jacobson 1996). These studies were unable to determine whether any of the outcomes were the result of hormonal disruption by PCBs.

In addition, epidemiological studies (ATSDR 1997) of workers involved in the production and use of PCBs have reported increased mortality from cancer, although results have not been consistent across studies. Data indicate that workers in capacitor manufacturing have increased incidences of liver, gallbladder, and biliary-tract cancer. Increased incidences of other specific cancers reported in exposed humans include gastrointestinal tract, malignant melanoma, lung, and brain cancers (Cogliano 1998). Recent studies have observed a significant association between non-Hodgkin's lymphoma and PCB concentrations in adipose tissue (Hardell et al. 1996; Rothman et al. 1997); however, another more recent study found no increase in tumor incidence (Kimbrough et al. 1999).

These epidemiological studies are complicated by several factors that make interpretation of the results difficult. These factors include exposure to other chemical contaminants, difficulty in accurately characterizing exposures, and few data on the types of congeners to which people are exposed. This concern regarding potential confounding factors was expressed recently in the report *Toxicological Effects of Methylmercury* (NRC 2000). The effects of prenatal PCB exposures are similar to those of methylmercury. The study population on which the NRC health recommendations for methylmercury were based also might have been affected by exposure to PCBs; therefore, exposures to PCBs had to be considered in the analysis of the data.

Due to the limitations of the available human data, animal data are used to assess the potential for health effects following PCB exposures (see Appendix G, Table G-1, for a summary of mammalian toxicity doses and endpoints). Animal studies have typically used commercial Aroclor mixtures for testing. However, as noted previously, these commercial mixtures differ substantially from the composition of PCBs typically found in sediments and from those to which humans are exposed through the consumption of contaminated fish and other foods. Thus, the toxicity characteristics of the Aroclors and sediment PCB mixtures are different, as are the possible interactions between the various congeners. In studies on mature animals, the most sensitive observed effect is induction of microsomal P-450 enzymes and liver enlargement for non-ortho-substituted and mono-ortho-substituted PCBs. However, in several developing animal species across vertebrate classes (birds, mammals), functional or structural effects on the nervous system appear to be at least as sensitive to dioxin-like PCBs as is induction of the P450 enzymes (Peterson et al. 1998). PCB exposure in animals has also been associated with a wasting syndrome, reduced body weight, immunotoxicity, vitamin A deficiency, and thyroid deficiency; reproductive effects in offspring include reduced birth weight, abnormal gonad development, slowed learning and memory loss, and other behavioral changes (Linet and Henshel 1998; Schantz et al. 1996).

The results of some studies indicate that PCBs containing 60% chlorine by weight are carcinogenic in animals. Data suggest that carcinogenic potency decreases as the percent chlorination decreases; but studies of PCBs containing less than 60% chlorine are few and have limitations (ATSDR and EPA 1998). As reported in the Second Annual Report on Carcinogens (NTP 1981), exposure to PCBs was considered to present sufficient evidence for carcinogenicity in rodents following bioassays of Aroclor 1254, Aroclor 1260, and Kanechlor 500. Following ingestion of diets containing Aroclor 1254, benign liver tumors were observed in male mice. Kanechlor 500 ingestion was associated with an increased incidence of hepatocellular carcinomas in male mice. Ingestion of Aroclor 1260 induced liver tumors in rats of both sexes and hepatocellular carcinomas and liver adenocarcinomas in female rats. In a more recent study (Brunner et al. 1996), Aroclors 1016, 1242, 1254, and 1260 induced liver tumors when fed to female rats, and Aroclor 1260 induced liver tumors when fed to male rats. Aroclor 1254 induced adenocarcinomas in the glandular stomach when fed to Fischer 344 rats. The consistent finding in these studies is the production of liver tumors that were benign in the majority of cases. Currently, EPA classifies PCBs as a probable human carcinogen (B2) based on data from Brunner et al. (1996) and suggestive, although inadequate, data in humans (IRIS 2000). The International Agency for Research on Cancer (IARC 1987) classifies PCBs as probably carcinogenic to humans (Group 2A), and the National Toxicology Program classified PCBs as reasonably anticipated to be a human carcinogen (NTP 1998).

The mechanism or mechanisms by which PCBs induce tumors in rodents remain unresolved. PCB mutagenicity studies, both in the presence and absence of a metabolic activation system, have generally been negative in bacterial, in vitro, and in vivo test systems. Therefore, PCBs and their metabolites do not appear to be mutagenic. It is possible that the induction of cellular enzymes, including cytochrome P-450s, could play a role in the carcinogenicity of PCBs, but exactly how that induction would lead to cancer has not been established. Although there is an overlap between the congeners present in sediments and those fed to rodents in the bioassay studies, there have been no animal studies conducted with congener mixtures actually found in sediments. Because the mixture of congeners present in sediment are site-specific, however, tests on individual congeners might be important to conduct so that the toxicity of a particular mixture could be extrapolated based on the toxicity of its component congeners. The number of toxicity tests required to do so might be reduced by the careful use of structure-activity relationships. PCBs, especially di-ortho-substituted congeners, can also inhibit cellular gap junction intracellular communication, thus promoting tumor growth. The AhR-active congeners form a receptor complex with the AhR nuclear translocator (ARNT) protein in the nucleus. ARNT is required by a number of other signal transduction pathways. Depletion of ARNT results in a wide range of pleiotropic effects.

The Potential Ecological Impacts of PCBs

Laboratory and field studies with wildlife have demonstrated a causal link between adverse health effects and PCB exposure (Giesy et al. 1994,a,b; Bowerman et al. 1995). The effects of PCBs have been previously reviewed (Poland and Knutson 1982; Safe 1984; Silberhorn et al. 1990; Barrett 1995). Chronic toxicity has been observed in fish, birds, and mammals; impacts include developmental effects, reproductive failure, liver damage, cancer, wasting syndrome, and death (Metcalfe and Haffner 1995). There was also some evidence that PCBs can affect the immune system of birds (Grasman et al. 1996) and marine mammals (de Swart et al. 1994, 1996; Ross et al. 1995) through diet.

Ecological exposure to PCBs is primarily an issue of bioaccumulation resulting in chronic effects rather than direct toxicity. PCBs bioaccumulate in biota by both bioconcentrating (being absorbed from water and accumulated in tissue to concentrations greater than those found in surrounding water) and biomagnifying (increasing in tissue concentrations as they go up the food chain through two or more trophic levels). At most contaminated sites, PCBs are predominantly bound to particles or strongly associated with an organic fraction. Therefore aquatic organisms are exposed to a combination of dissolved, sediment-associated, and food-associated PCBs. However, in terrestrial ecosystems, lower trophic level organisms are exposed to PCBs primarily through ingestion of soil and prey, although dermal absorption and inhalation might be important routes of exposure for certain species. At each higher trophic level, certain PCB congeners are selectively enriched or depleted because of selective metabolism and excretion of metabolites. As a result, organisms at the top of the food chain are generally at the greatest risk of adverse effects due to exposure to PCBs. However, foraging preferences, species sensitivity, and other site-specific factors can modify the magnitude of those risks.

Long-term studies of the effects of PCBs on aquatic ecosystems have not been conducted. The report *Hormonally Active Agents in the Environment* (NRC 1999a), recommends that population studies be conducted to assess the impacts of PCBs on alterations of population size, age structure, and dynamics. At present, many reports on the adverse effects of PCB exposures to wildlife are postmortem studies of individual animals (turtles and birds) found dead with high concentrations of PCBs in their bodies. Studies on a few specific populations of wildlife, such as mink, suggest that environmental contaminants including PCBs might be having a negative impact on the health of feral organisms. For example, Osowski et al. (1995) reported population declines in mink in Georgia, North Carolina, and South Carolina. Examination of liver tissues of mink indicated that concentrations of PCBs (0.2 milligrams per gram (mg/g)) and mercury (3.5 mg/g) were in the range known to cause reproductive impairment, growth deficits, and behavioral impairment. A significant correlation was observed between PCB body burdens in wild mink and in fish collected from their home ranges in New York, indicating that the food chain is being affected (Foley et al. 1988).

Studies of some bird species in the Great Lakes region show evidence of population decline, reproductive impairment, or both in several fish-eating species. Many of the declines might be caused by eating contaminated fish. Although declines were initially likely to be caused by DDE-induced toxicity, the populations still exhibit subtle effects, such as deformities that might be caused by dioxin-like PCBs (Giesy et al. 1994,a,b). It has been proposed that PCB concentrations in tree swallows nesting along the Hudson River might be responsible for their reduced reproductive success (Secord and McCarty 1997; Secord et al. 1999). In some locations, for example, the Niagara River, when wildlife health criteria values have been calculated and compared with the concentrations of PCBs found in local fish species, the amounts of PCBs in the fish have exceeded the criteria values (Newell et al. 1987).

The ecological impacts of PCBs on wildlife have also been assessed in the laboratory. Most of the laboratory studies on wildlife have confirmed the field work. However, these studies are problematic as controlled laboratory conditions cannot be used directly to predict effects in real-world populations because of changes in the concentrations and composition of PCBs as a function of space and time. Thus, the PCB mixture to which organisms are exposed at one time or at one location might be very different from that to which they are exposed in the laboratory or at other times or locations in the field. The pattern of relative proportions of PCBs in environmental mixtures is variable and does not resemble the composition of the original technical PCB mixtures that were released into the environment (Kannan et al. 1993; Corsolini et al. 1995a,b). Furthermore, the relative concentrations of various PCB congeners differ according to trophic level and species.

WEATHERING OF PCBs

The compositions of PCB congener mixtures that occur in the environment differ substantially from those of the original technical Aroclor mixtures released to the environment (Zell and Ballschmiter 1980; Giesy and Kannan 1998; Newman et al. 1998). As discussed previously, the difference is due to the changes in the composition of PCB mixtures over time after release into the environment because of several processes collectively referred to as "environmental weathering." The weathered multicomponent mixtures might have significant differences compared with Aroclor standards; the degree and position of chlorine substitution not only influences the physical and chemical properties of the PCB congeners but also their toxic effects. Weathering is a result of the combined effects of such processes as differential volatilization, solubility, sorption, anaerobic dechlorination, and metabolism, and results in changes in the composition of the PCB mixture over time and between trophic levels (Froese et al. 1998). Less-chlorinated PCBs are often lost rapidly due to volatilization and metabolism, whereas more-chlorinated PCBs are often resistant to degradation and volatilization and sorb more strongly to particulate matter. Bioaccumulation in the tissues of animals is greater for more-chlorinated PCBs than for less-chlorinated PCBs; therefore, more-chlorinated PCBs are more likely to biomagnify in food webs.

Microbial reductive dechlorination of PCBs is a process shown to occur in a variety of anaerobic environments (Bedard and Quensen 1995). This process does not remove all the chlorines and does not alter the basic structure of the biphenyl. The process results in a decrease in the concentrations of some congeners and an increase in others; therefore, the change in the total molar concentration of PCBs in sediments is generally not great. Reductive dechlorination occurs preferentially for chlorines in the meta and para positions, thereby selectively reducing the relative proportions of the PCBs that are laterally substituted. These congeners are also those that tend to have the greatest potency to cause AhR-mediated effects.

One of the most potent of the laterally substituted, non-ortho-substituted congeners is congener 126 (3,3′,4,4′,5-pentachlorobiphenyl). The absolute and relative concentration of this congener was reported to decrease by as much as 10- to 100-fold due to reductive dechlorination (Quensen et al. 1998). The total concentration of TCDD-toxicity equivalents (TEQ) in sediments (see Chapter 6 for a discussion of TEQ and toxicity equivalence factors (TEF)), either determined by the H4IIE bioassay or by application of TEFs to concentrations of individual congeners, was reduced 100-fold by reductive dechlorination (Quensen et al. 1998). Because the TEQ of the total PCB mixture has been shown to be the critical toxicant and the most predictive of toxicity of environmental mixtures of PCBs (Giesy and Kannan 1998; van den Berg et al. 1998), the TEQ reduction suggests that the toxicity of the total PCB mixture would be reduced by approximately 100-fold (Quensen et al. 1998). Furthermore, the bioavailability of the AhR-active congeners has been shown to be less than that for the di-ortho-substituted congeners. Thus, both processes, reductive dechlorination and selective sorption of coplanar PCB congeners, tend to reduce the toxicity of the mixture, relative to technical Aroclor mixtures, during the weathering process.

As was discussed above, the most accurate method of estimating the relative toxic potency of PCB mixtures is to measure the concentrations of individual congeners in tissues of receptors and correct their toxic potency by use of toxic potency factors. It is not appropriate to use thermodynamic models to predict the movement of total PCB or TEQ concentrations from one matrix or trophic level to another. The movement of individual congeners, or at least those with more similar partitioning characteristics, should be modeled and the congeners should be corrected for their toxic potency. Thus, to model the toxicity of the complex PCB mixture that is in sediments to higher trophic levels requires the application of both TEFs and congener-specific biomagnification factors (BMFs). When the toxicity of an example set of congener-specific concentrations in fish tissues to mink was estimated, it was found that the critical toxicant was the TEQ (Foley et al. 1988).

CONCLUSIONS AND RECOMMENDATIONS

PCBs are complex mixtures of chemicals that can have adverse effects on humans and wildlife. The committee's review of recent scientific information supports the conclusion that exposure to PCBs might result in chronic effects—such as cancer and immunological, developmental, reproductive, neurological effects—in wildlife, laboratory animals, and possibly humans. Therefore, the committee considers the presence of PCBs in sediments to pose potential long-term public health and ecosystem risks.

It must be understood by all affected parties that even if the risks at a site are managed such that a specific sediment concentration of PCBs is achieved, over time PCB concentrations will slowly change due to numerous factors including atmospheric inputs from other sources and biodegradation. Although considerable new information has become available in the past 2 decades and new information from field and laboratory studies is reported regularly, the committee finds that further research is particularly warranted in the following areas:

- Additional data are needed on the toxicological effects of exposure to multiple chemicals—PCBs plus PAHs, PCDDs, and metals—and to "real-world" mixtures of PCBs.
- To collect such data, further elucidation of the various mechanisms of toxic actions will be required.
- A better understanding of the contribution of PCB-contaminated sediments to the total global burden of PCBs is needed.
- The role of global cycling of PCBs in assessing the PCB problem at a specific site should be considered.

REFERENCES

Amaro, A.R., G.G. Oakley, U. Bauer, H.P. Spielmann, and L.W. Robertson. 1996. Metabolic activation of PCBs to quinones: reactivity toward nitrogen and sulfur nucleophiles and influence of superoxide dismutase. Chem. Res. Toxicol. 9(3):623-629.

Atlas, E., and C.S. Giam. 1981. Global transport of organic pollutants. Ambient concentrations in remote marine atmosphere. Science 211(4478):163-165.

ATSDR (Agency for Toxic Substance and Disease Registry). 1997. Toxicological Profile for Polychlorinated Biphenyls (Update). U.S. Department of Health and Human Services, Agency for Toxic Substance Disease Registry, Atlanta, GA.

ATSDR (Agency for Toxic Substances and Disease Registry). 2000. Toxicological Profile for Polychlorinated Biphenyls (Update). U.S. Department of Health and Human Services, Agency for Toxic Substance Disease Registry, Atlanta, GA. November.

ATSDR and EPA (Agency for Toxic Substance and Disease Registry and U.S. Environmental Protection Agency). 1998. Public Health Implications of Exposure to Polychlorinated Biphenyls (PCBs). Agency for Toxic Substance and Disease Registry, Public Health Service, U.S. Department of Health and Human Services. U.S. Environmental Protection Agency. [Online]. Available: http://www.atsdr.cdc.gov/DT/pcb007.html.

Ayotte, P., E. Dewailly, J.J. Ryan, S. Bruenau, and G. Lebel. 1997. PCBs and dioxin-like compounds in plasma of adult Inuit lining in Nunavik (Arctic Quebec). Chemosphere 34(5-7):1459-1468.

Barrett, J.C. 1995. Mechanisms for species differences in receptor-mediated carcinogenesis. Mutat. Res. 333(1-2):189-202.

Bedard, D.L., and J.F. Quensen. 1995. Microbial reductive dechlorination of polychlorinated biphenyls. Pp. 127-216 in Ecological and Applied Microbiology, Microbial Transformation and Degradation of Toxic Organic Chemicals, L.Y. Young, and C. Cerniglia, eds. New York: Wiley-Liss.

Benthe, C., B. Heinzow, H. Jessen, S. Mohr, and W. Rotard. 1992. Polychlorinated biphenyls indoor air contamination due to thiokol-rubber sealants in an office building. Chemosphere 25(7-10):1481-1486.

Bidleman, T.F., C.P. Rice, and C.E. Olney. 1976. High molecular weight chlorinated hydrocarbons in the air and sea: rates and mechanisms of air/ sea transfer. Pp. 323-351 in Marine Pollutant Transfer, H.L. Windom and R.A. Duce, eds. Lexington, MA: Lexington Books.

Bidleman, T.F., E.J. Christensen, W.N. Billings, and R. Leonard. 1981. Atmospheric transport of organochlorines in the North Atlantic gyre. J. Mar. Res. 39(3):443-464.

Bidleman, T.F., G.W. Patton, D.A. Hinckley, M.D. Walla, W.E. Cotham, and B.T. Hargrave. 1990. Chlorinated pesticides and polychlorinated biphenyls in the atmosphere of Canadian Artic. Pp. 347-372 in Long Range Transport of Pesticides, D.A. Kurtz, ed. Chelsea, MI: Lewis.

Birmingham, B., A. Gilman, D. Grant, J. Salminen, M. Boddington, B. Thorpe, I. Wile, P. Toft, and V. Armstrong. 1989. PCDD/PCDF multimedia exposure analysis for the Canadian population: detailed exposure estimation. Chemosphere 19(1-6):637-642.

Boon, J.P., and F. Eijgenraam. 1988. The possible role of metabolism in determining patterns of PCB congeners in species from Dutch Wadden sea. Mar. Environ. Res. 24(1-4):3-8.

Borlakoglu, J.T., and C.H. Walker. 1989. Comparative aspects of congener specific PCB metabolism. Eur. J. Drug Metab. Pharmacokinet. 14(2):127-132.

Bowerman, W.W., J.P. Giesy, D.A. Best, and V.J. Kramer. 1995. A review of factors affecting productivity of bald eagles in the Great Lakes region: implications for recovery. Environ. Health Perspect. 103(Suppl. 4):51-59.

Brunner, M.J., T.M. Sullivan, A.W. Singer, M.J. Ryan, J.D. Toft, R.S. Menton, S.W. Graves, and A.C. Peters. 1996. An Assessment of the Chronic Toxicity and Oncogenicity of Aroclor-1016, Aroclor-1242, Aroclor-1254, and Aroclor-1260 Administered in Diet to Rats. Battelle Study No SC920192.. Battelle, Columbus, OH.

Carpenter, D.O. 1998. Polychlorinated biphenyls and human health. Int. J. Occup. Med. Environ. Health 11(4):291-303.

Cogliano, V.J. 1998. Assessing the cancer risk form environmental PCBs. Environ. Health Perspect.. 106(6):317-323.

Connor, K., K. Ramamoorthy, M. Moore, M. Mustain, I. Chen, S. Safe, T. Zacharewski, B. Gillesby, A. Joyeux, and P. Balaguer. 1997. Hydroxylated polychlorinated biphenyls (PCBs) as estrogens and antiestrogens: structure-activity relationships. Toxicol. Appl. Pharmacol. 145(1):111-123.

Corsolini, S., S. Focardi, K. Kannan, S. Tanabe, A. Borrell, and R. Tatsukawa. 1995a. Congener profile and toxicity assessment of polychlorinated biphenyls in dolphins, sharks and tuna collected from Italian coastal waters. Mar. Environ. Res. 40(1):33-54.

Corsolini, S., S. Focardi, K. Kannan, S. Tanabe, and R. Tatsukawa. 1995b. Isomer-specific analysis of polychlorinated biphenyls and 2,3,7,8-tetrachlorodibenzo-p-dioxin equivalents (TEQs) in red fox and human adipose tissue from central Italy. Arch. Environ. Contam. Toxicol. 29(1):61-68.

de Swart, R.L., P.S. Ross, L.J. Vedder, H.H. Timmerman, S. Heisterkamp, H. Van Loveren, J. Vos, P.J.H. Reijnders, and A.D.M.E. Osterhaus. 1994. Impairment of immune function in harbor seals (Phoca vitulina) feeding on fish from polluted waters. Ambio 23(2):155-159.

de Swart, R.L., P.S. Ross, J.G. Vos, and A.D.M.E. Osterhaus. 1996. Impaired immunity in harbour seals (Phoca vitulina) exposed to bioaccumulated environmental contaminants: review of a long-term feeding study. Environ. Health Perspect. 104(Suppl. 4):823-828.

Dewailly, E., P. Ayotte, S. Bruneau, C. Laliberte, D.C. Muir, and R.J. Norstrom. 1993. Inuit exposure to organochlorines through the aquatic food chain in arctic Quebec. Environ. Health Perspect. 101(7):618-620.

Eisenreich, S.J., B.B. Looney, and G.J. Hollod. 1983b. PCBs in the Lake Superior Atmosphere 1978-1980. Pp. 115-125 in Physical Behavior of PCBs in Great Lakes, D. Mackay, S. Paterson, S.J. Eisenreich, and M.S. Simmons, eds. Ann Arbor, MI: Ann Arbor Science.

EPA (U.S. Environmental Protection Agency). 1994. Estimating Exposure to Dioxin-like Compounds, Volume II: Properties, Sources, Environmental Levels, and Background Exposures. EPA/600/6-88/005c. Office of Health and Environmental Assessment, U.S. Environmental Protection Agency, Washington, DC. June.

EPA (U.S. Environmental Protection Agency). 1999a. Draft Mercury, Polychlorinated Biphenyls, Alkyl Lead, and Benzo(a)pyrene and Hexachlorobenzene Reports published in reponse to the United States' committment in "The Great Lakes Binational Toxics Strategy; Canada-United States Strategy for the Virtual Elimination of Persistent Toxic Substances in the Great Lakes. Fed. Regist. 64(210):58841. (Nov. 1, 1999).

EPA (U.S. Environmental Protection Agency). 1999b. Use Authorization for, and Distribution in Commerce of, Non-liquid Polychlorinated Biphenyls; Notice of Availability; Partial Reopening of Comment Period; Proposed Rule. Fed. Regist. 64(237):69358-69364. (Dec. 10, 1999).

EPA (U.S. Environmental Protection Agency). 2000a. PCB Transformer Registration Database. Office of Pollution Prevention and Toxics, U.S. Environmental Protection Agency, Washington, DC. [Online]. Available: http:// www.epa.gov/ opptintr/pcb/xform.htm [Last updated: February 3, 2000].

EPA (U.S. Environmental Protection Agency). 2000b. National Toxics Inventory Maximum Achievable Control Technology (MACT) Report: United States NTI Air Pollution Sources—Polychlorinated Biphenyls (1996). AIRS Database. Office of Air Quality Planning and Standards, U.S. Environmental Protection Agency. [Online]. Available: http://oaspub.epa.gov/airsdata/nti. [March 12, 2001].

EPA (U.S. Environmental Protection Agency). 2000c. TRI On-site and Off-site Reported Releases (in pounds), By Chemical, U.S., 1998, All Industries. TRI Explorer Report. Office of Environmental Information, U.S. Environmental Protection Agency. [Online]. Available: http://www.epa.gov/triexplorer/ chemical.htm. [October 18, 2000].

Erickson, M.D. 1997. Analytical Chemistry of PCBs, 2nd. Ed. Boca Raton: CRC/ Lewis. 667 pp.

Evans, M.S., G.E. Noguchi, and C.P. Rice. 1991. The biomagnification of polychlorinated biphenyls, toxaphene and DDT compounds in a Lake Michigan offshore food web. Arch. Environ. Contam. Toxicol. 20(1):87-93.

Fein, G.G., J.L. Jacobson, S.W. Jacobson, P.M. Schwartz, and J.K. Dowler. 1984. Prenatal exposure to polychlorinated biphenyls: Effects on birth size and gestational age. J. Pediatr. 105(2):315-320.

Fischer, LJ., R.F. Seegal, P.E. Ganey, I.N. Peesah, and P.R. Kodavanti. 1998. Symposium overview: toxicity of non-coplanar PCBs. Toxicol. Sci. 41(1):49-61.

Fitzgerald, E.F., K.A. Brix, D.A. Deres, S.A. Hwang, B. Bush, G.L. Lambert, and A. Tarbell. 1996. Polychlorinated biphenyl (PCB) and dichlorodiphenyl dichloroethylene (DDE) exposure among Native American men from contaminated Great Lakes fish and wildlife. Toxicol. Ind. Health 12(3-4):361-368.

Foley, R.E, S.J. Jackling, R.J. Sloan, and M.K. Brown. 1988. Organochlorine and mercury residues in wild mink and otter: comparison with fish. Environ. Toxicol. Chem. 7(5):363-374.

Froese, K.L., D.A. Verbrugge, G.T. Ankley, G.J. Niemi, C.P. Larsen, and J.P. Giesy. 1998. Bioaccumulation of polychlorinated biphenyls from sediments to aquatic insects and tree swallow eggs and nestlings in Saginaw Bay, Michigan, USA. Environ. Toxicol. Chem. 17(3): 484-492.

Giam, C.S., H.S. Chan, and G.S. Neff. 1976. Concentrations and fluxes of phthalates, DDTs, and PCBs to the Gulf of Mexico. Pp. 375-386 in Marine Pollutant Transfer, H.L. Windom and R.A. Duce, eds. Lexington, MA: Lexington Books.

Giesy, J.P., and K. Kannan. 1998. Dioxin-like and non-dioxin-like toxic effects of polychlorinated biphenyls (PCBs): implications for risk assessment. Crit. Rev. Toxicol. 28(6):511-569.

Giesy, J.P., J.P. Ludwig, and D.E. Tillitt. 1994. Dioxins, dibenzofurans, PCBs, and colonial fish-eating water birds. Pp. 249-307 in Dioxins and Health, A. Schecter, ed. New York: Plenum Press.

Giesy, J.P., D.A. Verbrugge, R.A. Othout, W.W. Bowerman, M.A. Mora, P.D. Jones, J.L. Newsted, C. Vandervoort, S.N. Heaton, R.J. Aulerich, et al. 1994a. Contaminants in fishes from Great Lakes-influenced sections and above dams of three Michigan rivers. I: Concentrations of organo chlorine insecticides, polychlorinated biphenyls, dioxin equivalents, and mercury. Arch. Environ. Contam. Toxicol. 27(2):202-212.

Giesy, J.P., D.A. Verbrugge, R.A. Othout, W.W. Bowerman, M.A. Mora, P.D. Jones, J.L. Newsted, C. Vandervoort, S.N. Heaton, R.J. Aulerich, et al. 1994b. Contaminants in fishes from Great Lakes-influenced sections and above dams of three Michigan rivers. II: Implications for health of mink. Arch. Environ. Contam. Toxicol. 27(2):213-223.

Gladen, B.C., W.J. Rogan, P. Hardy, J. Thullen, J. Tingelstad, and M. Tully. 1988. Development after exposure to polychlorinated biphenyls and dichlorodiphenyl dichloroethene transplacentally and through human milk. J. Pediatr. 113(6):991-995.

Grasman, A., G.A. Fox, P.F. Scanlon, and J.P. Ludwig. 1996. Organochlorine-associated immunosuppression in prefledgling Caspian terns and herring gulls from the Great Lakes: an ecoepidemiological study. Environ. Health Perspect. 104(Suppl. 4):829-842.

Hardell, L., B. van Bavel, G. Linstrom, M. Frederikson, H. Hagberg, G. Liljegren, M. Nordstrom, and B. Johansson. 1996. Higher concentrations of specific polychlorinate biphenyl congeners in adipose tissue from non-Hodgkin's lymphoma patients compared with controls without a malignant disease. Int. J. Oncol. 9(4):603-608.

Hornbuckle, K.C., J.D. Jeremiason,C.W. Sweet, and S.J. Eisenreich. 1994. Seasonal variations in air-water exchange of polychlorinated biphenyls in Lake Superior. Environ. Sci. Technol. 28(8):1491-1501.

IARC (International Agency for Research on Cancer). 1987. Overall Evaluation of Carcinogenicity: An Updating of IARC Monograhps, Vol. 41, Supplement 7. Lyon, France: International Agency for Research on Cancer.

IRIS (Integrated Risk Information System). 2000. Integrated Risk Information System. National Center for Environmental Assessment, Office of Research and Development. U.S. Environmental Protection Agency. [Online]. Available: wysiwyg://4http://www.epa.gov/ncea/iris.htm [May 03, 1999].

Iwata, H., S. Tanabe, N. Sakal, and R. Tatsukawa. 1993. Distribution of persistent organochlorines in the oceanic air and surface seawater and the role of ocean on their global transport and fate. Environ. Sci. Technol. 27(6):1080-1098.

Iwata, H., S. Tanabe, K. Ueda, and R. Tatsukawa. 1995. Persistent organochlorine residues in air, water, sediments, and soils from the Lake Baikal Region, Russia. Environ. Sci. Technol. 29(3):792-801.

Jacobson, J.L., and S.W. Jacobson. 1996. Intellectual impairment in children exposed to polychlorinated biphenyls in utero. N. Engl. J. Med. 335(11):783-789.

Jacobson, J.L., S.W. Jacobson, and H.E. Humphrey. 1990. Effects of exposure to PCBs and related compounds on growth and activity in children. Neurotoxicol. Teratol. 12(4):319-326.

Jacobson, J.L., S.W. Jacobson, R.J. Padgett, G.A. Brumitt, and R.L. Billings. 1992. Effects of prenatal PCB exposure on cognitive processing efficiency and sustained attention. Dev. Psychol. 23(2):297-306.

Jensen, S. 1972. The PCB story. Ambio 1(4):123-131.

Kannan, K., S. Tanabe, A. Borrell, A. Aguilar, S. Focardi, and R. Tatsukawa. 1993. Isomer-specific analysis and toxic evaluation of polychlorinated biphenyls in striped dolphins affected by an epizootic in the western Mediterranean sea. Arch. Environ. Contam. Toxicol 25(2):227-233.

Kimbrough, R.D. 1980. Environmental pollution of air, water, and soil. Pp. 77-80 in Halogenated Biphenyls, Terphenyls, Naphthalenes, Dibenzodioxins, and Related Products, R.D. Kimbrough, ed. New York: Elsevier/ North-Holland Biomedical Press.

Kimbrough, R.D. 1987. Human health effects of polychlorinated biphenyls (PCBs) and polybrominated biphenyls (PBBs). Annu. Rev. Pharmacol. Toxicol. 27:87-111.

Kimbrough, R.D., M.L. Doemland, and M.E. LeVois. 1999. Mortality in male and female capacitor workers exposed to polychlorinated biphenyls. J. Occup. Environ. Med. 41(3):161-171.

Kodavanti, P.R.S., and H.A. Tilson. 1997. Structure-activity relationships of potentially neurotoxic PCB congeners in the rat. Neurotoxicology. 18(2):425-441.

Kodavanti, P.R.S., D.-S. Shin, H.A. Tilson, and G.J. Harry. 1993. Comparative effects of two polychlorinated biphenyl congeners on calcium homeostasis in rat cerebellar granule cells. Toxicol. Appl. Pharmacol. 123(1):97-106.

Linet, M.H., and D.S. Henshel. 1998. Histological effects of TCDD on gonadal development in chicken hatchlings. Toxicologist 42(1-S):256.

MacLeod, K.E. 1979. Sources of Emissions of Polychlorinated Biphenyls into the Ambient Atmosphere and Indoor Air. EPA 600/4-79-022. U.S. Environmental Protection Agency, Washington, DC. March.

Mavoungou, R., R. Masse, and M. Sylvestre. 1991. Microbial dehalogenation of 4,4'-dichlorobiphenyl under anaerobic conditions. Sci. Total. Environ. 101(3): 263-268.

McFarland, V.A., and J.U. Clarke. 1989. Environmental occurrence, abundance, and potential toxicity of polychlorinated biphenyl congeners: considerations for a congener-specific analysis. Environ. Health Perspect. 81:225-239.

Metcalfe, C.D., and G.D. Haffner. 1995. The ecotoxicology of coplanar polychlorinated biphenyls. Environ. Rev. 3(2):171-190.

Muir, D., F. Riget, M. Cleeman, J. Skaare, L. Kleivane, H. Nakata, R. Dietz, T. Severinsen, and S. Tanabe. 2000. Circumpolar trends of PCBs and organochlorine pesticides in the arctic marine environment inferred from levels in ringed seals. Environ. Sci. Technol. 34(12): 2431-2438.

Newell, A.J., D.W. Johnson, and L.K. Allen. 1987. Niagra River Biota Contamination Project: Fish Flesh Criteria for Piscivorous Wildlife. Technical Report 87-3. New York State Department of Environmental Conservation, Division of Fish and Wildlife, Bureau of Environmental Protection, Albany, NY.

Newhook, R.C. 1988. Polybrominated Biphenyls: Multimedia Exposure Analysis. Contract report to the Department of National Health and Welfare, Ottawa, Canada.

Newman, J.W., J.S. Becker, G. Blondina, and R.S. Tjeerdema. 1998. Quantitation of aroclors using congener-specific results. Environ. Toxicol. Chem. 17(11):2159-2167.

NRC (National Research Council). 1979. Polychlorinated Biphenyls, Washington, DC: National Academy of Science.

NRC (National Research Council). 1999a. Hormonally-active Agents in the Environment. Washington, DC: National Academy Press.

NRC (National Research Council). 1999b. National Symposium of Contaminated Sediments—Coupling Risk Reduction with Sustainable Management and Reuse. Washington, DC: National Academy Press.

NRC (National Research Council). 2000. Toxicological Effects of Methylmercury. Washington, DC: National Academy Press.

NTP (National Toxicology Program). 1981. Second Annual Report on Carcinogens. Research Triangle Park, NC: U.S. Dept. of Health and Human Services, Public Health Service, National Toxicology Program.

NTP (National Toxicology Program). 1998. Eighth Report on Carcinogens, 1998. Full report. 8th Ed. Research Triangle Park, NC: U.S. Dept. of Health and Human Services, Public Health Service, National Toxicology Program.

Oakley, G.G., U. Devanaboyina, L.W. Roberson, and R.C. Gupta. 1996. Oxidative DNA damage induced by activation of polychlorinated biphenyls (PCBs): implications for PCB-induced oxidative stress in breast cancer. Chem. Res. Toxicol. 9(8):1285-1292.

Osowski, S.L., L.W. Brewer, O.E. Baker, and G.P. Cobb. 1995. The decline of mink in Georgia, North Carolina, and South Carolina: the role of contaminants. Arch. Environ. Contam. Toxicol. 29(3):418-423.

Panshin, S.Y., and R.A. Hites. 1994a. Atmospheric concentrations of polychlorinated biphenyls at Bermuda. Environ. Sci. Technol. 28(12):2001-2007.

Panshin, S.Y., and R.A. Hites. 1994b. Atmospheric concentrations of polychlorinated biphenyls at Bloomington, Indiana. Environ. Sci. Technol. 28(12):2008-2013.

Peterson, R.E., H.M. Theobald, and G.L. Kimmel. 1998. Developmental and reproductive toxicity of dioxins and related compounds: cross-species comparisons. Crit. Rev. Toxicol. 23(3):283-335.

Poland, A., and J.C. Knutson. 1982. 2,3,7,8-tetrachlorodibenzo-p-dioxin and related halogenated aromatic hydrocarbons: examination of the mechanism of toxicity. Ann. Rev. Pharmacol. Toxicol. 22:517-554.

Quensen, J.F, 3rd ., M.A. Mousa, S.A. Boyd, J.T. Sanderson, K.L. Froese, and J.P.

Giesy. 1998. Reduction of aryl hydrocarbon receptor-mediated activity of polychlorinated biphenyl mixtures due to anaerobic microbial dechlorination. Environ. Toxicol. Chem. 17(5):806-813.

Rice, D.C. 1995. Neurotoxicity of lead, methylmercury, and PCBs in relation to the Great Lakes. Environ. Health Perspect. 103(suppl.9):71-87.

Rice, D.C., and S. Hayward. 1998. Lack of effect of 3,3'4,4',5-pentachlorobiphenyl (PCB 126) throughout gestation and lactation on multiple fixed interval-fixed ratio and DRL performance in rats. Neurotoxicol. Teratol. 20(6):645-650.

Risebrough, R.W., P. Rieche, D.B. Peakall, S.G. Herman, and M.N. Kirven. 1968. Polychlorinated biphenyls in the global ecosystem. Nature 220(172):1098-1102.

Rogan, W.J., and B.C. Gladen. 1991. PCBs, DDE, and child developmnet at 18 and 24 months. Ann. Epidemiol. 1(5):407-413.

Ross, P.S., R.L. de Swart, P.J.H. Reijnders, H. Van Loveren, J.G. Vos, and A.D.M.E. Osterhaus. 1995. Contaminant-related suppression of delayed-type hypersensitivity and antibody response in harbor seals fed herring from the Baltic Sea. Environ. Health Perspect. 103(2):162-167.

Rothman, N., K.P. Cantor, A. Blair, D. Bush, J.W. Brock, K. Helzisouer, S.H. Zahm, L.L. Needham, G.R. Pearson, R.N. Hoover, G.W. Comstock, and P.T. Strickland. 1997. A nested case-control study of non-Hodgkin's lymphoma and serum organochlorine residues. Lancet 350(9073):240-244.

Safe, S. 1984. Polychlorinated biphenyls (PCBs) and polybrominated biphenyls (PBBs): biochemistry, toxicology, and mechanism of action. Crit. Rev. Toxicol. 13(4):319-395.

Safe, S. 1994. Polychlorinated biphenyls (PCBs): environmental impact, biochemical and toxic responses, and implications for risk assessment. Crit. Rev. Toxicol. 24(2):87-194.

Schantz, S.L., A.M. Sweeney, J.C. Gardiner, H.E. Humphrey, R.J. McCaffrey, D.M. Gasior, K.R. Srikanth, and M.L. Budd. 1996. Neuropsychological assessment of an aging population of Great Lakes fisheaters. Toxicol. Ind. Health 12(3-4):403-417.

Secord, A.L., and J.P. McCarty. 1997. Polychlorinated Biphenyl Contamination of Tree Swallows in the Upper Hudson River Valley, New York. Effects on Breeding Biology and Implications for Other Species. U.S. Fish & Wildlife Service, Cortland, NY.

Secord A.L., J.P. McCarty, K.R. Echols, J.C. Meadows, R.W. Gale, and D.E. Tillitt. 1999. Polychlorinated biphenyls and 2,3,7,8-tetrachlorodibenzo-p-dioxin equivalents in tree swallows from the upper Hudson River, New York State, USA. Environ. Toxicol. Chem. 18(11):2519-2525.

Silberhorn, E.M., H.P. Glauert, and L.W. Robertson. 1990. Carcinogenicity of polyhalogenated biphenyls: PCBs and PBBs. Crit. Rev. Toxicol. 20(6):439-496.

Subramanian B.R., S. Tanabe, H. Hidaka, and R. Tatsukawa. 1983. DDTs and PCB isomers and congeners in antarctic fish. Arch. Environ. Contam. Toxicol. 12(6):621-626.

Suter, G.W. 1993. Ecological Risk Assessment. Boca Raton: Lewis.

Tanabe, S. 1988. PCB problems in the future: foresight from current knowledge. Environ. Pollut. 50(1-2):5-28.

Tanabe, S., M. Kawano, and R. Tatsukawa. 1982. Chlorinated hydrocarbon in the Antarctic, Western Pacific, and Eastern Indian Oceans. Trans. Tokyo Univ. Fish. 5:97-109.

Tanabe, S., H. Hidaka, and R. Tatsukawa. 1983. PCBs and chlorinated hydrocarbon pesticides in Antarctic atmosphere and hydrosphere. Chemosphere 12(2):277-288.

Taylor, P.R., C.E. Lawrence, H.L. Hwang, and A.S. Paulson. 1984. Polychlorinated biphenyls: influence on birthweight and gestation. Am. J. Public Health 74(10):1153-1154.

Taylor, P.R., J.M. Stelma, and C.E. Lawrence. 1989. The relation of polychlorinated biphenyls to birth weight and gestational age in the offspring of occupationally exposed mothers. Am. J. Epidemiol. 129(2):395-406.

Tilson, H.A., P.R. Kodavanti, W.R. Mundy, and P.J. Bushnell. 1998. Neurotoxicity of environmental chemicals and their mechanism of action. Toxicol. Lett. 102-103:631-635.

USCG (U.S. Coast Guard). 2001. FOIA Requests. National Response Center Database. [Online]. Available: http://www.nrc.uscg.mil/foia.html. [January 18, 2001].

Van den Berg, M., L. Birnbaum, B.T.C. Bosveld, B. Brumström, P. Cook, M. Feely, J.P. Giesy, A. Hanberg, R. Hasegawa, S.W. Kennedy, T. Kubiak, J.C. Larsen, F.X. R. van Leeuwen, A.K. Djien Liem, C. Nolt, R.E. Peterson, L. Pollinger, S. Safe, D. Schrenk, D. Tillitt, M. Tusklind, M. Younes, F. Waren and T. Zacharewski.. 1998. Toxic equivalency factors (TEFs) for PCBs, PCDDs, PCDFs for humans and wildlife. Environ. Health Perspect. 106(12):775-792.

van Leeuwen, C.J., and J.L.M. Hermens, eds. 1995. Risk Assessment of Chemicals: An Introduction. Dordrecht: Kluwer. 374p.

WHO (World Health Organization). 1976. Polychlorinated Biphenyls and Terphenyls. Environmental Health Criteria 2. Geneva: World Health Organization.

WHO (World Health Organization). 1992. Polychlorinated Biphenyls and Terphenyls. 2nd Ed. Environmental Health Criteria 140. Geneva: World Health Organization.

Zell, M., and K. Ballschmiter. 1980. Baseline studies of the global pollution. III. Trace analysis of polychlorinated biphenyls (PCB) by glass capillary gas chromatography in environmental samples of different trophic levels. Fresenius Z. Anal Chem. 304:337-347.

3

The Framework

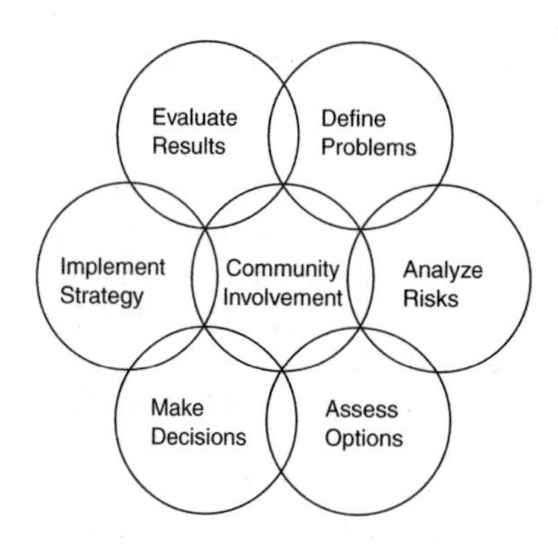

Each PCB-contaminated site is unique in terms of level and extent of contamination, the types of communities and ecosystems that are located near or affected by the site, the site history, the effects of the contamination on human and ecological health, and the applicable regulations. During the public sessions held by the committee, it became apparent that the processes used by the regulatory agencies, such as EPA and state environmental agencies, and by the industries—generally termed potentially responsible parties at Superfund sites—to deal with each site were neither uniform and consistent nor clear to outside parties. At present, the approaches appear to vary considerably from site to site or even at specific locations within a larger site. Data collection, applicable models, and interpretation of the results were not always presented in a manner to make the risks at a site, particularly the relative risks of alternative actions at a site, evident to all. The processes used by the federal and state agencies responsible for remediating contaminated sites should be broadened to address some of the concerns raised during the public sessions. EPA's development of the *Contaminated Sediment Management Strategy* (EPA 1998a) and *Remediation Guidance Document for Contaminated Sediments* (in preparation) will, over time, assist in managing risks at contaminated sediment sites in a more consistent manner. The available EPA strategy document, however, provides only the broad principles that should guide agency response to contaminated sediment sites. In addition, the second goal of the EPA strategy is to reduce the volume of existing contaminated sediment, a goal that does not necessarily reduce risk.

Therefore, in consideration of what the committee heard during the public sessions, particularly with respect to the need for a collaborative and iterative process to manage a broad range of risk, the committee determined that a framework was necessary to provide a consistent structure to assess the risks from PCB contamination and the risks associated with the various technologies and options that might be used to clean up a site. The framework should be risk-based and applicable to a variety of sites, where risks can range from obvious and short term to less evident and decades long. The framework should also be able to build upon many of the methods and procedures developed by the regulatory agencies, but it should go beyond the traditional human-health and ecological risk-assessment processes, such as those in the EPA's *Risk Assessment Guidance for Superfund* (RAGS) (EPA 1989).

As discussed in Chapter 1, the Committee on Remediation of PCB-Contaminated Sediments was given the task of selecting and refining such a scientific risk-based framework for assessing the remediation alternatives for exposure of humans and other organisms to PCBs in contaminated sediments.

RISK-MANAGEMENT FRAMEWORK

PCB-contaminated sediments threaten ecosystems, marine resources, and human health and can have significant impacts on affected communities. Any decision regarding the specific choice of a risk management strategy for a contaminated sediment site must be based on careful consideration of the advantages and disadvantages of available options and a balancing of the various risks, costs, and benefits associated with each option.

The committee appreciates that more than one risk-assessment and risk-management framework has been developed and can be effective in assessing and managing a variety of activities (e.g., NRC 1983, 1994, 1996; EPA 1998b, 1999). Furthermore, some of these assessment procedures are mandated by various laws and regulations. The committee reviewed several of these frameworks and found that, although useful for specific situations, generally they are not applicable to the variety of sites and situations where PCB-contaminated sediments are an issue. One of the earliest risk-assessment frameworks was that developed by the National Research Council (NRC) in 1983 in *Risk Assessment in the Federal Government: Managing the Process*. This risk-assessment and risk-management paradigm (Figure 3-1) institutionalized risk assessment for many government regulatory agencies. This paradigm begins with field and laboratory research and data collection on health effects, extrapolation methods, exposure estimates, and population characterization. These

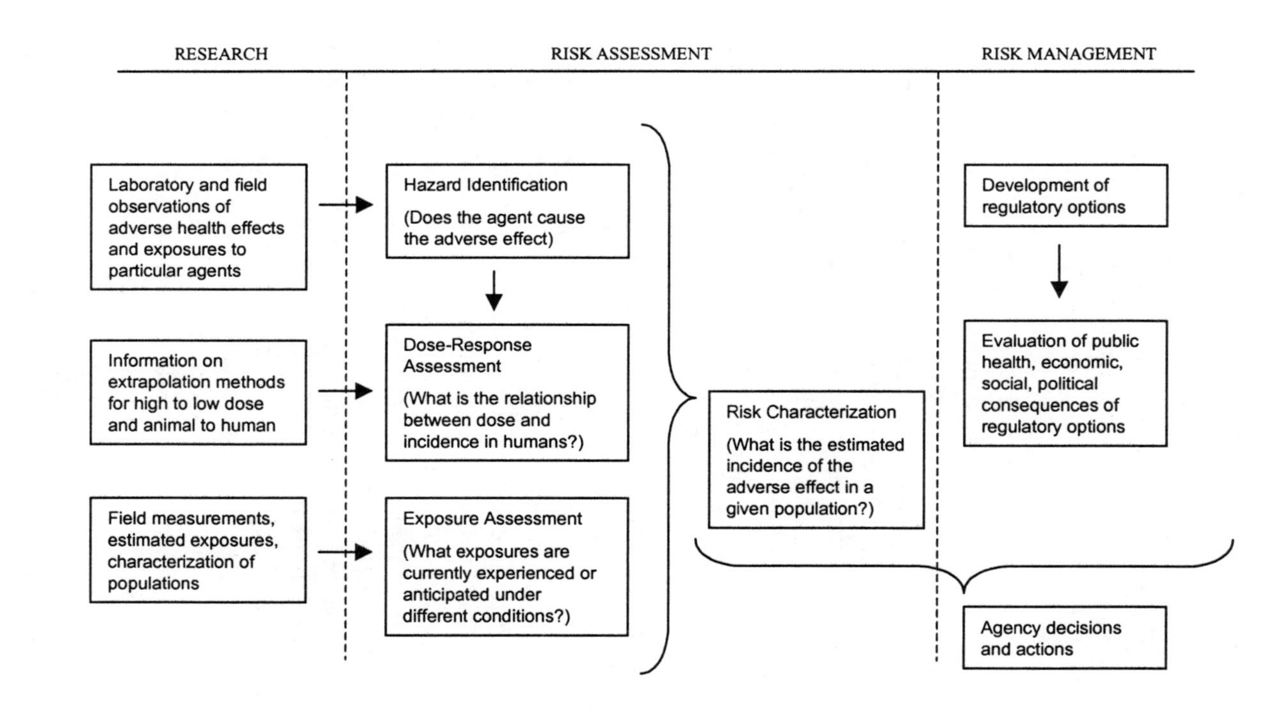

FIGURE 3-1 Elements of risk assessment and risk management. Source: NRC (1983).

data and methods are then used to conduct the risk assessment, which consists of three steps: hazard assessment, dose-response assessment, and exposure assessment. The three assessments are then combined into a risk characterization, which describes the nature of the adverse effects and the strength of the evidence and gives an estimated incidence of the adverse effect in a specified population with some expression of the uncertainties. The risk characterization is used to inform the risk-management step, where the regulatory alternatives are weighed and chosen. This decision step also "entails consideration of political, social, economic, and engineering information with the risk-related information" (NRC 1983) to develop the regulatory options. This risk-assessment approach was further developed a decade later in the NRC (1994) report *Science and Judgment in Risk Assessment.*

EPA also developed a risk-assessment paradigm, particularly for human health (EPA 1989), and, more recently, a modified paradigm for ecological risk assessment (Figure 3-2) (EPA 1998b). This paradigm builds on the NRC framework but puts the risk assessment into more of a regulatory framework. This EPA risk-assessment paradigm begins with a dialogue between the risk manager and the risk assessor to plan the risk-assessment process. On the basis of that dialogue, the problem-formulation stage identifies available information for the risk characterization, the assessment endpoints, a conceptual model of the problem (a risk hypothesis of the responses of the ecological entities and the ecosystem processes that act upon them), and an analysis plan for determining the effects and exposures. In the risk-analysis phase, the paradigm closely follows the NRC model by incorporating measures of exposure, ecosystem and receptor characteristics, and effect into any exposure profile and a stressor-response profile. From the risk-analysis phase, a risk characterization is developed that includes a risk estimation and a risk description. The risk characterization is then communicated by the risk assessor to the risk manager, who is responsible for the risk-management decisions. During this entire process, data are acquired as necessary, the various stages are iterated as necessary, and the results are monitored. EPA risk-assessment guidance also indicates that "all involved parties" contribute to the problem-formulation phase of the risk assessment. These parties include the remedial project manager, risk-assessment team, regional Superfund biological technical assistance group, potentially responsible parties, natural resource trustees, and stakeholders in the natural resource at issue, such as local communities and state agencies.

These and other similar frameworks are adequate for the general approach to assessment and management of risk. In fact, for sites that have been designated as a National Priority Site under the Comprehensive Environmental Response, Compensation, and Liability Act (CERCLA), there is specific guidance and procedures for conducting the mandatory remedial investigative and feasibility studies (see Figure 3-3) (EPA 1989; NRC 2000). In general,

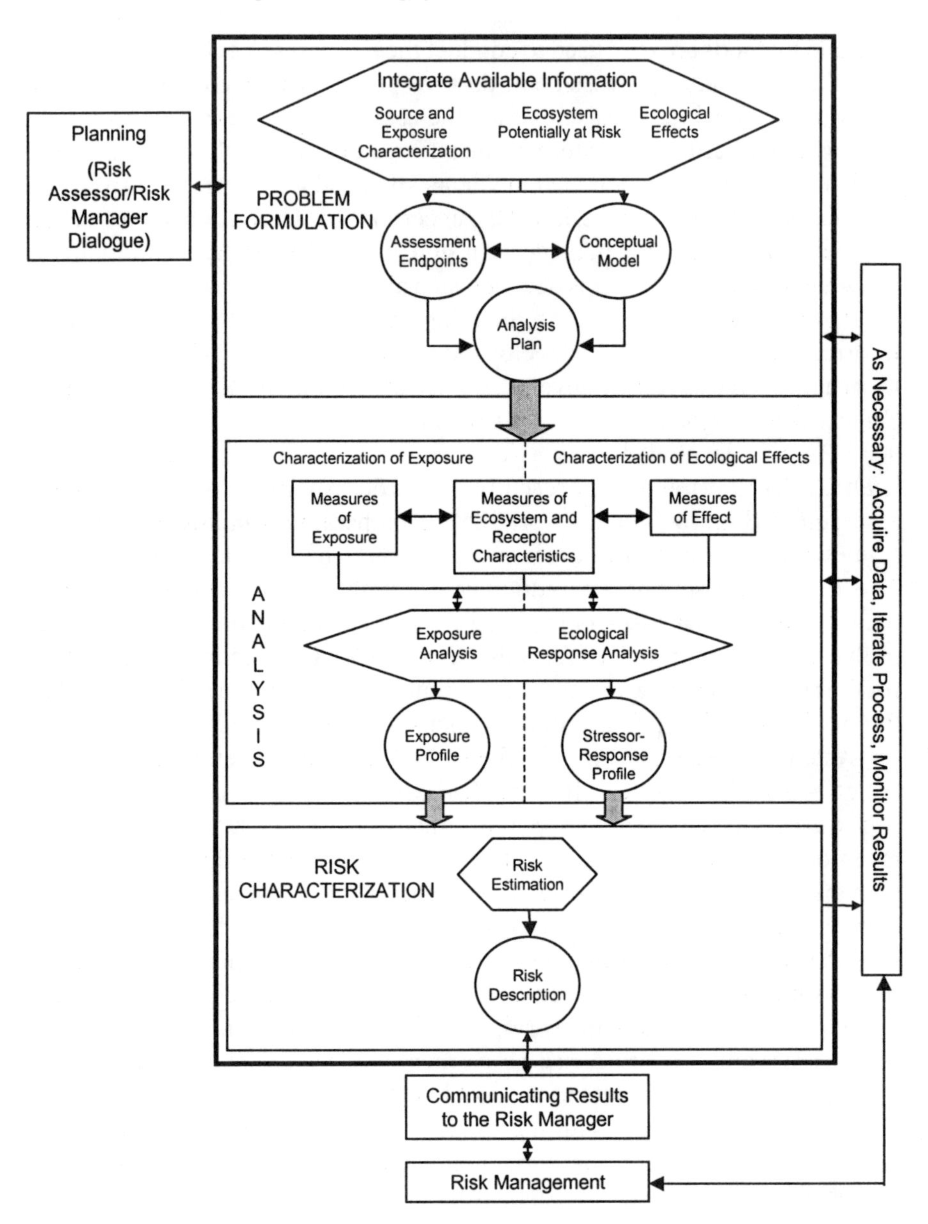

FIGURE 3-2 The ecological risk assessment framework, with an expanded view of each phase. Within each phase, rectangular boxes designate inputs, hexagon-shaped boxes indicate actions, and circular boxes represent outputs. Source: EPA (1998b).

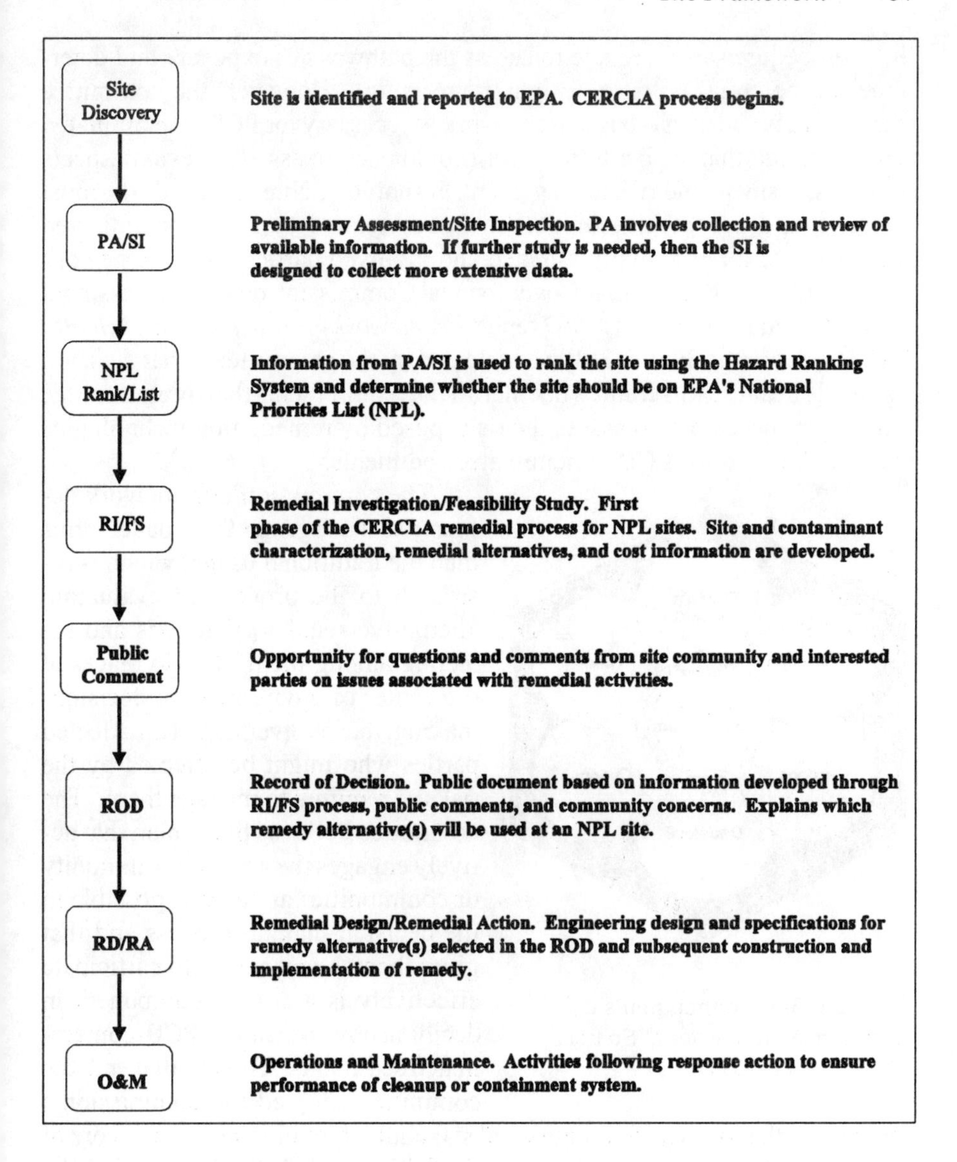

FIGURE 3-3 Steps in the Comprehensive Environmental Response, Compensation, and Liability Act (CERCLA) remedial process. As shown, opportunities for community review of potential remedies occur only after a list of remedial alternatives has been developed. The public's opportunity to help identify possible alternatives or plan the site evaluation is limited. As shown, the process unfolds sequentially, with little opportunity to go back to earlier stages. Source: NRC (2000).

these procedures are adequate to assess the pathways of exposure and determine the potential risks to designated receptors. However, the committee found that a broader risk-based framework is necessary for PCB-contaminated sediments and that such a framework should encompass all relevant aspects of risk assessment and risk management. It should be able to provide the most effective and acceptable solution to issues related to PCB-contaminated sediments for all potentially affected and involved parties. The framework developed by the Presidential/Congressional Commission on Risk Assessment and Risk Management in its 1997 report *Framework for Environmental Health Risk Management* (Figure 3-4) was selected by the committee. That decision was made because this framework met all the requirements that the committee found were necessary to assess the risks posed by remediation technologies that may be used for PCB-contaminated sediments.

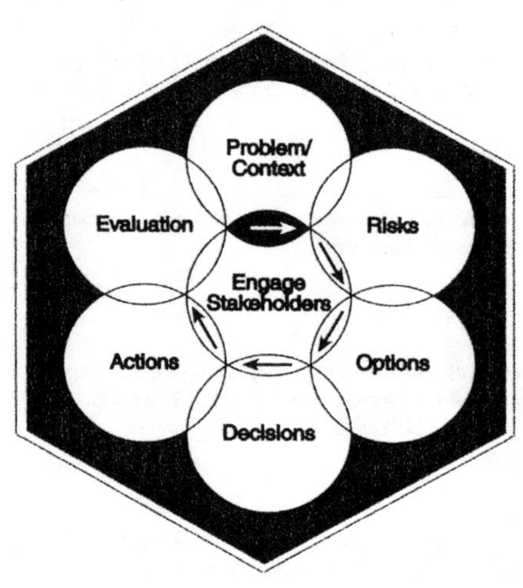

FIGURE 3-4 Commission's risk-management framework. Source: PCCRARM (1997).

The commission's framework defines risk management in broader terms than the traditional usage, which is restricted to the process of evaluating alternative regulatory actions and selecting among them. It also places at the center of environmental decision-making, the involvement of all affected parties who might be affected by the risks or required to manage them. The committee believes that a plan that actively engages the affected community or communities as early as possible in the decision-making process and that gives them the resources to participate effectively is a critical component in deciding how to clean up PCB-contaminated sediments. Toward that end, the committee adopted the commission's framework that places community and stakeholder involvement at the core of the decision-making process. The goal of this process should be to build a partnership and an effective working relationship among the regulatory agencies, site owners, affected community members, and other interested parties (Box 3-1).

The commission's framework is designed to help all types of risk managers—government officials, private-sector businesses, individual members of the public—make good risk-management decisions. The framework has six stages (Figure 3-4):

BOX 3-1 Advantages of the Commission's Framework

- The framework is all encompassing from problem definition to long-term evaluation after risk management.
- All affected parties are involved in all stages of the risk management process.
- All types of risks can be considered in the framework.
- The framework is iterative, and any stage can be revisited as new information becomes available.
- The framework can be applied at new sites and at sites where the management process is ongoing.
- The framework can incorporate existing methods and procedures (e.g., EPA ecological risk-assessment guidelines) if they are agreed upon by all affected parties.

1. Define the problem in a public-health and ecological context and establish risk-management goals.
2. Analyze the risks associated with the problem.
3. Examine options for addressing the risks.
4. Make decisions about which options to implement.
5. Take actions to implement the decisions.
6. Conduct an evaluation of the actions' results.

The framework is conducted as follows:

- In collaboration with all affected parties.
- With the use of more than one iteration of the framework if new information becomes available that changes the nature of the problem or the need for risk assessment or risk-management decisions.

The framework involves all affected parties from the earliest problem definition stage through the decision-making and decision-evaluation stages. Affected parties contribute to decision-making, rather than just providing feedback about decisions made by others. The importance of community involvement in all stages of the risk-management process was made clear to the committee at each of its public meetings. However, the committee recognizes that, depending on their situations, affected parties will have differing and often opposing views and goals. In those situations, federal, tribal, state, and local authorities must be given the authority to make decisions.

As indicated in Figure 3-4, the framework involves affected parties through all stages of the process. Note that the stakeholders or affected parties are at the very center of the framework and that the initial problem/context stage is highlighted with a bold overlap and larger arrow in the hexagon in the figure. The framework is designed so that any stage of it may be revisited as appropriate, giving risk managers and the affected public the flexibility to re-visit earlier stages of the process when new findings made during later stages shed light on earlier deliberations and decisions. The framework is intended to supplement, not supplant, the CERCLA remedial process mandated by law for Superfund sites.

It is often in the application and interpretation of the results of the assessments that opinions diverge and polemics ensue. These differences of opinion are a function of perspective, mandate, and political and economic goals. The commission's framework was selected by the committee because it was the most complete and flexible framework and deals with contaminated sediment situations from the identification of the site through risk assessment, implementation of the risk management strategy, and long-term evaluation.

Other frameworks considered by the committee deal primarily with risk assessment, which is critical to the selection of a risk-management strategy and is part of a risk-based framework but not all of it. Although the other frameworks provide for input from parties other than the risk assessor or risk manager, this input is not an integral part of the decision-making process. Furthermore, in each of the frameworks considered by the committee, only health and ecological risks assessments are delineated. Risks such as economic, social, and cultural risks are not considered, nor are risks to humans and ecosystems beyond the immediate problem area. Finally, the other frameworks do not involve an iterative review that includes the risk-reduction goals and management strategy—that is, the other frameworks are linear.

RATIONALE FOR AND ADVANTAGES OF THE FRAMEWORK

The committee understands that the conduct and uses of risk assessment are not without critics. A few of the criticisms that have appeared in the literature (NRC 1994; O'Brien 2000) are cited here, because the committee believes that the commission's framework is responsive to the concerns raised in these criticisms. The committee emphasizes that our citation of these criticisms does not mean that we believe them to be valid, nor is the order of their listing meant to suggest our opinion regarding their possible importance.

1. Some experts have noted that risk assessment often neglects important aspects of risk, including synergistic interactions from multiple chemical

exposures, risk aggregation from combinations of multiple chemicals, individual variability to chemical exposures, and noncancer risks.

2. Some people believe that credible risk assessment might be impossible to obtain with the existing state of science. These people believe that there is not sufficient knowledge about the cancer and noncancer effects caused by exposure to most chemicals to make accurate risk estimates.

3. Some people believe that it is nearly impossible to separate *risk assessment* (the scientific and objective process of determining hazard, exposure, susceptibility, and risk) from *risk management* (the procedures used to reduce risks). These people believe that risk assessors typically are hired by those who desire a certain outcome and that risk assessments can be manipulated too easily to favor that outcome.

Those who reject *risk assessment* may favor *alternatives assessment*. An alternatives assessment "involves consideration of the pros and cons of a decent range of options" and "includes the public whenever they might be harmed by activities considered in the assessment" (O'Brien 2000). Its method of disclosure and analysis is based on the National Environmental Policy Act (NEPA) regulations, which govern the actions of federal agencies. Alternatives assessment includes risk assessment but relegates risk assessment to looking at the disadvantages of each alternative. If applied to PCB-contaminated sediments, an alternative assessment might include the following elements:

1. Presentation of a full range of options for responding to the PCB-contaminated sediments. The options would include those that seem to promise the least adverse impact on environmental and social health and those that seem to promise the greatest environmental and social advantages.

2. Presentation of the potential adverse and beneficial effects of each alternative. The types of adverse and beneficial consequences considered for each option would vary but might include aesthetic, cultural, economic, and social effects. The presentation would also consider direct, indirect, and cumulative effects.

The public would be involved in "scoping" reasonable alternatives and "commenting" on the presentation and evaluation of their potential beneficial and adverse effects.

The commission's framework is responsive to the concerns raised by critiques of the risk-assessment process:

- The framework provides an integrated, holistic approach to solving public-health and environmental problems in context. (It offers a broad vehicle for incorporating all aspects of risk.)

- The framework ensures that decisions about the use of risk assessment and economic analysis rely on the best scientific evidence and are made in the context of risk- management alternatives. (It focuses on best available science, places risk assessment within the confines of risk-management decision-making, emphasizes consideration of alternatives and recommends peer review of both technical and economic analyses.)
- The framework allows for evaluation of the pros and cons of remedial alternatives under consideration.
- The framework emphasizes the importance of collaboration, communication, and negotiation among all affected parties so that public values can influence risk-assessment and risk-management strategies. (It involves active participation by affected parties.) As a result, it should produce risk-based decisions that are more likely to be successful than decisions made without adequate and early involvement of affected parties.
- The framework accommodates critical new information that might emerge at any stage of the process. (It is iterative.)

The commission's framework is also general enough to be applicable to a variety of situations, including decisions related to the risk management of PCB-contaminated sediments. As noted in the commission report (PCCRARM 1997), "the level of effort and resources invested in using the Commission's framework can be scaled to the importance of the problem, potential severity and economic impact of the risk, level of controversy surrounding it, and resource constraints." This flexibility makes the framework ideal for providing a foundation for the committee's risk-based approach to managing PCB-contaminated sediments.

No two PCB-contaminated sites are identical; therefore, no one risk management technology or strategy is appropriate for all sites. The affected parties must agree on the risk-management goals, the likely immediate short-term and long-term risks, the available and reasonable risk-management options for consideration, and the pros and cons of the risk-management options. The risks to be considered are those associated with the present PCB contamination, those that might occur while risk management is undertaken, and the short- and long-term risks that remain after risk management has taken place. Every attempt should be made to quantify risks to the extent feasible and in a timely manner, but there will always be some element of value judgment incorporated into the risk assessment.

OUTLINE OF THE RISK-MANAGEMENT FRAMEWORK

The committee refined the commission's framework as follows for application to the risk management of PCB-contaminated sediments. The committee's framework used in concert with the EPA guidance for ecological risk assessment (EPA 1998b) and the *Risk Assessment Guidance for Superfund: Human Health Evaluation Manual* (EPA 1989) should provide sufficient guidance to perform valid risk assessments of PCB-contaminated sediments that can be used in examining options for risk management of PCB-contaminated sediments.

At the center of the framework is community involvement (Chapter 4), deserving of its own chapter in recognition of the commission's and the committee's conclusion that risk management decisions made with adequate and early involvement of affected parties are more likely to be successful than decisions without such involvement. The risk-management framework promotes participation of affected parties at each stage of the risk-assessment and risk-management process.[1]

The *first* stage in the framework is defining the problem and setting management goals (Chapter 5). The problem-definition stage is the most important step in application of the risk-management framework, because it characterizes the problem from public health, ecological, social, economic, and cultural perspectives; determines general risk-management goals; and identifies affected parties and decision-makers. These goals may need to be modified as new information is gained about the risks as the site and the best strategy for managing them. Problem characterization should be performed in collaboration with all affected parties.

The intent of managing PCB-contaminated sediments is to reduce present and future risks to human health and the environment from exposure to PCBs. In the problem-definition phase, affected parties are identified, a site assessment is begun, and risk-reduction goals are set. Problem definition should include identification of the perceptions of affected parties (particularly the directly affected community or communities) on the risks posed by the PCB-contaminated sediments problem. Affected parties often propose explicit questions that provide useful guidance for the risk- assessment stage, matters that would be neglected in the standard technical risk assessment (PCCRARM 1997). Involvement of the affected parties in the risk-management framework requires active outreach efforts by the regulatory agencies to engage the affected parties, including the community and industry as partners. This

---

[1] As noted in Chapter 4, the committee prefers to use the terms "interested" and "affected parties" instead of "stakeholders" in this report, following the guidance of the NRC Committee on Risk Characterization (NRC 1996). The committee makes this distinction to emphasize the need to identify and include the most affected members of the community, some of whom may be the least powerful and most socially disadvantaged.

involvement must be understood to be a learning process by all affected parties. Furthermore, in many cases, technical assistance might be required to enable the affected parties to participate fully in the risk-management process.

The *second* stage in the framework is analyzing the risks (Chapter 6). The risk-analysis stage is the risk assessment–risk characterization step of the process in which the specific risks posed by the PCB-contaminated sediments are assessed and placed into their multisource, multimedia, multichemical, and multirisk contexts and in which the gathered information is combined into a characterization of the risk of the PCB-contaminated sediments to humans and the environment.

During the analysis phase, assessment endpoints are selected, conceptual models are developed, and exposure to stressors and the relationship between stressor levels and human and ecological effects are evaluated. Assessment endpoints, specific expressions of the actual value that is to be protected, are the ultimate focus in risk characterization, and act as a link to the risk-management process (such as policy goals). Effect measures or measurement endpoints are responses that might be more easily measured than assessment endpoints but are related quantitatively or qualitatively to the assessment endpoints. The analysis phase involves collection and integration of information on toxicity of the PCBs, PCB concentrations and spatial distribution (at a more detailed level than during the problem-definition stage), and exposure conditions (temporal and spatial patterns), as well as observations or predictions of adverse effects. In the risk-characterization step, risk is estimated through integration of the exposure and stressor-response profiles and is described by lines of evidence for human and ecological effects.

The *third* stage in the framework is examining management options (Chapter 7). This stage of the risk-management process involves identifying risk-management options and evaluating their effectiveness, feasibility, costs, benefits, unintended consequences, and habitat, cultural and social impacts. In this stage, the pros and cons of an appropriate range of options are considered. Options include socioeconomic and institutional controls, source control, in situ sediment-management technologies, and removal and ex situ storage, disposal, and/or treatment, and natural attenuation. Often a combination of technologies will be needed at a site. Affected parties, particularly the directly affected community or communities, can and should play an important role in all facets of identifying and evaluating options.

The *fourth* stage in the framework is making a decision (Chapter 8). During this stage of the framework, decision-makers review the information gathered during the analyses of risks and options to select the most appropriate risk management strategy. The strategy selected should be one that actually reduces overall risk, not merely transfers risk to another site or another affected population. The decision-making process necessary to arrive at an optimal management strategy is complex and likely to involve numerous site-specific considerations.

Management decisions must be made, even when information is imperfect. There are uncertainties associated with every decision that need to be weighed, evaluated, and communicated to affected parties. Imperfect knowledge must not become an excuse for not making a decision.

Once again, involving the affected parties is key to a successful outcome. Involving affected parties and incorporating their recommendations where possible reorients the decision-making process from one dominated by regulators and responsible parties to one that includes those who must live with the consequences of the decision. As noted in the commission's report, this fosters successful implementation.

The *fifth* stage in the framework is implementing the management strategy (Chapter 9). This stage is crucial to the success of the management strategy. It is important in this step to remember that no two sites are identical, and therefore the risk-management strategy will vary from site to site. It is also important to keep in mind that a single option will rarely suffice at a given site. It is much more common that a combination of options will have to be chosen to reduce the risk due to contamination at a site.

The *sixth* stage in the framework is evaluating results (Chapter 10). At this stage of risk management, decision-makers and affected parties review what risk-management actions have been implemented and how effective they have been in meeting the management goals established in the first stage. Evaluation of the remedy is an essential part of the process. It is only after the effectiveness of the remedy has been evaluated that the parties can decide whether more actions are needed at the site and whether implementation of the selected options actually accomplished the desired result. As with other stages of the risk-management process, the evaluation will benefit if affected parties are involved. In general, to gain credibility with all affected parties at multiple sites, objective evaluation is needed to ensure that the risk-assessment process and the required investments do generate the protection desired and do so with approximately the financial costs and technical and social complications expected from the action plan. This process is iterative; for instance, a pilot project might be performed to determine the effectiveness of a proposed risk-management option. The process should be sufficiently flexible to assimilate the information as it becomes available, and the actions can be adjusted.

Environmental monitoring—that is, sampling and analysis of environmental media, including biota—during and after the risk-management action is implemented is imperative to evaluate whether the risk-management goals have been met. Failure to achieve the management goals could necessitate a reexamination of those goals, in addition to selection of new management measures and their commensurate risks. Reviews of ongoing or completed management projects are of great value when planning for future management efforts. The lack of information on past actions can limit management efforts at future sites, particularly when a relatively new remediation technology is being considered for use.

New information might emerge during evaluation that is of sufficient importance to indicate that parts of the framework should be reviewed and repeated. The committee emphasizes that the risk-management process must be iterative and adaptive as important new information, ideas, and perspectives as well as new risk-management options come to light. Each iteration might provide additional certainty and information to support further risk-management decisions, or it might require a course correction. The commission's risk-management framework provides that flexibility. Each of the stages in the framework, including the need for community involvement in risk management strategy, is discussed in greater detail in the following chapters.

CONCLUSIONS AND RECOMMENDATIONS

Based on the public sessions and the committee's experience, the following recommendations are made with regard to the selection of a framework for assessing the risks associated with PCB-contaminated sediments and the possible remediation technologies and strategies that may be used at a contaminated site:

- All risk-management decisions regarding PCB-contaminated sediments should continue to be made within a risk-management framework.
- All risk-management decisions must include active and continuing involvement from affected parties (particularly the directly affected community or communities) and the use of all available sources of information.
- The framework should be a refinement of the 1997 Framework for Environmental Health Risk Management developed by the Presidential/Congressional Commission on Risk Assessment and Risk Management (PCCRARM 1997). The commission's framework provides for an iterative and collaborative process involving all affected parties. Furthermore, it is applicable to both new contaminated sites and sites where substantial work has already been done. Therefore, when the site evaluation or risk-management process is already under way, the commission's framework can still be used to reassess available data, involve affected parties, and make, implement, and evaluate management decisions.

REFERENCES

EPA (U.S. Environmental Protection Agency). 1989. Risk Assessment Guidance for Superfund, Volume 1. Human Health Evaluation Manual, Part A. Interim Final. EPA/540/1-89/002. Office of Emergency and Remedial Response, U.S. Environmental Protection Agency, Washington, DC. December. [Online]. Available: http://www.epa.gov/superfund/programs/risk/ragsa/index.htm. (Last updated May 12, 1999).

EPA (U.S. Environmental Protection Agency). 1998a. EPA's Contaminated Sediment Management Strategy. EPA-823-R-98-001. Office of Water, U.S. Environmental Protection Agency, Washington, DC. April.

EPA (U.S. Environmental Protection Agency). 1998b. Guidelines for Ecological Risk Assessment. Final. EPA-630-R-95-0026F. Risk Assessment Forum, U.S. Environmental Protection Agency, Washington, DC. [Online]. Available: http://www.epa.gov/ncea (Last updated May 03, 1999).

EPA (U.S. Environmental Protection Agency). 1999. Risk Assessment Guidance for Superfund, Vol. 3, Part A. Process for Conducting Probabilistic Risk Assessment. Draft. EPA 000-0-99-000. OSWER 0000.0-000. PB99-000000. Office of Solid Waste and Emergency Response, U.S. Environmental Protection Agency, Washington, DC. December. [Online]. Available: http://www.epa.gov/superfund/pubs.htm.

NRC (National Research Council). 1983. Risk Assessment in the Federal Government: Managing the Process. Washington, DC: National Academy Press. 191pp.

NRC (National Research Council). 1994. Science and Judgment in Risk Assessment. Washington, DC: National Academy Press.

NRC (National Research Council). 1996. Understanding Risk: Informing Decisions in a Democratic Society, P.C. Stern and H.V. Fineberg, eds. Washington, DC: National Academy Press.

NRC (National Research Council). 2000. Natural Attenuation for Groundwater Remediation. Washington, DC: National Academy Press.

O'Brien, M. 2000. Making Better Environmental Decisions, An Alternative to Risk Assessment. Cambridge, MA: The MIT Press.

PCCRARM (Presidential/Congressional Commission on Risk Assessment and Risk Management). 1997. Framework for Environmental Health Risk Management. Vol. 1. Washington, DC: The Commission.

4

Community Involvement

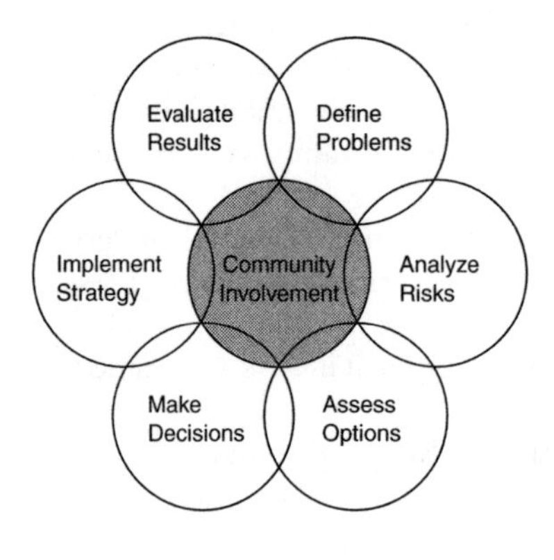

Involving members of a community affected by a contaminated site is critical to the successful completion of a risk-management project. Decisions are less likely to be accepted by a community if the people directly affected by the decisions are not included in the decision-making process. The risk-management framework developed by the Presidential/Congressional Commission on Risk Assessment and Risk Management (1997) emphasizes that the "stakeholders"[1] are "active partners so that different technical perspectives, public values, perceptions, and ethics are considered." This framework goes beyond the classic EPA model of public participation, which is largely a passive model in which the public provides feedback to the regulatory agencies about decisions made and reports written in closed sessions (EPA 1992).

The active public participation model of the commission's report is supported by a broad range of research on public participation efforts (NRC

[1]Stakeholders are defined by the Presidential/Congressional Commission as "parties who are concerned about or affected by the risk management problem." Because the term "stakeholder" has been used by others to mean only the legally involved parties (i.e., the potentially responsible parties (PRPs) and the regulatory agencies negotiating and overseeing the remediation action), the committee chose not to use the term "stakeholder" throughout this document, preferring to use the clearer phrase "affected parties" or "members of the affected community."

1992; Kunreuther et al. 1993; Ashford and Rest 1999; English et al. 1993; Lynn 1988; NRC 1996; Rich et al. 1995; Hance et al. 1988). Although many important lessons are derived from this research, an essential component for successful implementation of a risk-based framework is the direct and early involvement of all affected parties, including the affected public, as full partners in all parts of the risk-management strategy, including goal setting, evaluating options, setting priorities, evaluating different risk-management strategies, and making a final decision.

This chapter summarizes concerns raised at the public meetings held by the committee during its deliberations. The chapter then discusses the need for community involvement, the benefits, how the community is defined to include both interested and affected parties, and ways to identify and involve the interested and affected parties in the decision-making process. In doing so, the committee presents results from the social science literature and the conclusions drawn by the committee based on the public meetings and on the wealth of information provided to the committee by community and environmental groups; state, tribal, and federal government agencies; and private companies and interests (see Appendix C for a list of these materials).

SPECIFIC COMMUNITY CONCERNS

During the public sessions held in Washington, DC; Green Bay, WI, where the committee visited the Fox River; and Albany, NY, where the committee visited the Hudson River, the committee heard from grassroots community organizations, environmental groups, government agencies, commercial and industrial concerns, and the general public. Some of the following messages were conveyed to the committee at these meetings:

- The public was dissatisfied with its level of involvement in the management efforts. Involvement was largely limited to commenting on draft documents provided by regulatory agencies. Little active involvement was observed. Participants made it clear that the regulatory agencies have to communicate with the public during all phases of the decision-making process, not just after a decision is made and the public is asked to comment on a completed plan.
- There was concern about the lack of consideration of indigenous community issues. One speaker from the Mohawk community of Akwesasne described the cultural destruction and community breakdown suffered by Mohawk people, because the PCB contamination along the St. Lawrence River had resulted in fishing advisories, which forced his tribe to stop fishing.

That separated the tribe from its traditional life style of fishing for food and for bartering for other goods from farmers. He further stated that it is difficult for some people to understand and respect these concerns and to factor them in when deciding how to clean up the river.

- Some participants expressed a lack of trust in the government agencies involved at various sites, because the agencies were perceived to be secretive.
- There were contradictory statements regarding the need to balance short-term risks with long-term benefits. One commenter said that the risks of active remediation (dredging) far outweighed the risks of no remediation, while others said that it was necessary to accept some short-term risks to gain long-term benefits. Other commenters felt that active remediation would not provide any long-term benefits.
- Farmers expressed concern about the economic impact on their farms if long-term storage sites for dredged PCB-contaminated sediments were located near their property. Concerns focused on whether property values would be reduced and whether their agricultural products would be labeled as "tainted" by PCBs.
- Businesses were concerned about reduced business opportunities and property values due to adverse publicity associated with remediation in general and with the siting of a long-term storage facility to hold dredged PCB-contaminated sediments. There was also concern about how long it would take to complete the cleanup.
- One presenter hoped the committee would provide "a critical and unbiased review" of the reports prepared by the government agencies and responsible parties, because he no longer knew what to believe.
- Some participants criticized the use of natural attenuation, saying that it was a "do-nothing" approach.

Community groups expressed a number of concerns about natural attenuation, particularly that it was a "do-nothing" approach. They stated that natural attenuation is being used or proposed at many sites, because it is inexpensive, not because it provides adequate protection of public health (NRC 2000). Other community groups, such as farmers and marina owners, were concerned that the natural attenuation option be evaluated thoroughly as a viable option for some areas. Natural attenuation processes will be a part of any risk-management strategy, because complete removal of PCB-contaminated sediments cannot be reasonably achieved with present technologies.

Those views were expressed largely by the members of the affected communities. Different views and comments were raised by the regulatory agencies and by the responsible parties. Several of the responsible parties supported the use of natural recovery and natural attenuation at sites where they were involved and wanted to see cost considerations factored into the risk-management process. There was some disagreement among the regulatory agencies and the responsible parties about the appropriate data, models, and methodologies to be used at the sites.

The public meetings were very important in providing the committee with a first-hand view of the varied and complicated issues surrounding the management of PCB-contaminated sediments. The many different perspectives and concerns of the various interested parties were presented along with a great deal of experiential and technical information. These meetings were an integral and invaluable aid to the committee's understanding and in its deliberations on how to best reduce the risks associated with managing PCB-contaminated-sediment sites.

DEFINING THE COMMUNITY

Community involvement is a process by which individuals and groups come together in some way to communicate, interact, exchange information, provide input around a particular set of issues, problems, or decisions and share in decision-making to one degree or another (Ashford and Rest 1999). Communities are generally defined as a group of people affiliated by geographic proximity, special interest, or similar situations to address issues affecting the well-being of those people (CDC 1997). The community surrounding a contaminated site includes the people who live closest to the site and whose health may be at risk and/or whose property, property values, or economic welfare is adversely affected by the contamination; local business owners; elected officials; local government agency representatives; workers at the site; and others who live farther from the site but who are indirectly affected. The views of these different constituencies might vary substantially, depending on whether their health is at risk and/or their property, property values, or economic welfare is or would be adversely affected by the contamination or the proposed risk-management strategy. In directly affected communities, the members of local organized community efforts do not generally include the regulators or the companies responsible for the contamination as affected parties, at least not in the same way as they include themselves as affected (Ashford and Rest 1999).

Potentially affected parties might include individuals that live by or use the waterway, tribal groups, subsistence or sports fishers who might live near or only occasionally use the water resources, and commercial businesses that rely on fishing, recreational boating, or tourism for subsistence or economic survival. In the case of PCB-contaminated-sediment sites, identifying who is directly affected is complicated by the fact that many of these sites are large river or estuarine areas that cover many miles. In these instances, communities and businesses downriver may also be affected by contamination occurring upriver. For example, migratory fish contaminated by PCBs at one site might be eaten by people living hundreds of miles away.

Several of these interested and affected parties have different and conflicting values. For example, there are clear differences in perspectives, goals, and objectives between those who are legally responsible for the contamination and its cleanup; those with regulatory oversight and responsibility; those whose health, economic well-being, or quality of life have been or will be affected by the contamination; those who speak on behalf of purely environmental or ecological considerations; those who might be adversely affected by the remediation options; and those who strive to protect the interests of future generations (English et al. 1993).

"Stakeholders" Versus "Affected Parties"

Some studies of public participation and involvement have introduced the notion of a stakeholder (English et al. 1993; Ashford and Rest 1999). Stakeholders are defined broadly as parties with legitimate interest (or stake) in the issues or impending decisions about the contamination (English et al. 1993) and generally consist of representatives of organized constituencies or institutions—industry, labor, environmental groups, government regulators, local government (Ashford and Rest 1999). The president/congressional commission defines stakeholders as "parties who are concerned about or affected by the risk management problem." Stakeholders may include responsible parties, government regulators, industry and business, elected officials, unions, environmental advocacy groups, consumer-rights organizations, education and research institutions, trade associations, religious groups, and others. The directly affected community may be one of these many parties or may be part of several, but not all, of the many subsets of stakeholders.

In its ideal form, stakeholder involvement treats all stakeholders as equals and attempts to break down the "us/them" notion inherent in public participation (English et al. 1993). However, in its practical form the goals of inclusiveness and representativeness are not always met (Ashford and Rest 1999). The stakeholder approach may be dominated by stakeholders with the most powerful vested interests, effectively diluting the voices of the least powerful members of the community. This concern is supported by the environmental justice literature, which documents that it is sometimes the people who are most affected by the contamination who have the weakest voice in the remedy (Sexton et al. 1993; IOM 1998) and the least influence on the decision-making process.

Clearly, government and agency officials, responsible parties, and the business community have many resources on which to draw, including the ability to access other powerful interests (Ashford and Rest 1999). Other interested parties, such as national environmental organizations, may also have some resources and a degree of influence, but the grassroots neighborhood groups in the contaminated community generally do not. Nor do members of these groups usually have the time, experience, or resources that they need to participate as equal partners in a stakeholder process. Small-business owners have similar constraints.

The committee distinguishes between members of the community and stakeholders. We prefer the term "affected parties" following the guidance of the NRC Committee on Risk Characterization (NRC 1996). We make this distinction to emphasize the need to identify and include the most affected members of the community, who often are the least powerful and most socially disadvantaged.

The Need for Community Involvement

There is a long tradition of community involvement and public participation in government decision-making going back to the town meetings held in New England over 150 years ago (Folk 1991). During the early 1970s, federal regulations, notably the National Environmental Policy Act, were passed that provided for public participation (O'Brien 2000). The role of public and community involvement changed dramatically during the late 1970s as grassroots community-based organizations (see Box 4-1) formed to represent the interests of the people most directly affected by exposure to chemicals leaking from contaminated sites (Gottlieb 1993; Dowie 1995). Perhaps the most dramatic and well-known of these sites is the Love Canal landfill located in Niagara Falls, NY. More than 900 families were evacuated from this community in 1978 and 1980 following strong organized community efforts (Gibbs 1998). Since that time, literally thousands of similar grassroots organizations have been formed across the country. The database of the Center for Health, Environment, and Justice lists more than 8,000 such groups (Gibbs 1998).

These neighborhood organizations provide a voice in decisions that affect the health and well-being of the community. They provide a focal point for involving interested participants by providing scientific information to the

BOX 4-1 Grassroots Community-Based Organizations

Grassroots community-based organizations are local neighborhood groups that have formed with the primary purpose of addressing a local contamination problem. Thousands of these organizations have been formed across the country in response to contamination problems. Grassroots organizations do not exist at every PCB-contaminated site, but at some sites, there may be more than one such group. These organizations provide a good base for community involvement in site decision-making. Two grassroots organizations that addressed the committee during visits to the Hudson River area were the Housatonic River Initiative from Pittsfield, Massachusetts, and the Arbor Hill Environmental Justice Corporation from Albany, New York. These organizations have been very active in addressing PCB-contaminated sediments issues affecting their communities.

community, a place to meet, emotional support, a sense of empowerment, and a mechanism for getting community concerns relayed back to other involved parties (Unger et al. 1992; Edelstein 1988). Grassroots groups typically, but not exclusively, consist of people who do not see themselves as environmentalists in a traditional way (Gibbs 1998). Most grassroots community groups are formed not because they choose to get involved in environmental issues but because these issues have intruded upon their lives. It is important, however, not to exclude members of the community who do not participate in or belong to these community-based organizations in the risk-management process. Surveys have found a relatively high level of public trust in these organizations. For example, McCallum et al. (1991) found that survey respondents had five times as much trust in environmental groups (whether local or national) as in chemical industry officials and nearly four times as much trust in these groups as in local and federal officials. Grassroots organizations can provide valuable information that professionals not familiar with the local environment and community might overlook (Ashford and Rest 1999; Brown and Mikkelsen1997; ATSDR 1996). They might know who is sick and with what disease, have valuable first-hand historical knowledge of past practices that might have led to the contamination, and be familiar with local environmental conditions.

With time, many of these groups lose faith and trust in the institutions that they once believed were established to help protect their interests (Ozonoff and Boden 1987; McCallum et al. 1991; Edelstein 1993; Rich et al. 1995;

Brown and Mikkelsen 1997). Many of these groups also become distrustful of scientific and technical experts from both government and private institutions who often are viewed by the community as being slow to acknowledge hazards, quick to minimize risks, and often prefer to wait for more scientific evidence before taking action to protect public health (Ashford and Rest 1999). These groups—and the public in general—have experienced that science and technical expertise can be frequently politicized (Nelkin 1975; Fiorino 1989; Eden 1996). Furthermore, the interpretation of scientific evidence cannot be completely isolated from personal experiences, social background, and values of any individual providing the interpretation (Ashford 1988). These factors have led to increased transparency of the process and more interaction and participation by these groups in institutional decision-making (Yosie and Herbst 1998; Ashford and Rest 1999).

Benefits of Community Involvement

There are many benefits to involving communities in the risk-management process. Participation makes the process more democratic, lends legitimacy to the process, educates and empowers the affected communities, and generally leads to decisions that are more accepted by the community (Fiorino 1990; Folk 1991; NRC 1996). The affected community members can contribute essential community-based knowledge, information, and insight that is often lacking in expert-driven risk processes (Ashford and Rest 1999). Community involvement can also assist in dealing with perceptions of risk and helping community members to understand the differences between different types and degrees of risk.

When the community is involved in decision-making it is possible that the public will reject or wish to modify a cleanup plan favored by a government agency. However, by participating in the process and being aware of the issues that affect a choice of remediation alternatives, the community will be more likely to accept the compromises that invariably have to be made. A community that has participated in the risk-management plan is less likely to protest the plan after the decision has been made. An example of that benefit is Waukegan Harbor, where the Citizens Advisory Group was an integral part of the remedial action plan (RAP), assisting in the review of the RAP, obtaining cooperation and access from local businesses, and identifying an expanded study area to prevent additional contamination of the harbor. Thus, although getting and keeping the community involved requires additional time and effort from the beginning and throughout the process, the net time spent from problem formulation to final closure can actually decrease.

Several studies have shown that community involvement is likely to result in decreased legal liability, the possibility that new alternatives will emerge, and avoidance of conflicts that can consume resources (Rich et al. 1995; English et al. 1991; Hance et al. 1988). For example, English et al. (1991) found that incorporating the interests of the community can reduce total management costs. Responsible parties, regulators, and community members are likely to benefit from following the basic principles of early community involvement, providing the community with a voice in decision-making, and building an effective relationship with community members.

Another important benefit is that community members can provide the institutional and historical memory about the area that needs to be included in the management plan. Court documents, records of land use, aerial photographs, and official company histories do not always provide full and accurate information. Community members can also identify potentially affected areas by pointing out where changes have occurred in the animal or plant communities or in the physical environment, especially if the changes occurred relatively suddenly.

INVOLVING THE COMMUNITY

As indicated in the risk-based framework presented in Chapter 3, community involvement is an integral component of any strategy to clean up PCB-contaminated sediments, but no one-size-fits-all community involvement plan will work at every site. Each approach must be tailored to the site and its particular social and institutional setting (Renn et al. 1995; English et al. 1993; Kasperson 1986). The process needs to involve all interested and affected parties, who should be considered equals (Box 4-2).

A number of examples of successful community involvement have been documented (Lynn 1988; ATSDR 1996; Ashford and Rest 1999). In reviewing successful case studies, Ashford and Rest (1999) concluded that successful community involvement is a process and that the process should be designed to improve communication with the community, education of community members, and specific participation in information review and decision-making. (See Box 4-3 for an example of a successful community involvement program used by the U.S. Department of Energy.)

Three key principles that have emerged from studies of community involvement are (1) involve the community from the beginning; (2) provide the community with the resources they need to participate effectively in the decision-making process; and (3) build an effective working relationship with the community.

BOX 4-2 Community Involvement Principles

- Provide community with resources to participate.
- Build an effective ongoing relationship with community.
- Establish trust and respect among all parties.
- Involve community as a respected partner.
- Empower community representatives.
- Communicate.
- Make the commitment of time and attention.
- Ensure that the process is open and transparent from the start.
- Take responsibility for community outreach and education.

BOX 4-3 Public Participation Programs at the Department of Energy

The U.S. Department of Energy (DOE) has designed a public participation program that involves the appointment of Site-Specific Advisory Boards (SSABs). The SSABs may include representatives of local governments, tribal nations, civic groups, environmental firms, and interested individuals. DOE has assigned obligations to the local site boards that include keeping the boards informed about key issues and upcoming decisions; requesting recommendations well in advance of DOE deadlines; considering and responding in a timely manner to all board recommendations; and providing adequate funding for administrative and technical support.

Organized citizen participation has been successful at DOE's Los Alamos National Laboratory. In one program to characterize site-wide groundwater, an External Advisory Group (EAG) was appointed to advise the responsible managers at the Los Alamos Laboratory concerning the implementation of the hydrogeologic work plan. At the same time, the EAG meets independently with all individual citizens, representatives of citizens groups, tribal nations, and state and federal regulatory representatives. The EAG has effectively acted as facilitator between the external parties' interest and the managers of the effort to develop the hydrogeologic work plan. During the last 2 years, many of the concerns have been mediated effectively by this process.

Numerous research studies at contaminated sites have concluded that the best way to build public trust in the selected risk-management plan is to involve the public in the selection process as early as possible rather than involving them after a plan is chosen (Hance et al. 1988; Ashford et al. 1991; English et al. 1991, 1993; Mitchell 1992; Rich et al. 1995; ATSDR 1996; NRC 1996; PCCRARM 1997; Ashford and Rest 1999; Chess and Purcell 1999). These studies support the presidential/congressional commission's framework approach that community involvement should begin with the initial discovery of contamination or adverse health effects and continue throughout the framework process. The community must be involved in defining the problems and goals for risk management, setting priorities, assessing the risks associated with the contamination, identifying and evaluating different remediation options (such as in the remedial investigation/feasibility studies process), selecting the final risk-management strategy, and evaluating the efficacy of the management strategy in meeting the goals. Early involvement helps take community members out of a reactive role and offers more meaningful engagement in discussion of options, tradeoffs, and consequences of the issues that need to be addressed (PCCRARM 1997).

Current practice under Superfund and the Resource Conservation and Recovery Act (RCRA) limits formal community involvement to commenting on key decisions already made by EPA. For example, as shown in Figure 3-3, community involvement in the Superfund process does not formally begin until after the government and the responsible parties have completed the RI/FS phase.

More recently, however, the EPA Superfund Community Involvement Program has been advocating greater involvement by the affected communities (EPA 1999b) and has developed updated guidance for community involvement during the remedial investigation and feasibility study process at Superfund sites. It must be remembered that not all PCB-contaminated-sediment sites are Superfund sites, nor is EPA necessarily the lead regulatory agency involved at the site. This factor might make the implementation of a community-involvement plan less consistent. The agency has identified a number of lessons that they have learned about involving the affected community at Superfund sites (see Box 4-4). These lessons emphasize that a more active community-involvement process leads to a risk-management strategy that is more acceptable to a broader cross-section of the community. Table 4-1 illustrates how early and active community involvement can assist in the management process. In addition, several states and other federal regulatory agencies have public participation guidance that encourages strong, active, and early community involvement in various forms of environmental management activities (McLoud et al.1999; Hance et al. 1988).

BOX 4-4 Lessons Learned About Superfund Community Involvement

The following is a summary of EPA Superfund staff comments on how community involvement has helped to clean up sites.

- Community involvement improves decision quality and increases public confidence in the project remediation plan.
- Community members can make a "significant contribution" to cleanup efforts. Ask for help from the community members.
- Agency personnel should

  1. Be persistent, visible, and sincere in interactions with community members and be available to answer questions.
  2. Make sure the public understands the remediation and cleanup process and goals and the implications of those goals (i.e., reduction of risk).
  3. Be sure the public has the knowledge to understand the technical issues and access technical documents; help provide technical assistance if necessary (e.g., Technical Assistance Grants).
  4. Involve the community proactively and early in the remediation process.
  5. Respect the values, priorities, and knowledge of the community members.
  6. Plan ahead. Take the time to learn about the citizens, their thoughts, their values, their priorities, and their concerns before you develop the plan of action. Anticipate their likely reactions to the plan.
  7. Be willing to shed your own preconceptions and to listen and learn from your critics.
  8. Share ownership of the project with the community. Share responsibility, work and credit.
  9. Partner with community leaders to engage the public.
  10. Maintain frequent, open, and honest communication. Do not try to deceive the community or keep them ignorant of plans or facts.
  11. Make sure the shared information is written or spoken in language that the community understands.
  12. Foster trust and cooperation, not antagonism, and act in good faith.
  13. Go beyond the traditional community relations approach. The minimal approach required by CERCLA and the National Environmental Policy Act does not work to engage the community or to gain their trust and cooperation. Adapt your own style and activities to the specific needs of the community.

Source: Adapted from EPA (1999a).

The committee is encouraged to see the recent emphasis on community involvement as evidenced by materials on EPA's Superfund website (EPA 1999a) and recent EPA draft and final guidance (EPA 1999b, 2000). The application of this new emphasis at various sites should, within the framework

TABLE 4-1 A Comparison Between Two Remediation Projects

Tacoma (Commencement Bay)	New York (Hudson River)
Community views might have been divergent (though apparently not exceedingly polarized) at first, but a long, open, consensus-based process with active community participation during the discussion and planning phases led to a more coherent consensus.	Divergent community views were geographically based. Upper HR – Fewer citizens appeared to support dredging, and more citizens focused on the disruptions of dredging. Lower HR – More citizens appeared to support dredging. More focus on the potential beneficial health and economic outcomes of unrestricted fish consumption.
Citizens demanded and became actively involved in the process early. The citizens obtained the necessary education and technical expertise, and provided science-based opinion to the negotiations.	Citizens were limited to providing passive feedback at EPA-defined points in the process. Feedback is permitted on nearly completed draft documents and nearly finalized decisions.
Committee Observation: The community-involvement process allowed the citizens to interact with regulators and potentially responsible parties and each other over time. Trust built up as differences were discussed and sorted out, mutual solutions were introduced, and compromises were made.	Committee Observation: The community-involvement process does not allow the citizens to interact in a fact-based environment, and maintains a divergent community with little communication between the community groups.
Distribution of contamination is relatively small and relatively discrete.	Widespread contamination is at a relatively high level with continuing leakage of PCBs into the river system.
Highest contamination concentration is moderate (25 parts per million (ppm)).	Highest contamination concentration is severe (>1,000 ppm).
Community involvement was successful.	Community involvement was unsuccessful.

recommended by the committee, result in a more effective process for risk management at PCB-contaminated sites (see Box 4-5)

Traditionally, funding for community-based groups was limited to individual contributions and local fund-raising efforts. In more recent years, community-based organizations have received funding from small local foundations, government grants, and in some instances from private companies. In addition, federal programs now provide funds through the EPA's Technical Assistance Grants (TAG) for Superfund projects and through environmental justice grants, the Technical Assistance for Public Participation Program (TAPP) for Department of Defense sites, and the Environmental Justice Partnership for Communication grants provided by the National Institute for Environmental Health Sciences.

The EPA Technical Assistance Grants program provides $50,000 to grassroots community-based groups so that they can hire their own technical advisor to help community members understand scientific and technical information. As of February 2001, more than 230 TAGs totaling more than $18 million had been issued (Lois Gartner, EPA Office of Solid Waste and Emergency Response, personal communication to Stephen Lester, Feb. 15, 2001). Although these grants have assisted many community groups, the program has limitations. One problem is that TAGs are not provided early enough in the process. In practice, they have often been awarded after the site management plan has been selected. Other limitations are that they are available only for communities with Superfund sites (Ashford et al. 1991), they cannot be used to evaluate health studies (CHEJ 1990), and only one TAG is permitted per Superfund site.

BOX 4-5 Public Participation at the St. Paul Waterway Project

The St. Paul Waterway Project in Commencement Bay, Washington State, is an example of the difference that community support can make to the timely and successful completion of a sediment management project. The Tacoma community was consulted early and openly and given a voice in shaping the cleanup strategy to be chosen for the site. They came together early in the process and agreed that capping the contamination in-water was the most cost-effective and environmentally protective solution. With community support, ranging from tribes and environmental groups to labor and industry, Simpson (the company conducting the remediation activities) was able to complete its planning and obtain the necessary construction permits and approvals for the project within less than 2 years.

The number of TAGs awarded per site is a problem that the committee believes should be addressed. Awarding only one TAG per site might work for a relatively small site, but in situations such as the Hudson River that covers hundreds of miles, it does not. At these large sites, many communities are affected, and their interests and needs might vary greatly. Selecting one group to represent all the voices along a river seems difficult if not impossible. At a minimum, consideration should be given to allowing the broader communities with a diversity of perspectives to have access to the resources they need to participate effectively in any discussions about PCB-contaminated-sediment remediation and management. Another option is to provide more than one TAG per site for sites that cover large areas and have communities with distinctly different interests.

The DOD also awards limited technical assistance funds to communities through the Technical Assistance for Public Participation (TAPP) program that began in 1998. TAPP provides $25,000 to Restoration Advisory Boards (RABs) to obtain technical assistance and enable them to provide more educated feedback about the issues related to remediation of the DOD sites of concern (CPEO 1998).

EPA also supports a Technical Outreach Services to Communities program administered by universities associated with the Hazardous Substances Research Centers (HSRCs). These university research centers are located around the country and offer communities no-cost, non-advocate technical assistance and education on the hazardous substance issues they face. Although these programs cannot serve as advocates for the community, they provide a technical resource to assist community participation in the decision-making process. Additional information can be found by contacting the HSRC program director with the EPA Office of Research and Development or by accessing www.tosc.org.

These programs are included in what EPA terms "building-capacity" programs in communities to increase a community's knowledge about technical and legal aspects of the management effort, thus making them more effective partners. Building-capacity programs may include seminars and courses about technical aspects or other aspects of a problem or possible solutions, training in the processes used (such as risk assessment), or enabling community access to technical outreach support through such mechanisms as the TAG grants. Even though a building-capacity program increases the knowledge base of the community members, it does not guarantee the community a place at the negotiating table or provide the community with any decision-making power.

Industry is another potential source of funding for community-involvement efforts (see the Asarco Tacoma Smelter case study in Appendix D).

Responsible parties can help provide technical and logistic support for community groups, if asked. Such foundations as the Hudson River Foundation may also assist communities by sponsoring educational seminars and conferences, supporting research on the water bodies, and being involved in educational outreach.

The National Institute of Environmental Health Sciences (NIEHS) funds basic scientific research in areas that may pertain to PCB-contaminated sediments and their management. Three types of centers are supported by NIEHS: (1) 21 environmental health science centers that conduct research on toxicology, epidemiology, occupational health, and prevention and community outreach; (2) five marine and freshwater biomedical sciences centers that focus on the development of alternative marine and freshwater models for toxicological research and the study of human health impacts of seafood-borne toxins; and (3) a developmental center to study health problems of underserved and underrepresented human populations (NIEHS 2001). The community outreach efforts of most of these centers are not designed to deal with the broad range of PCB issues encountered by communities and other affected parties. With time and experience, however, these centers might provide more resources.

Build a Working Relationship with the Community

In a 1996 study of community-involvement programs, the Agency for Toxic Substances and Disease Registry (ATSDR) found that to build an effective working relationship, community involvement should be viewed as a dynamic and developing relationship between community members and agencies and not as something an agency "does to a community." Community leaders interviewed as part of the ATSDR study emphasized the importance of getting to know citizens as "real people," seeking community input in designing outreach and education materials, keeping community members updated on new developments, and being forthcoming with information rather than withholding it.

Establish Trust and Respect Between All Parties

Trust and respect are the fundamental underpinnings of a nonadversarial working relationship. Although building trust and mutual respect takes time and may initially slow the decision-making process, the eventual solution is more likely to satisfy all partners and engender less resentment and protest within and from the affected community. Communities often exhibit an absence of trust in the public trustees (i.e., the regulatory agencies), and a deep mistrust of the PRPs; overcoming this lack of trust can result in a more satisfactory outcome to all parties.

Involve the Community As a Respected Partner

The community needs to be considered a respected partner in the decision-making process. That means affording them respect for their opinions and knowledge—technical, ecological, social, and historical—which may not always mesh with what the regulators and managers "know." Efforts should be made to create a partnership between all the interested and affected parties. Full partnership in the decision-making process may not always be easy to accomplish, given certain legal and logistic constraints. However, the more the affected community is able to participate in the decision-making process, the more likely they are to accept the outcome, even when the outcome includes some compromise. For example, the current practice of community involvement at most Superfund and RCRA sites brings the community in so late that the government is essentially asking the public to ratify an agency decision rather than asking them for input. As a result, the community is not given a chance to help plan the characterization of a site or evaluate all the possible remediation alternatives.

Empowerment

Successful community involvement requires empowerment of the community representatives by the community and by the other participants in the management and decision-making process. That means that the rest of the participants (the regulatory representatives and the PRPs) must recognize the community as full participants in the process and they must be given the resources to enable their participation.

Communication

Effective communication strategies are critical to the success of a project. Communication must be in a form that people can access, use, and understand. It is important to be aware of and use the languages spoken by the different members of the community; native speakers should be used as interpreters if necessary. Communication between agencies or government-sponsored entities is also important and can include agencies at different levels of government (local, state, and federal).

Commitment

Involving the community takes a commitment of time and attention on the part of the other parties, especially the regulatory agencies. Involving the public takes time, planning, preparation, and preliminary work to ensure involvement and acceptance by the community and affected public.

Process Transparency

The process must be open and transparent from the start. An open process allows a new person to come into the process at any time. A transparent process makes it clear how decisions are made, what the roles and responsibilities of the participants are, and what information is available, and how it is used in making decisions. Having a historical record that traces decisions, process, and requests from the community as well as providing the facts of the situation will help newcomers become effective participants more quickly. A certain amount of responsibility rests on a new person to learn the history and facts, although it is expected that other involved participants would help in this initiation and training process.

COMMUNITY OUTREACH AND EDUCATION

There will be times when there is little or no apparent community interest at a site and when no grassroots community organization exists. In some places, what seems to be an uninterested community may actually be a community that is not aware of the issue, unclear about the health risks posed by the problem, and unsure about the implications of the existence of the site and of the proposed management plan for themselves and their families. It takes time to develop an awareness of an issue within a community, and it takes time for the community to build activity around the site management effort into their daily schedule.

In these instances, the involved regulatory agencies should take the responsibility for reaching out to the affected community. The extent of necessary outreach and ultimately the level of community involvement, will depend on the magnitude of the contamination problem, the number of people who are affected, and the interest of community members. According to ATSDR's study of community-involvement programs, community outreach is most effective when it begins with an effort to learn as much as possible about the community, including its culture, diversity, geography, and political relationships (ATSDR 1996).

Ways to accomplish this effort include prominent placement of public notices and feature articles with meeting and contact information highlighted in the item in commonly read newspapers or local magazines. In communities with a potentially affected non-English-speaking population, these articles and notices should be placed in native language newspapers, radio, and television stations. The non-English notices and articles should be placed early and throughout the process to ensure that the non-English speaking population does not feel disenfranchised and therefore resistant to the outcome.

EPA's National Environmental Justice Advisory Committee (NEJAC), formed to advise the agency on issues of environmental justice, suggests reviewing correspondence files and media coverage about the site. NEJAC also suggests identifying key individuals who represent different interests in the community and learning as much as possible about these people and their concerns (EPA 1996). This preliminary outreach step can be accomplished by personal contact, telephone, or letters.

Educating the affected community is also a critical part of outreach. To have effective community involvement, the community needs to be educated about the technical, economic, and political aspects of the problem. Studies of community outreach programs have suggested the following guidelines for community education (EPA 1996; ATSDR 1996; CDC 1997; English et al. 1993):

- Educational materials provided to the community should be culturally sensitive, relevant, and translated when necessary.
- Materials should be readily accessible, written in a manner that is easy to understand, and timely.
- Unabridged materials should be placed in accessible repositories, such as public libraries.
- Meetings should be scheduled to make them accessible and user friendly. They should be held at times that do not conflict with work schedules, dinner hours, and other commitments at facilities that are local and convenient and that represent neutral space, taking into account child-care concerns and time and travel expenses.
- Meetings should be advertised in a timely manner in the print and electronic media, and notices should provide a phone number and/or address for people to contact about the meeting.

- Agency staff working on the outreach effort should be trained in cultural, linguistic, and community-outreach techniques.

Education encourages interest in the project, increases everyone's ability to communicate with each other, helps keep the public involved in the decision-making process, and helps all affected parties set realistic goals for the management project. Education can help the public understand not just the science, but the economics and the physical, chemical, and structural limitations of the remediation options. Education allows the public, who often have historical knowledge of the area, to provide helpful insights into the site, the weather patterns and site responses to the weather, the local environment, and what the habitat was once like and might be like again, if restoration to a similar habitat is a goal of the management project.

Once involved and informed about the project, the community partners should identify their priorities and goals for the process and educate the regulators and other partners about them. Identification of cultural or ethnic priorities or mores might affect what the interested and affected parties of the group think about the project and how they set goals for it.

When steps have been taken to educate the affected parties, ATSDR suggests using a community-guided approach to determine the appropriate level of community involvement. Under this approach, the agency works with community members to develop a community-involvement plan that meets community needs as well as agency requirements. To make this plan work, the agency must view community involvement as a central pillar of its work, not as an add-on (ATSDR 1996).

The community is not the only party that needs to be educated. The regulatory agencies and PRPs also need education. The regulatory agencies may want to increase community participation but not know how or be aware of good risk communication and facilitation techniques. The state and federal agencies should recognize the value of training personnel so that they can become and remain more aware of community needs and issues and learn better ways to work with the affected and interested community groups and individuals.

THE ROLE OF REGULATORY AGENCIES

The state and federal agencies also have a responsibility in the community-involvement process. Agencies need to ensure the existence of institutional memory within the agency. Community members may not always have the time, finances, or ability to stay with the process throughout its duration, and the agencies also change personnel, sometimes frequently. In that situation, there can be a lack of continuity in personnel and in "memory" of what has happened at the site through the lengthy negotiation or management processes, and resources are wasted as negotiations are repeated about points settled long ago. In addition, site information already obtained (including sampling and tracer information) can be forgotten and lost in the accumulation of documents. Promises made to the community can also be forgotten, as they are made orally all too often at a meeting and not written down.

Agencies (local, federal, and state) can help to ameliorate this problem by requiring a running historical record that continues outside the multitudes of legal and scientific documents that are created during any management process (i.e., beyond the administrative record). This record should be organized and readily accessible and, where possible, available via the Internet and as printed material at local libraries. Most important of all, all promises and commitments made in meetings need to be immediately incorporated into the main section of the time-line, possibly through the use of meeting minutes including action items. This listing of commitments and promises then becomes a point of reference for the communities and the project managers. The creation of such a record must be a high priority for regulatory agencies. The time-line needs to be publicly available from the start and distributed to all meeting participants as well as others who have indicated their interest. The public is thus informed and can provide corrections as needed.

MECHANISMS FOR INVOLVING THE COMMUNITY

A variety of methods are available for involving a community in an open decision-making process. These methods include public meetings and hearings, citizen advisory boards, workshops, citizen surveys, citizen juries and review panels, focus groups, alternative dispute resolution, mediation, and negotiation processes (English et al. 1993; Renn et al. 1995; NRC 1996; Ashford and Rest 1999). Although there is no agreement on the best method, two of the more common methods used are public hearings and meetings and citizen and community advisory boards. Both methods are briefly discussed below.

Public hearings and meetings are the most traditional and familiar form of public participation (Ashford and Rest, 1999). Hearings are often required by law and have been used by agencies to present information and defend their decisions. Public meetings are similar to public hearings in format, but they are not required by law. Hearings and meetings are easy to convene, open to everyone, and provide the opportunity for community members to present their views and possibly affect decisions (NRC 1996). Potential disadvantages include their tendency to occur late in the decision-making process (NRC 1996) and the possibility that they might be dominated by organized interests, the most outspoken critics in the community, and individuals most at ease with public speaking in the community (Ashford and Rest 1999). The NRC (1996) described public hearings as being best suited for presenting alternative views, information, and concerns and less useful for dealing with imbalances, creating trust, or promoting dialogue. There is rarely a decision or consensus at the end of such meetings, allowing agencies great latitude to ignore any comments and thus create further mistrust.

Community advisory groups are formed at many contaminated sites where there is high community interest. These groups are typically formed to examine one or more issues, provide ongoing advice to an agency or organization, and make recommendations on specific issues (Ashford and Rest 1999). Advisory groups with well-defined charges, adequate resources, and neutrally facilitated processes have been found to have significant policy impacts (Lynn and Busenberg 1995). The NRC (1996) found that advisory groups could overcome many of the deficiencies of public hearings and be extremely productive, if provided the "information that members want, access to appropriate agency personnel, or an independent technical advisor as needed." They have the advantage of meeting over time, allowing for more in-depth examination of issues, creating relationships, and developing mutual understanding and respect for differing views. Disadvantages are limited inclusiveness and representativeness, a high level of commitment required of members, a need for technical expertise, and questions about whether the agency will use the group's recommendations (NRC 1996; Ashford and Rest 1999).

Concerns about the make-up of advisory groups, especially those proposed by responsible parties, have been raised (Lewis et al. 1992; Renn et al. 1995). The primary concerns are that members are often hand-picked by the government or institution seeking advice and that membership often consists of all major interested parties, rather than only those who are directly affected by the contamination. Also, unempowered groups and those with views that diverge too far from those of the agency are often excluded, undermining the legitimacy and credibility of the approach (NRC 1996) and providing few opportunities to address issues outside the charge of the group. In those cases, advisory groups are perceived as vehicles for accomplishing a predetermined agenda rather than as mechanisms for facilitating true community involvement.

Several federal agencies use the advisory-group approach. EPA uses citizen advisory groups (CAGs) at Superfund sites with environmental justice concerns (EPA 1995), although the degree to which these groups can influence decision-making appears to be quite small. EPA anticipates that these CAGs "will serve primarily as a means to foster interaction among interested members of an affected community, to exchange facts and information, and to express individual views of CAG participants while attempting to provide, if possible, consensus recommendations from the CAG to EPA" (EPA 1995).

ATSDR uses citizen advisory panels in conducting health assessments at contaminated sites (ATSDR 1996). The Department of Defense (DOD) uses restoration advisory boards (RABs) in the cleanup of DOD installations (CEQ 1995). RABs are established when (1) installation closure involves the transfer of property to the community; (2) at least 50 citizens petition for an advisory board; (3) the federal, state, or local government requests formation of an advisory board; or (4) the installation determines the need for an advisory board (CEQ 1995).

Conclusions And Recommendations

Risk management of PCB-contaminated sediment sites should include early, active, and continuous involvement of all affected parties and communities as partners. Although the need for involvement of the affected communities has often been recognized, it has not been implemented on a consistent basis. Community-involvement efforts at most PCB-contaminated sediment sites have not been effective. At communities visited by the committee, the committee found an active, involved, and educated public that is eager to participate in the site evaluation and remedy selection process. Despite the presence of a concerned public at these sites, the committee found that the opportunities for public involvement generally had been limited to prescribed times in the cleanup process and that the process was dominated by the PRPs and the agencies with little communication among affected parties. Significant distrust developed at some of these sites among the regulatory agencies and between the regulatory agencies, the site owners, the affected communities, and other interested parties. This distrust led to a gridlock at many sites, with extensive delays or no forthcoming management decision. Involvement by all affected parties in the entire decision-making process may help avoid the gridlock and expedite the management process.

The framework developed by the Presidential/Congressional Commission on Risk Assessment and Risk Management established a critical role for all affected parties by integrating them into the risk management process from the initial problem-definition stage through all of the remaining stages. Community involvement will be more effective and more satisfactory to the community if the community is able to participate in or directly contribute to the decision-making process. Passive feedback about decisions already made by others is *not* what is referred to as community or stakeholder involvement. The committee's interpretation of the commission's report (PCCRARM 1997), and the committee's site visits and discussions with agency personnel, industry representatives and community members indicate that passive feedback does not appear to be an effective community-involvement strategy compared with an active participation model.

There are many benefits to community involvement including giving communities a rightful place at the table and in the process, educating the community on the issues, and providing a means for communication and collaboration. Community involvement at PCB-contaminated sediment sites should include representatives of all those who are potentially at risk due to contamination, although special attention should be given to those most at risk. People want to be treated with respect, recognized for their knowledge of their community, and considered equals in the process.

No one strategy for involving the public at PCB-contaminated sediment sites is appropriate in every case. At sites where grassroots organizations formed to address the problem already exist, these groups can provide the basis for public-involvement plans. At other sites, regulators and site owners have a responsibility to educate the affected public and determine the level of interest in participating in the decision-making process.

The affected communities need to be given the resources to effectively participate.

The committee makes the following recommendations for improving the involvement of affected parties at PCB-contaminated-sediment sites:

- Federal, and state environmental regulations and guidelines for managing contaminated sites that affect communities immediately adjacent to these sites should be changed to allow community involvement as soon as the presence of contamination is confirmed at these sites.
- The key lessons identified in the 1999 EPA report *Lessons Learned from Community Involvement at Superfund Sites* should be carefully reviewed by regulatory agencies. The report encourages a more active community role in planning and implementation of a risk-management strategy. The report emphasizes that a more active community-involvement process results in a site-management solution that is more acceptable to a broader cross-section of the community.
- EPA, state regulators, and responsible parties should ensure that interested community groups can obtain independent technical assistance to help understand the health risks and potential remedies being considered to manage PCB-contaminated-sediment sites. The availability of this assistance should be timely and provided by an objective source.
- EPA should provide more than one Technical Assistance Grant (TAG) for sites that cover large distances, as is the case with most PCB-contaminated-sediment sites, and have more than one affected community with distinctly different interests. Awarding only one TAG per site might work for

a conventional site, but it does not work for situations such as the Hudson River that covers hundreds of miles. In these situations, many communities are affected, and the interests and needs of the different communities might vary greatly.

REFERENCES

Ashford, N.A. 1988. Science and values in the regulatory process. Stat. Sci. 3(3):377-383.

Ashford, N.A,. and K.M. Rest. 1999. Public Participation in Contaminated Communities, Center for Technology, Policy, and Industrial Development, Massachusetts Institute of Technology, Cambridge, MA.

Ashford, N.A., C. Bregman, D.E. Hattis, A. Karmali, C. Schabacker, L.J. Schierow, and C. Whitbeck. 1991. Monitoring the Community for Exposure and Disease: Scientific, Legal, and Ethical Considerations. Center for Technology, Policy, and Industrial Development, Massachusetts Institute of Technology. Rep. No. U60/CCU/100929-02. Agency for Toxic Substances and Disease Registry, Atlanta, GA.

ATSDR (Agency for Toxic Substances and Disease Registry). 1996. Learning from Success: Health Agency Effort to Improve Community Involvement in Communities Affected by Hazardous Waste Sites. Through a cooperative agreement with Boston University School of Public Health and Henry S. Coles & Associates. Agency for Toxic Substances and Disease Registry, Atlanta, GA.

Brown, P., and E.J. Mikkelsen. 1997. No Safe Place: Toxic Waste, Leukemia, and Community Action. Berkeley, CA: University of California Press.

CDC (U.S. Centers for Disease Control and Prevention). 1997. Principles of Community Engagement. Committee on Community Engagement, Office of Public Health and Science, Centers for Disease Control and Prevention, Atlanta, GA.

CHEJ (Center for Health, Environment & Justice). 1990. Report on a Meeting Between ATSDR and Community Representatives, Washington. DC. June 30.

CEQ (Council of Environmental Quality). 1995. Improving Federal Facilities Cleanup: A Report of the Federal Facilities Policy Group. Council of Environmental Quality, Office of Management and Budget, Washington, DC. October.

Chess, C., and K. Purcell. 1999. Public participation and the environment: Do we know what works? Environ. Sci. Technol. 33(16):2685-2692.

CPEO (Center for Public Environmental Oversight). 1998. Report of the National Stakeholders Forum on Monitored Natural Attenuation. Center for Public Environmental Oversight, San Francisco Urban Institute, San Francisco State University. October. [Online]. Available: http://www.cpeo/org/pubs/narpt html.

Dowie, M. 1995. Losing Ground, American Environmentalism at the Close of the Twentieth Century. Cambridge, MA: The MIT Press.

Eden, S. 1996. Public participation in environmental policy: considering scientific, counter-scientific and non-scientific contributions. Public Understand. Sci. 5(3):183-204.

Edelstein, M.R. 1988. Contaminated Communities: the Social and Psychological Impacts of Residential Toxic Exposure. Boulder, CO: Westview Press.

Edelstein, M.R. 1993. When the honeymoon is over: environmental stigma and distrust in the siting of a hazardous waste disposal facility in Niagara Falls, New York. Research in Social Problems and Public Policy 5:75-95.

English, M.R., D. Counce-Brown et all. 1991. The Superfund Process: Site-Level Experience. Knoxville, TN: Waste Management Research and Education Institute.

English, M.R., A.K. Gibson, D.L. Feldman, and B.E. Tonn. 1993. Stakeholder Involvement: Open Processes for Researching Decisions About the Future Uses of Contaminated Sites. Waste Management Research and Education Institute, University of Tennessee, Knoxville. September.

EPA (U.S. Environmental Protection Agency). 1992. Community Relations in Superfund: A Handbook. EPA/540/R-92/009. OSWER 9230.0-03C. PB92-963341. Office of Solid Waste and Emergency Response, Environmental Protection Agency, Washington, DC. January.

EPA (U.S. Environmental Protection Agency). 1995. Guidance for Community Advisory Groups at Superfund Sites. EPA 540-R-94-063. OSWER-9230.0-28. Office of Emergency and Remedial Response, Environmental Protection Agency, Washington, DC. December.

EPA (U.S. Environmental Protection Agency). 1996. The Model Plan for Public Participation. EPA 300-K-96-003. Public Participation and Accountability Subcommittee, National Environmental Justice Advisory Council, Office of Environmental Justice. November.

EPA (U.S. Environmental Protection Agency). 1999a. Lessons Learned about Superfund Community Involvement, Features Partnerships at Waste Inc, Superfund Site, Michigan City, Indiana. Environmental Protection Agency, Region 5.

EPA (U.S. Environmental Protection Agency). 1999b. Risk Assessment Guidance for Superfund: Vol. 1. Human Health Evaluation Manual Supplement to Part A: Community Involvement in Superfund Risk Assessments. EPA 540-R-98-042.. OSWER 9285.7-01E-P. PB99-963303. Office of Solid Waste and Emergency Response, Washington, DC. [Online]. Available: http://www.epa.gov/oerrpage/superfund/programs/risk/ragsa/ci_ra.pdf . March.

EPA (U.S. Environmental Protection Agency.). 2000. Draft Public Involvement Policy. Fed. Regist. 65(250):82335-82345. (December 28, 2000).

Gibbs, L.M. 1998. Love Canal: The Story Continues . . . Stoney Creek, CT: New Society Pub.

Gottlieb, R. 1993. Forcing the Spring: The Transformation of the American Environmental Movement. Washington, DC: Island Press.

Fiorino, D.J. 1989. Environmental risk and democratic process: a critical review. Columbian Journal of Environmental Law 14:501-547.

Fiorino, D.J. 1990. Citizen participation and environmental risk: a survey of institutional mechanisms. Science, Technology and Human Values 15(2):226-243.

Folk, E. 1991. Public participation in the Superfund cleanup process. Ecology Law Quarterly 18(1):173-221.

Hance, B.J., C. Chess, and P.M. Sandman. 1988. Improving Dialogue with Communities: A Risk Communication Manual for Government, Environmental Communication Research Program, New Jersey Agricultural Experimental Station. Trenton, NJ: Division of Science and Research.

IOM (Institute of Medicine). 1998. Toward Environmental Justice: Research, Education, and Health Policy Needs. Washington, DC: National Academy Press.

Kasperson, R.E. 1986. Six propositions on public participation and their relevance for risk communication. Risk Anal. 6(4):275-281.

Kunreuther, H., K. Fitzgerald, and T. Aarts. 1993. Siting noxious facilities: a test of the facility siting credo. Risk Anal. 13(3):301-318.

Lewis, S.J. et al. 1992. The Good Neighbor Handbook. A Community-based Strategy for Sustainable Industry. Waverly, Mass: Good Neighbor Project .

Lynn, F.M. 1988. Citizen Involvement in Hazardous Waste Sites: Two North Carolina Success Stories. Environmental Impact Assessment and Review 7(4):347-361.

Lynn, F.M. and G.J. Busenberg. 1995. Citizen advisory committees and environmental policy: What we know, what's left to discover. Risk Anal. 15(2):147-162.

McCallum, D.B., S.L. Hammond, and V.T. Covello. 1991. Communicating about environmental risks: how the public uses and perceives information sources. Health Education Quarterly 18(3):349-361.

McLoud, S., H. Stanbrough, and C. Stern, eds. 1999. Watershed Action Guide for Indiana, IDEM Watershed Management Section. Indianapolis, IN: Indiana Department of Environmental Management (IDEM). [Online]. Available: http://www.state.in.us/idem/owm/planbr/wsm/. (Feb. 20, 2001)

Mitchell, J.V. 1992. Perception of risk and credibility at toxic sites. Risk Anal. 12(1):19-26.

NIEHS (National Institute of Environmental Health Sciences). 2001. Community Outreach and Education Program (COEP). [Online}. Available: http://www.-niehs.nih.gov/centers/coep/coepcver.htm (Last updated July 27, 2000).

NRC (National Research Council). 1992. Assessment of the U.S. Outer Continental Shelf Environmental Studies Program: III. Social and Economic Studies. Washington, DC: National Academy Press. 164 pp.

NRC (National Research Council). 1996. Understanding Risk, Informing Decisions in a Democratic Society, P.C. Stern and H.V. Fineberg, eds. Washington, DC: National Academy Press. 264 pp.

NRC (National Research Council). 2000. Natural Attenuation for Groundwater Remediation. Washington, DC: National Academy Press. 292 pp.

Nelkin, D. 1975. The political impact of technical expertise. Social Stud. Sci. 5:34-54.

O'Brien, M. 2000. Making Better Environmental Decisions, An Alternative to Risk Assessment. Cambridge, MA: The MIT Press. 286 pp.

Ozonoff, D., and L. I. Boden. 1987. Truth and consequences: health agency responses to environmental health problems. Sci. Technol. Hum. Values 12(3&4): 70-77.

PCCRARM (Presidential/Congressional Commission on Risk Assessment and Risk

Management). 1997. Framework for Environmental Health Risk Management: Final Report. Washington, DC: The Commission.

Renn, O., T. Webler, and P. Wiedemann. 1995. A need for discourse on citizen participation: Objectives and structure of the book. Pp. 1-16 in Fairness and Competence in Citizen Participation: Evaluating Models for Environmental Discourse. Dordrecht: Kluwer Academic.

Rich, R.C., M. Edelstein, W.K. Hallman, and A.H. Wandersman. 1995. Citizen participation and empowerment: the case of local environmental hazards. Am. J. Community Psychol. 23(5):657-676.

Sexton, K., K. Olden, and B. Johnson. 1993. "Environmental Justice": the central role of research in establishing a credible scientific foundation for informed decision making. Toxicol. Ind. Health 9(5):685-728.

Unger, D.G., A. Wandersman, and W. Hallman. 1992. Living near a hazardous waste facility: Coping with individual and family distress. Am. J. Orthopsychiatry 62(1):55-70.

Yosie, T.F., and T.D. Herbst. 1998. Using Stakeholder Processes in Environmental Decision-Making. An Evaluation of Lessons Learned, Key Issues, and Future Challenges. Washington, DC: Ruder Finn.

5

Defining the Problem and Setting Management Goals

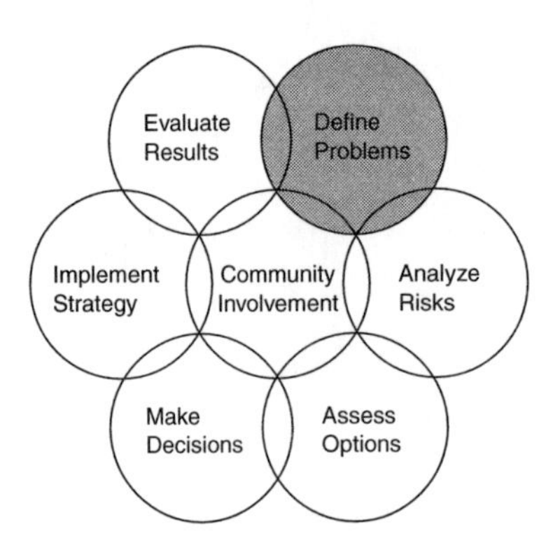

Applying the Presidential/Congressional Commission on Risk Assessment and Risk Management's framework to manage risks at a PCB-contaminated site begins with defining the problems associated with PCBs in the sediments and putting them into their public-health, environmental, socioeconomic, and cultural contexts (PCCRARM 1997). The presence of PCBs in the sediments of a water body can cause a number of problems for the human and ecological communities that live near or in the water. Depending on the specific site, in addition to possible adverse human health effects, PCB contamination might result in a variety of problems, such as fish consumption advisories, limitations on navigable dredging, and adverse effects in wildlife. If the problems associated with the PCB-contaminated sediments are not identified appropriately at the outset of the management process, subsequent risk-assessment and management activities might be conducted, but the risks might not be wholly addressed or they might be transferred to a different community or ecosystem. Consequently, the risk-management process should begin by clearly defining the problems thought to be associated with the contamination at a particular site. Each site will be different in terms of the human health, ecological, cultural, economic, and social contexts in which the PCB problem occurs and needs to be managed (Box 5-1).

This chapter presents an approach that can be used at a PCB-contaminated site to recognize and define the problems associated with the contaminated sediments, determine the extent of the contamination and its sources, set risk-management goals to address the human health and environmental problems that affect the communities, assess other concerns, such as loss of income or cultural traditions, arising from the PCB contamination, and set priorities for risk management (Box 5-2). Some of the potential obstacles to implementing a risk-management strategy at the site are also briefly examined.

BOX 5-1

Risk management is the process of identifying, evaluating, selecting, and implementing actions to reduce risk to human health and ecosystems. The goal of risk management is to take scientifically sound, cost-effective, integrated actions that reduce or prevent risks while taking into account social, cultural, ethical, political, and legal considerations (PCCRARM 1997).

DEFINING THE PROBLEM

The committee prepared this report based on the assumption that the sites at which this risk-management framework would be applied have already been identified as having PCB contamination. PCB contamination can be identified by a variety of mechanisms, such as fish sampling, national pollution dis-

BOX 5-2 Defining the Problem

What are the problems caused by the PCB-contaminated sediments?

Are there other potential sources of the health or ecological problem?

Who is affected by the problems?

Who might be affected by the risk-management strategy?

How big is the problem?

What are the sources of the contamination?

Are there other chemicals of concern?

What are the risk-management goals?

What are the obstacles to risk management?

charge elimination system (NPDES) permitting, bulk sediment analysis required as part of dredging or construction of shoreline structures, or monitoring for Clean Water Act section 303(d) to identify affected water bodies. In some instances, a site with known PCB contamination might become a concern if the criteria for designating sediments as "contaminated" or "hazardous" are lowered, and the contaminant levels at the site now exceed those criteria. Also, PCBs might not be the major cause of the immediate problems. For example, a decline in a fish population might not be the result of PCB contamination but rather overfishing. Such a possible cause should be included in the initial list of problems even if subsequent risk analysis indicates that overfishing is not the underlying cause of the decline in fish. At newly identified PCB-contaminated sites, the committee expects that the framework can be more easily applied when the problems associated with the contamination have not been determined yet and the differences among the affected parties are less pronounced. At newly identified sites, the extent of the contamination and the presence of other chemicals might not have been determined. One of the first steps in assessing the problems at a new site is conducting a preliminary site assessment to determine the extent of the contamination, the presence of any confounding factors, such as other contaminants, and sources of contamination.

At many sites, contamination has been identified and work is under way to characterize the site or develop or implement a management strategy. At these ongoing sites (i.e., sites where the management process has begun), the general problems associated with the presence of PCBs in the sediments might have been identified, although their extent and severity might be disputed by some of the affected parties. Some of the problems that might be identified by the affected parties include PCB levels in fish or wildlife that have resulted in fish or wildlife consumption advisories, prohibitions against swimming, and lack of navigational dredging because of restrictions on dredged material disposal. Table 5-1a,b presents an overview of some of the information that has been collected at selected PCB-contaminated-sediment sites during the management process, including some basic site characteristics such as site size, the management goals, the levels of PCBs that have been found at the site, the presence of co-contaminants, the selected management option chosen and in some cases implemented at the site, estimated costs of the management action, and the source of the contamination.

Once the affected parties have been identified (see Chapter 4), their knowledge, understanding, and concerns about the site and the problems associated with the PCB contamination should be canvassed. This information should be stated in general terms and might focus on both immediate concerns (e.g., not eating fish) or long-range concerns (e.g., restoration of a habitat) as well as problems with the site in its present condition or problems that are anticipated to arise during or subsequent to its management, regardless of the risk-management strategy chosen. Furthermore, it is incumbent upon the regulators to explain to the other affected parties, the regulatory ramifications of the contamination and any regulatory processes that must, by law, be applied at the site. For example, if the site is on the National Priorities List (i.e., it is a Superfund site), the EPA RI/FS process must be applied (EPA 1999a).

In the sections below, the committee identifies and provides examples of the problems and impacts that stem not only from the presence of PCBs in the sediments at a site but also those that might arise during and after risk management. The problems and impacts include not only public-health and ecological concerns but also social, cultural, and economic impacts. Not every problem is applicable to every site and risk-management strategy; furthermore, each site will have problems that cannot be anticipated. Table 5-1a,b provides an overview of selected PCB-contaminated-sediments sites. From this table, it is evident that the motivation behind the management at the majority of these sites stems from human health or ecological concerns. For instance, at Commencement Bay, the risk-management goals were the reduction of contaminant concentrations in sediments to concentrations that would support a healthy marine environment and protect the health of people eating seafood from the bay.

Public-Health Problems

Food chains are the most likely source of exposure for higher trophic organisms, particularly via fish consumption. As a result, consumption of contaminated fish might be perceived as the major problem on a river by the affected parties, particularly by subsistence fishermen or those for whom fishing is a cultural value. In addition to fish, consumption advisories have also been issued for other wildlife. The Massachusetts Department of Public Health issued a provisional waterfowl consumption advisory in August 1999 on the basis of PCB concentrations found in wood ducks and mallards collected near the Housatonic River (EPA 1999c). The U.S. Food and Drug Administration (FDA) limit for PCBs in poultry is 3 ppm, and the ducks had average concentrations of 7.1 ppm in breast tissue (EPA 1999c). In general, PCB concentrations in sediments have not adversely affected activities like boating because of the lack of direct contact between boaters and PCBs.

TABLE 5-1a Overview of Selected Sites with PCB-Contaminated Sediments

Name of Site	Size of Site (Area of Concern)	Site Characteristics	PCB Levels	Other Major Contaminants
Commencement Bay, WA (Projects 1, 2, and 3)	Project 1: 2.5 miles; Project 2: 3,000 ft wide by 750 ft long (Sitcum Waterway) and (5,200 ft long by 300 ft wide (Milwaukee Waterway); Project 3: 1.2 miles long by 600 ft wide (GE 1999)	Intertidal areas, 7 inland waterways (GE 1999)	Maximum PCB concentration at Superfund site, 25 ppm (EPA 1999b)	Arsenic, cadmium, copper, lead, zinc, PAHs, beryllium, chromium, mercury, nickel selenium, silver, zinc, anilines (GE 1999)
Fox River, WI (Projects 1, 2, and 3)	Project 1: 9 acres; Project 2: 3 acres; Project 3: 39 miles (GE 1999)	Impounded by 12 dams and 17 locks (GE 1999)	35 soft sediment deposits from Lake Winnebago to DePere (32 miles) with an estimated PCB mass of 4,200 kg (4.63 tons); dispersed sediment deposition in the lower 7 miles with estimated PCB mass of 26,500 kg (29.2 tons) (GE 1999)	Metals (mercury), PAHs (GE 1999)
General Motors Central Foundry (Messena, NY)	27 acres (GE 1999)	Nearshore area; St. Lawrence River, Raquette River, Turtle Creek (GE 1999)	Raquette River (flows along the southern boundary of the site), maximum PCB levels of 390 ppm (EPA 1999b)	Mainly PCBs (GE 1999)

Housatonic River, MA (Projects 1 and 2)	Project 1: 500 ft; Project 2: 0.5 mile reach (GE 1999)	Project 1: hot spot: Upper Housatonic River; Project 2: Housatonic River starting in Pittsfield, MA (GE 1999)	Sediment volume, 6,000 yd^3 (Scenic Hudson 2000)	Mainly PCBs (GE 1999)
Hudson River, NY	43 miles (GE 1999)	17 major tributaries drain 13,365 square miles of land located in eastern NY, VT, MA, and CT. The lower river is a tidal estuary (EPA 1999b)	Approximately 1 million lb dispersed throughout entire Hudson River system south of Fort Edward; after dredging, 498,000 to 656,000 lb remain in river (EPA 1999b)	Mainly PCBs (EPA 1999b)
Lake Hartwell, SC	730 acre - Sangamo Weston/ Twelve Mile Creek/ Lake Hartwell (EPA 1999e)	12-Mile Creek, Lake Hartwell (GE 1999)	Between 1955 and 1977, 400,000 lb PCBs discharged into Town Creek, a major tributary of Lake Hartwell (EPA 1999e)	VOCS, PCE, TCE, other organics (EPA 1999b)
Manistique Harbor, MI	Drains 1,450 square miles; area of concern, 1.7 miles (Fox River Group 2000)	Manistique Harbor, Manistique Bay, Lake Michigan (GE 1999)	In 1999, concentrations of 990 ppm and 620 ppm in the dredging areas of the Harbor and the North Bay (Fox River Group 2000)	Mainly PCBs (GE 1999)
New Bedford Harbor, MA (Projects 1 and 2)	18,000 acres (EPA 1999d)	Tidal estuary/harbor/bay (EPA 1999d)	In 1989, hot spot 5 acre area containing 45% of PCBs; concentrations ranging from 4,000 ppm to over 200,000 ppm (EPA 1992)	Heavy metals with concentrations from below detection to 4,000 ppm (EPA 1992)

TABLE 5-1a *(Continued)*

Name of Site	Size of Site (Area of Concern)	Site Characteristics	PCB Levels	Other Major Contaminants
Palos Verdes, CA	17-27 square miles (GE 1999)	Estuary (GE 1999)	Highest levels reported in the 3 square miles of sediments on the Palos Verdes Shelf (GE 1999)	DDT (GE 1999)

TABLE 5-1b Overview of Selected Sites with PCB-Contaminated Sediments

Name of Site	Selected Remediation Action	Remediation Goals	Cost of Remediation	Source of Contamination
Commencement Bay, WA (Projects 1, 2, and 3)	Dredging, excavation, and treatment of all contaminated soils (GE 1999)	Goal is reduction of contaminant concentrations in sediments to levels that will support a healthy marine environment and will protect the health of people eating seafood from the bay; Sediment Quality Objectives (SQOs) used to achieve goal; goal established to allow a diverse range of uses in the bay, including industrial, commercial, navigation, fisheries, and recreation (GE 1999)	Project 1: $39.1 million; Project 3: $36.9 million (GE 1999)	Storm water runoff, contaminated groundwater, and other industrial operations (Asarco Tacoma Smelter) (GE 1999)

Fox River, WI (Projects 1, 2, and 3)	Hydraulic dredging (GE 1999)	Demonstration dredging project, with ultimate target of 1 ppm for PCBs (GE 1999)	Project 1: $2.5 million for removal, dewatering, and water treatment of 55,000 yd^3; Project 2: $4 million (GE 1999)	10 or more paper mills discharged during production of carbonless copy paper, recycling of wastepaper, in addition to other sources (GE 1999)
General Motors Central Foundry (Messena, NY)	Hydraulic dredging, wet excavation, capping (GE 1999)	Reduce PCB levels in fish (GE 2000)	$10 million total (GE 2000)	General Motors Powertrain facility emitted PCBs in the hydraulic fluids used in diecasting machinery. Evidence of historical discharges from an outfall pipe (GE 1999)
Housatonic River, MA (Projects 1 and 2)	Dry/wet excavation; commercial landfill (GE 1999)	Some removal prior to PCB sampling, with additional removal based on agency field decisions regarding (1) whether remaining PCB levels pose an imminent hazard, (2)whether removal is technically practicable, and (3) whether additional removal will contribute to a long-term remedy; removal to at least 5-foot depth with potential for 8-foot based on the three criteria; unofficially, 1 ppm residual PCBs was the target (GE 1999)	$4.5 million total (GE 2000)	Implosion at General Electric transformer manufacturing facility (GE 1999)

TABLE 5-1b *(Continued)*

Name of Site	Selected Remediation Action	Remediation Goals	Cost of Remediation	Source of Contamination
Hudson River, NY	In-place containment, reinforcement of banks (EPA 1999b)	Not identified	Estimated cost of full-scale hot spot dredging $34,000,000 and $55,000,000 (EPA 1999b)	General Electric (GE) discharged PCBs into the river from 1947 until 1977 (GE 1999)
Lake Hartwell, SC	Extraction and treatment by air stripping and/or carbon adsorption and contaminated groundwater; discharge of treated water; excavation of materials contaminated with >1 ppm of PCBs (EPA 1999b)	Reduce PCB levels in fish to <2 ppm FDA limit by natural recovery (GE 2000)	Thermal separation, $47,900,000-63,300,000 (EPA 1999b)	Site historically used for capacitor manufacturing involving dielectric fluids; waste disposal process included land-burial of off-specification capacitors and wastewater treatment sludge at site and satellite disposal facilities (EPA 1999e)

Manistique Harbor, MI	Diver-assisted hydraulic dredging, commercial landfill (Fox River Group 2000)	Reduce PCB in fish levels, reduce carcinogenic and noncarcinogenic risks to $<10^{-4}$ and <1, respectively; exceptions for high-end subsistence and some high-end recreational exposure due to fish consumption (GE 2000)	$10.2 million (1994) plus additional $4.24 million due to 6/7/95 Action Memorandum (GE 1999)	Numerous sources, including Edison Sault Electric, Warshawsky Brothers Iron and Metal, Manistique Papers, Inc. (GE 1999)
New Bedford Harbor, MA (Projects 1 and 2)	Dredging, development of confined disposal facilities (CDFs) (EPA 1999d)	Remove PCB mass at an optimal residual concentration-to-volume removed ratio, and reduce PCB flux to the water column (GE 2000)	Project 1: Record of Decision cost estimate $12.2 million; Project 2: $120 to $130 million (GE 1999)	Factories located along the Acushnet River discharged industrial processed wastes from the 1940s to the late 1970s (GE 1999)
Palos Verdes, CA	Natural attenuation proposed for biodegradation of DDT, based on 1998 University of Michigan study, capping (GE 1999)	Not identified	Not identified	Transport of DDT through the Los Angeles sanitary sewer system; PCBs discharged from a Westinghouse Electric facility (or facilities) (EPA 1992)

Ecological Problems

The presence of PCBs in sediments can also have impacts on wildlife populations, particularly threatened and endangered species that are already considered to be at risk. For example, declines in mink populations in the southeastern United States have been attributed to PCB contamination (Osowski et al. 1995).

Although increases in PCB concentrations in fish are generally considered to pose a human health risk to people that consume them, PCB exposure might also result in reproductive impairments in freshwater fish and in other aquatic organisms. Reduced numbers and species of aquatic organisms have been observed below seep areas in a PCB-contaminated marsh at Sullivan's Ledge in New Bedford, Massachusetts (EPA 1999b).

Social Impacts

The presence of contaminated sediments at a site and their management are not without social impacts on a community. The presence of contaminated sediments in a community might prevent the use of the water body for recreational activities, such as boating or swimming. Playgrounds near the Housatonic River in Massachusetts required the removal of contaminated soils before children could use them.

The affected community might perceive that their drinking water is contaminated or unsafe that their community is unhealthy as a result of the contamination. They also might be more fearful for their own health and that of their children.

The presence of contamination might also lead to divisiveness among neighbors, some believing that the contamination is of concern and others believing that it is not. Divisiveness might also occur between those who are concerned about the effects of the contamination on wildlife and those who do not consider the ecosystem to be at risk.

Because some management efforts at PCB-sediment sites involve the use of heavy construction equipment over long periods of time (many years in some cases), the community might be subjected to increased noise pollution caused by the remediation (e.g., dredging) and disrupted by the trucks hauling sediments and the digging of holding ponds. Increased air pollution can result from a greater number of vehicles, road dust, and site operations, and new roads or other facilities might have to be built causing further disruption to the community. There might be fears of contamination leaking from holding and dewatering areas, spills from the trucks that transport the sediment, or leaks from landfills.

Cultural Impacts

The most significant impacts of PCB-contaminated sediments and risk-management efforts on culture occur in those populations whose identity is linked in some way to fishing. For American Indian and other groups for whom fishing is an integral cultural value, the main concerns include the health of families, impacts on the health of aquatic and terrestrial wildlife, and severe disruptions in their way of life. For populations who are not American Indian, impacts are most likely to be felt by communities and individuals whose livelihood and way of life revolve around commercial fishing and consumption of seafood. For American Indian populations, the bans on consumption of PCB-contaminated fish, turtles, and shellfish have produced serious disruptions in the normal pattern of life on some reservations (Carpenter 1995). American Indians who once were able to subsist at least in part by fishing have been forced to obtain food from sources outside their communities. Cultural practices that involve fishing traditions over generations in some cases can no longer be practiced.

The nearness of PCB-contaminated sites to Indian lands and landfill disposal of PCB sediments near tribal lands have had a disruptive effect on the way of life for some American Indians. The inability of those Indian nations to have a decision-making role with regard to PCB management has left some with a sense of alienation and inability to control an important aspect of their future (Akwesasne Research Advisory Committee 1999).

Other less obvious impacts on cultural practices include possible avoidance of breast-feeding in populations exposed to PCB-contaminated fish. In the United States, human breast milk from individuals in the general population commonly contains PCBs. At least a quarter of samples tested had PCBs concentrations at or above 1.5 ppm, the FDA action level for food-grade commercial milk (21 CFR Section 109.30). Health-care workers faced with counseling women about breast-feeding have a difficult dilemma, because the impact of avoiding breast-feeding might be more detrimental to infant health than the possible negative impacts due to PCB exposure. Women who have a cultural, social, or economic inclination toward breast-feeding might find recommendations to avoid the practice unacceptable.

Economic Impacts

The economic implications of PCB contamination and subsequent risk-management efforts can affect commercial activities in several ways. The presence of PCB-contaminated sediments might have impacts on nearby and even distance communities. Individuals and businesses whose income is dependent on commercial fishing might feel other direct economic impacts. For example, PCB contamination of sediments in Buzzards Bay, Massachusetts, has disrupted commercial fishing in the area (Farrington et al. 1985). In addition, PCB contamination of sediments can be a costly problem for maintenance of navigational channels and development of ports and harbors. On a microeconomic level, fishing bans that restrict or prevent consumption of fish and shellfish from contaminated waters can have a financial impact on families who depend on subsistence fishing to provide a portion of their diet.

PCB contamination and subsequent management activities can affect commercial, recreational and subsistence fisheries. In New Bedford Harbor, Massachusetts, total PCB-related losses to the commercial lobster fishery alone were estimated at $2.0 million (McConnell and Morrison 1986), and the loss to recreational fishing was estimated to be approximately $3.1 million (McConnell 1986). Those figures might understate losses, as they include only the cost of traveling to more distant sites to avoid contamination. Other costs, such as any potential reduction in catch, lost recreation days, or any impacts associated with fishers going out of business, are not included in these estimates.

If the water body was navigable prior to contamination, costs associated with continued navigational dredging can increase as a result of disposal of contaminated sediments. Costs include dumping fees, transportation (frequently out of state) and other activities, such as dewatering. Concerns about remobilization of contaminants or the high costs of handling contaminated sediments can restrict maintenance dredging (NRC 1997). Lack of maintenance dredging might result in restrictions on vessel traffic, delays in cargo handling and shipment, and the likelihood of accidents. The resultant economic loss in shipping could be substantial at a national level. For example, in 1999, over $670 billion in imports and exports traveled through U.S. ports (DOT 2000). A loss that amounts to even a small fraction of the annual value of cargo transportation could imply large losses over time for the nation as a whole.[1]

PCB contamination can result in recreational losses. The total loss to recreational beach users in the New Bedford Harbor area was estimated to be between $8.3 and $11.4 million (McConnell 1986). Again, that amount includes only the losses associated with selecting alternative sites that are less desirable and more distant, thereby requiring additional travel costs. Other impacts, such as a reduction in the number of recreation days, are not included in these estimates.

---

[1]The loss excludes both the transport of domestic commercial products and all noncommercial transportation.

If the contamination has resulted in a change in diet from a fish-based protein source to a less healthful diet, there might be health impacts, and thus economic costs, associated with the dietary changes such as increased risk for heart disease.

PCB contamination can also result in reduction in property values for home owners living adjacent to contaminated waters and for those living near sediment disposal sites. Mendelsohn (1986) estimates that PCB contamination in New Bedford Harbor resulted in losses totaling $27.3 to $39.7 million to nearby home owners. These and other impacts can result in large economic losses to the nation as a whole.[2] Concerns regarding reduced property values have been expressed by farmers who also face potentially reduced demand for their agricultural products because of consumer fears of contamination (Borden 1999).

In addition, unforseen problems associated with contamination might occur. The committee learned from one meeting participant that he was unable to dispose of the zebra mussels that infested his water front because they contained such high concentrations of PCBs and were considered hazardous waste.

There are also economic impacts associated with any risk-management strategy. One significant impact is on potentially responsible parties who face having to finance management activities. The large amounts of sediments that might have to be managed at a site means that the financial impact on responsible parties might be extremely high. When responsible parties are no longer capable of paying for remediation efforts, the costs for cleanups might be borne by local, state, or federal governments.

In addition to the actual cost of the strategy itself, the strategies can have impacts on the broader community. For example, dredging can have economic impacts on the community beyond the cost of the dredge and crew. Increased costs can be incurred for road maintenance as a result of the transport of the sediments to hazardous waste landfills and for obtaining access to the contaminated areas if the adjacent shorelines are privately owned. Some businesses might be adversely affected by management of the sediments if the implementation of the strategy curtails access to the river (e.g., use of silt curtains) or prevents use of a navigable channel in the water because of the placement of a cap or location of a dredge. Management activities might have

---

[2]Note that total losses cannot be determined simply by adding these numbers. Additional categories of losses might be associated with health or other effects. Furthermore, these numbers might include common elements resulting in double counting of some damages. For example, some of the losses in housing values might reflect the loss of nearby opportunities for recreational amenities.

a significant economic impact on marinas and other businesses that rely on recreational or commercial use of the waterway.

Other costs associated with risk-management strategies include the possible need to site a hazardous waste landfill in the area and the cost of monitoring the site before, during, and after the risk-management strategy has been implemented.

SITE ASSESSMENT

For sites with known or suspected PCB contamination, existing information should be reviewed as part of a preliminary assessment to help identify present or historical sources of contamination, the spatial extent of contamination, approximate levels of contamination (e.g., in sediments and/or fish), evidence of possible effects to humans and the environment, and the existence of co-contaminants at the site. Although the available information oftentimes provides an incomplete characterization of the site, this preliminary assessment provides a useful starting point for the risk-management process, particularly in identifying affected parties and in assisting the affected parties to prioritize the problems and set general risk-management goals. For example, a larger site might pose more concerns, as more communities, both human and ecological, can be affected by the contamination and therefore might have more varied perceptions of the problems at a site. A small site might have fewer problems associated with it and only a few communities that are affected.

Sources of PCBs

In a given region, historical uses of PCBs (e.g., in the manufacture of electrical capacitors, carbonless copy paper, etc.) have often served as good indicators of potential PCB contamination. In many instances, the manufacture and use of PCBs have resulted in their release to the environment through permitted discharges or inadvertent release with subsequent sediment contamination. For example, leaking transformers may release PCBs to nearby water bodies.

Although virtually all uses of PCBs in the United States were banned in 1977, PCBs continue to be released at many sites and are adding directly to PCB inventories in contaminated sediments and to elevated PCB concentrations in fish. Existing sources of PCBs might include leaching of PCB oils and contaminated groundwater from industrial sites, erosion of PCB contaminated soils from the watershed, and bank erosion. In addition, low concentrations of PCBs from high-volume discharges such as municipal wastewater treatment plant discharges, combined-sewer overflows, and storm drains might also serve as significant sources of PCBs to aquatic systems. Because of the dispersal of PCBs in the atmosphere, atmospheric deposition of PCBs by wet and dry deposition and gaseous exchange might also be important at certain sites.

The distinction between historical and current sources of PCBs can have a major impact on defining the extent of the contamination problems and setting risk-management goals. For example, remediation efforts at the General Electric Plant on the Hudson River at Fort Edward, New York, were complicated when PCBs were found to have entered the bedrock along the river and, in spite of the removal of tons of contaminated sediments, seepage from the bedrock continues to release many pounds of PCB into the river. Efforts to remove the PCBs from the bedrock are still in progress.

Characterization of Present Contamination

Sediment problems associated with PCBs occur at a variety of different types of site, for example, nearly dry sites to deep ocean, rural to metropolitan areas, and highly commercial to subsistence living conditions (see Table 5-1 for a description of selected sites). At a given site, horizontal and vertical variations in PCB concentrations are common and are dependent on the history of PCB loadings and on the temporal and spatial deposition patterns of fine- and coarse-grained sediments. For example, at the Raisin River in Michigan, PCB sediment surface concentration ranged from 11 to 28,000 ppm and subsurface concentrations ranged from 0.78 to 29,000 ppm prior to remediation (GE 2000). At the Reynolds Metals Company Superfund site in Massena, New York, on the St. Lawrence River, PCBs were detected at concentrations up to 690 ppm.

In addition to PCB-sediment-sampling results, fish monitoring data can provide further information on PCB-contamination levels. For preliminary assessment, PCB concentrations in fish might serve as a more appropriate indicator of contamination since they provide a more direct measure of PCB effects on wildlife and of PCB exposures to humans consuming contaminated fish. EPA has developed national guidance for states to determine whether humans are at risk and what the local fish consumption advisories should be (EPA 2000).

Preliminary review of existing information can also be used to assess the geographical extent of the problem and set risk-management goals. In a situation such as the Hudson River or Commencement Bay, an appreciation of the extent of the contaminated site will help determine who should participate in setting the risk-management goals. Communities and industries above or below the actual site of major contamination might need to be involved if they are directly or indirectly affected by the contamination, or if they contribute to the problem for others.

Co-contaminants

The presence of co-contaminants at a site complicates the problem definition. Depending on the initial problem at the site (e.g., fish advisories or recontamination of a flood plain) and the co-contaminants at a particular site, the nature of the problem can change substantially. For example, if a site contains mercury and PCBs, as do some sites around the Great Lakes, fish advisories might reference both chemicals. However, as is evident from some recent controversy regarding the consumption of fish contaminated with PCBs and mercury by pregnant women, it might be difficult to determine whether the developmental impairments seen in their children result from exposure to the PCBs, mercury, or both chemicals, since both chemicals have been implicated as developmental toxicants (Renner 2000).

The majority of PCB-contaminated sites contain other contaminants, such as polycyclic aromatic hydrocarbons (PAHs), dichlorodiphenyltrichloroethane (DDT), and heavy metals, such as mercury. The presence of co-contaminants might have a significant impact on the options for appropriate risk management, the management outcomes, and on whether a waterway is "clean" after remediation. Although cleanup at a PCB site might also reduce contamination from other materials, cleanup of all PCB sites will not necessarily render the fish edible in all parts of a watershed, because more sites are contaminated with DDT rather than PCBs. At some sediment sites that are primarily contaminated with heavy metals, PCBs are a secondary contaminant.

SETTING RISK-MANAGEMENT GOALS

Once the general problems have been identified for a particular site and agreed upon by the affected parties, the next step is to set priorities and determine the risk-management goals (Box 5-3). At most sites, human health concerns are the first priority, as is reflected in the number of fish consumption advisories. At some sites, wildlife might be a second priority, particularly if the wildlife are consumed by humans; at other sites, economic issues might

BOX 5-3 Possible Risk-Management Goals

- Minimize or eliminate exposure to PCBs, site co-contaminants, and remediation-related byproducts.
- Reduce or eliminate exposure and adverse effects due to long-term storage of PCB-contaminated sediments.
- Reduce or eliminate impacts on American Indian culture and life style.
- Mitigate habitat impacts.
- Prevent economic hardship to the community.
- Maintain or improve property values.
- Remove contaminated sediments.

be most important. Loss of real estate values might be a concern if dredged sediments are to be disposed of on nearby land. Because there can be so many priorities, it is important that the affected parties be included in the priority-setting phase and that their values be reflected in the risk-management goals that are set.

The goals might require revision as the risk-management strategy progresses, the cost and effectiveness of the strategy becomes more apparent, and trade-offs come into play. The risk-management goals should be identified early in the framework to help guide the next stage in the framework—analyzing risks. These goals should initially be defined in general (nonquantifiable) terms and relate back to the concerns and problems identified during the information-gathering stage of this process. It should be emphasized that the problems might also need to be revised as the risk assessment indicates the severity of the effects of the PCBs to human health, wildlife, or other concerns. Some possible goals might be one or more of the following: being able to eat fish from the water body, preventing recontamination of the river bank or flood plain, and being able to conduct navigational dredging. The appropriate goals depend in part upon the judgment of the affected parties, who must be involved in establishing the risk-management goals. For instance, reducing concentrations in sediments or biota might be appropriate if it is believed that conditions are likely to remain stable. However, if it is generally believed that eventual remobilization of all buried materials is inevitable, an appropriate goal might be mass removal.

Knowing what the ecosystem is like now and what the affected parties want the ecosystem to be like after implementation of the risk-management strategy can help all parties develop a mutually acceptable vision for the site.

It is important to recognize throughout the process that the system's ability to recover naturally should be measured against both in situ and ex situ remediation technologies to ascertain the benefits of such measures.

Placing a number of constraints on achieving the management goals (e.g., specific concentrations to be achieved in fish tissue) might lead to reduced flexibility in controlling risks, thereby limiting the extent to which risks can be reduced. At this point in the framework, the actual risks at the site are unknown. The stated goals must be flexible to allow for feedback from the risk-assessment process and subsequent analysis of the management options so that, if necessary, affected parties can redefine the management goals. Risk-management goals should also be realistic. All risk-management strategies will require some level of financial investment; a degree of willingness by all affected parties to be educated about the problems; an appreciation that the process is lengthy, detailed, and at times tedious; and revisiting various stages of the framework more than once. It might be unrealistic for a community to hold out for a goal of restoring a reach of the river to preindustrial conditions, for a potentially responsible party to expect to avoid paying some or all of the costs for a problem they caused, or for a regulatory agency to expect all affected parties to immediately appreciate the legal constraints under which it operates.

During the risk-management process, it might also become evident that the goals will need to be adjusted if it appears that the risk-management strategy is not achieving the initial goals. The technical options chosen might be inadequate to deal with the contamination problem, and other options will have to be explored. Furthermore, site specific processes can be very disruptive to the community, causing problems that were not initially obvious. For example, if the risk assessment indicates that the extent of the PCB contamination is greater than originally thought, the amount of sediment to be removed, if dredging is the chosen management option, can result in the treatment and transport of greater volumes of sediment, an increased number of trucks to transport the sediment, and a greater level of noise and activity for the neighboring community. Because the sediments will have to be disposed of at an appropriate site, the community near the disposal site might be concerned about the increased volume of sediment and truck traffic. That community cannot be neglected in developing the risk-management goals for a site. They will also have long-term problems and will be asked to commit to dealing with the contamination and the associated risks as well.

As the framework process proceeds, it will be necessary to establish further goals, possibly of a more technical nature, and modify the general goals identified in this initial framework stage. A detailed goal that can be developed later in the framework is, for example, the level of residual PCBs that will be acceptable (if at all) after remediation.

One approach to ensure that the risk-management process is fair and understandable to all affected parties is to develop a decision plan before assessments begin. In these plans, a series of decision points are described in a decision tree. In some cases, the thresholds and statistical methods to be applied to data are agreed to before the risk assessment is begun.

If the science is not used in an objective and ethical manner, the entire process can be subverted. It is not appropriate for any interested party to slant the results of scientific investigations to meet political, economic, or social needs. It is absolutely critical that scientists keep the results of their studies free of bias, regardless of their affiliations.

CONCLUSIONS AND RECOMMENDATIONS

The committee found that the impacts of PCB-contaminated sediments extend beyond traditional human health and ecological risks considered by EPA and other regulatory agencies. The committee emphasizes that societal, cultural, and economic impacts should also be considered when developing risk-management goals for the contaminated-sediment sites. All affected parties should be involved in setting these goals. Among the impacts that might affect communities are restrictions on commercial and recreational fishing that can impact local communities, such as occurred in New Bedford Harbor where PCB-contaminated sediments resulted in economic losses to the commercial lobster fishery. Cultural impacts can result when subsistence use of a resource is lost, affecting such traditions as sharing among the community or passing on indigenous knowledge to younger generations, as occurred among the Mohawk Community of Akwesasne on the St. Lawrence River. Marine transportation can be affected by restrictions on dredging due to the need to handle contaminated sediments.

Preliminary site assessments should be used to identify present and historical sources of PCB releases. The spatial extent of the contamination, the concentrations of PCBs in the sediments, possible effects on humans and the environment, and the presence of co-contaminants should be considered when determining the problems at the site and possible risk-management goals.

REFERENCES

Akwesasne Research Advisory Committee. 1999. Superfund Clean-up Akwesasne: A Case Study in Environmental Justice. Akwesasne Task Force on the Environment Research Advisory Committee, Hogansburg, NY.

Borden T.A. 1999. Testimony to NRC on PCB Contaminated Sediments. Washington County Farm Bureau, NY. November 8.

Carpenter, D.O. 1995. Communicating with the public on issues of science and public health. Environ. Health Perspect. 103(Suppl.6):127-130.

DOT (U.S. Department of Transportation). 2000. U.S. Foreign Waterborne Transportation Statistics. Office of Statistical & Economic Analysis, U.S. Maritime Administration, U.S. Department of Transportation. [Online]. Available: http://www.marad.dot.gov/statistics/usfwts/pr_final1999.html [July 2, 2000].

EPA (U.S. Environmental Protection Agency). 1992. EPA Proposes Cleanup for Second Portion of New Bedford Harbor Superfund Site. EPA Environmental News January 22.

EPA (U.S. Environmental Protection Agency). 1999a. Risk Assessment Guidance for Superfund: Vol. 1. Human Health Evaluation Manual Supplement to Part A: Community Involvement in Superfund Risk Assessments. EPA 540-R-98-042. OSWER 9285.7-01E-P. PB99-963303. Office of Solid Waste and Emergency Response, U.S. Environmental Protection Agency, Washington, DC. March. [Online]. Available: http://www.epa.gov/oerrpage/superfund/programs/risk/ragsa/ci_ra.pdf .

EPA (U.S. Environmental Protection Agency). 1999b. Superfund Public Information System (SPIS). EPA 540-C-97-003. Office of Emergency and Remedial Response, U.S. Environmental Protection Agency, Washington, DC. September. Available as CD ROM.

EPA (U.S. Environmental Protection Agency). 1999c. Waterfowl Samples from Housatonic River Show Elevated Levels of PCBs; State Department of Public Health Issues Duck Consumption Advisory [Press Release]. August 27, 1999. U.S. Environmental Protection Agency, New England. [Online]. Available: http://www.epa.gov/region01/pr/files/082799a.html. [October 13, 2000]

EPA (U.S. Environmental Protection Agency). 1999d. New Bedford Harbor Superfund Site Update. U.S. Environmental Protection Agency, Region 1. [Online]. Available: http://www.epa.gov/oerrpage/superfund/sites/ [August 23, 1999].

EPA (U.S. Environmental Protection Agency). 1999e. Record of Decision (ROD) Abstract. Site: Sangamo Weston/Twelve-Mile/Hartwell PCB; Location: Pickens, SC. EPA/ROD/R04-94/178. [Online]. Available: http://www.epa.gov/oerrpage/superfund/sites/query/rods/r0494178.htm [January 14, 1999].

EPA (U.S. Environmental Protection Agency). 2000. Guidance for Assessing Chemical Contaminant Data for Use in Fish Advisories. Vol. 2. Risk Assessment and Fish Consumption Limits. 3rd Ed. EPA 823-B-00-008. Office of Water, U.S. Environmental Protection Agency, Washington, DC.

Farrington, J.W., R.W. Tripp, A.C. Davis, and J. Sulanowski. 1985. One view of the role of scientific information in the solution of enviro-economic problems. Pp. 73-102 in Proceedings of the International Symposium on Utilization of Coastal Ecosystems: Planning, Pollution and Productivity, 21-27 Nov. 1982, Rio Grande, Brazil, Vol.1. N.L. Chao and W. Kirby-Smith, eds. Rio Grande, RS, Brasil: Editora da FURG.

Fox River Group. 2000. Dredging-Related Sampling of Manistique Harbor - 1999 Field Study. Technical Report. Prepared by Blasland, Bouck & Lee, Inc., Syracuse, NY. June. [Online]. Available: http://www.foxrivergroup.org/pdf_files/Man_2000.pdf.

GE (General Electric Company). 1999. Major Contaminated Sediment Site Database. Available: http://www.hudsonwatch.com/mess/.

GE (General Electric Company). 2000. Environmental Dredging: An Evaluation of Its Effectiveness in Controlling Risks. Prepared by Blasland, Bouck & Lee for General Electric Company, Albany, NY. August.

McConnell, K.E. 1986. The Damages to Recreational Activities from PCBs in New Bedford Harbor. Rockville, MD: National Oceanic and Atmospheric Administration.

McConnell, K. E. and B. G. Morrison. 1986. Assessment of Economic Damages to the Natural Resources of New Bedford Harbor: Damages to the Commercial Lobster Fishery. Prepared for NOAA Oceanic Assessment Division, Rockville, MD. December.

Mendelsohn, R. 1986. Assessment of Damages by PCB contamination to New Bedford Harbor Amenities Using Residential Property Values. Unpublished report prepared for NOAA Oceanic Assessment Division, Rockville, MD. November.

NRC (National Research Council). 1997. Contaminated Sediments in Ports and Waterways: Cleanup Strategies and Technologies. Washington, DC: National Academy Press.

Osowski, S.L., L.W. Brewer, O.E. Baker, and G.P. Cobb. 1995. The decline of mink in Georgia, North Carolina, and South Carolina: the role of contaminants. Arch. Environ. Contam. Toxicol. 29(3):418-423.

PCCRARM (Presidential/Congressional Commission on Risk Assessment and Risk Management). 1997. Framework for Environmental Health Risk Management: Final Report. Washington, DC: The Commission.

Renner, R. 2000. PCBs may mar results of in utero mercury testing. Environ. Sci. Technol. 34(19):410A-411A.

Scenic Hudson. 2000. Accomplishments at Contaminated Sediment Cleanup Sites Relevant to the Hudson River: An Update to Scenic Hudson's Report Advances in Dredging Contaminated Sediments, Poughkeepsie, NY. September.

6

Analyzing Risks

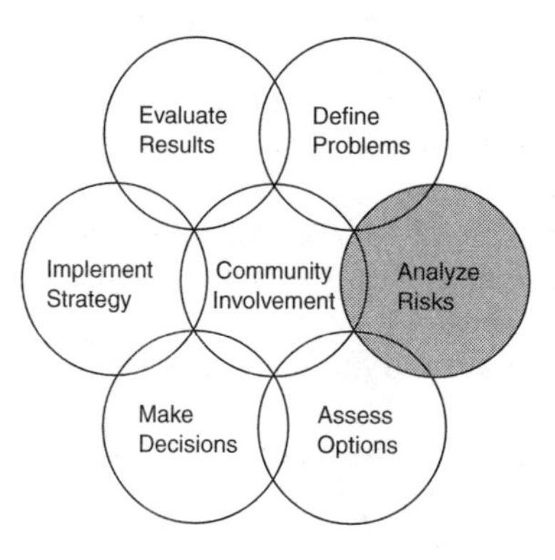

The primary objective for managing PCB-contaminated sediments is the reduction of risk. A critical part of this evaluation is the characterization of existing and potential risks to affected parties. In analyzing risks from PCB-contaminated sediments, the primary focus has been on human health and ecological effects from exposure, the emphasis being on the bioaccumulation of PCBs through the aquatic food web and the human health effects associated with the consumption of contaminated seafood. Risks associated with water consumption and inhalation near contaminated sites are also considered. In addition, PCB contamination might result in economic, social, or cultural impacts to affected parties. These impacts might include loss of a commercial or recreational fishery, decline in property values, reduced commercial opportunities, increased health risks associated with changes from a fish-based to a possibly less healthy diet, and loss of cultural traditions (e.g., the passage of fishing and hunting rites from one generation to the next).

Analyzing all the risks associated with PCB-contaminated sediments is a complicated, multifaceted task and is best addressed within a prescribed, methodical framework of an environmental risk assessment (ERA). Although a number of frameworks are available for conducting ERAs (as described in Chapter 3), the committee endorses the application of the general framework developed by the Presidential/Congressional Commission on Risk Assessment and Risk Management (1997). The U.S. Environmental Protection Agency

(EPA) guidance for human health and ecological risk assessment (EPA 1997b,1999) is generally consistent with the commission's framework and is commonly used in conducting ERAs at PCB-contaminated-sediment sites. This guidance is limited to human health and ecological risk assessment and needs to be extended to explicitly include social, cultural, and economic impacts.

This chapter provides an overview of the ERA process and discusses the use and limitations of scientific information in each of the steps of a risk assessment of PCB-contaminated sediments. The steps that are described include exposure assessment to PCBs; ecological effects and human health effects from PCB exposure; PCB risk characterization; social, cultural, and economic impacts of PCB contamination; and comparative risk assessment. A summary of findings and specific recommendations for conducting ERAs at sites with PCB-contaminated sediments are given at the end of this chapter.

ENVIRONMENTAL RISK ASSESSMENT

ERA provides a process to evaluate the probability that adverse effects are occurring or might occur in the future because of the presence of contamination (see Box 6-1). The framework for this assessment is designed to follow a flexible, tiered approach beginning with a screening-level assessment followed by more detailed evaluations of the site. In this approach, initial or screening-level assessments are used to identify the issues and possibly rebut the presumption of risk. This assessment is typically based on minimal data and very protective assumptions. Three outcomes are possible from this screening-level assessment. First, the screening assessment might indicate that the degree or extent of contamination is sufficiently small to pose no significant risk. Second, the risk might be predicted to be relatively great, but the extent of contamination is sufficiently small to make effective management technically feasible and relatively cost-effective. In such cases, the decision to initiate a particular risk-management strategy or not may be taken without further refinement of the risk assessment. Third, if potential risks cannot be rebutted and the extent of contamination is such that a rapid and effective risk-management strategy cannot be easily identified and applied, a more refined ERA should be conducted. A refined ERA should begin with a baseline assessment to quantify the existing and potential risks associated with PCB contamination as described in this chapter. The baseline risk assessment should be followed by an examination of potential risk-management options (Chapter 7), the development of a risk-management strategy (Chapter 8), the implementation of the risk-management strategy (Chapter 9), and a short- and

BOX 6-1 Issues in Environment Risk Assessment

The analysis of risk for PCB-contaminated sediments is best addressed within the systematic structure of an ERA. The discussions of risk assessment throughout this chapter largely focus on the following questions:

1. What are the risks posed by PCB-contaminated sediments before and after remediation?
2. What are the populations potentially at risk?
3. What are appropriate assessment endpoints?
4. What are appropriate measurement endpoints?
5. What are the primary exposure pathways of PCBs to the receptors of concern?
6. What site characteristics are most important in affecting PCB exposure?
7. What methods are available for predicting future PCB exposure?
8. What information is available on human and ecological health effects from PCB exposure?
9. How reliable are risk estimates for human and ecological health?
10. How can social, cultural, and economic impacts be incorporated into risk assessments?

long-term evaluation of the risk-management strategy to determine if the management goals have been achieved (Chapter 10).

The ERA process consists of three steps (problem formulation, analysis, and risk characterization). Problem formulation involves defining the specific contaminants of concern, delineating the areas of concern, and identifying populations potentially at risk and their size. The analysis phase for human health and ecological risk assessments includes an identification of exposure pathways, a characterization of exposures, and an assessment of the relationship between exposures and effects. Finally, the risk-characterization phase involves quantifying overall risks to humans and wildlife. Each of these steps is discussed in further detail below.

In this discussion, problem formulation, analysis, and risk characterization are presented sequentially, but the committee emphasizes that the ERA should be considered an iterative process. Information obtained during the analysis or risk characterization can lead to a reevaluation of the problem formulation, new data collection or analysis, or even a reevaluation of the initial problem definition and risk-management goals. Similarly, selection of a management option or the evaluation of risk-management results can lead to a reformulation of the problem and might require additional data collection and analysis, such as congener-specific measurements of PCBs in key receptors or media. The extent to which additional site-specific information is collected should be balanced with the costs of conducting the ERA and the costs of risk management. The committee stresses that references to cost refer not only to monetary costs or risks but also to social and political costs of actions or lack of actions in a timely manner.

PROBLEM FORMULATION

The problem-formulation stage for PCB-contaminated-sediment sites involves discussions among the various affected parties to identify the specific geographic areas of concern, all possible risks to humans and wildlife from immediate and long-term exposure to PCBs and from remedial activities, the identification and size of the populations potentially at risk, and the possible presence of co-contaminants at the site. This information is used to identify clearly the assessment endpoints, select measurement endpoints, and develop a conceptual model for the site. At most sites with PCB-contaminated sediments, human health assessment endpoints include both carcinogenic and noncarcinogenic effects (e.g., children born with learning dysfunctions). Special consideration should be given to certain sensitive subpopulations, such as women of child-bearing age, pregnant women, and young children. Other populations who eat fish from contaminated water ecosystems on a regular basis might also be at increased risk. Such populations include many American Indian tribes, immigrants from fishing cultures, such as Southeast Asia, and subsistence fishers who rely upon fish as a major source of protein. For ecological assessment endpoints, reproductive success and population sustainability of resident fish, piscivorous and other predatory birds, and marine mammals are often considered.

Assessment endpoints are used to select measurement endpoints, for which indirect effects, sensitivity and response time, diagnostic ability, and practicality issues are considered. Measurement endpoints are responses (e.g., litter size in mink) that can be measured more easily than assessment endpoints (e.g., reproductive success in mink) but are related quantitatively or qualitatively to the assessment endpoints. Whenever practical, multiple measurement endpoints should be chosen to provide additional lines of evidence for each assessment endpoint. For example, for humans, it might be possible to measure PCB concentrations in food and in human tissues. For predatory fish, birds, and mammals, it might be possible to measure concentrations of PCBs in prey and in predator tissue. Additional measurement endpoints should be selected to assess effects from other chemicals, from non-chemical stressors (e.g., habitat alterations), and from the proposed remedial actions. If feasible, measurement endpoints should be compared with a reference site that has many of the same characteristics as the study area.

Wildlife and humans can be exposed to PCBs either directly from abiotic media, such as sediments, water, or air, or indirectly through diet. As an example, PCBs can enter the food chain by accumulating in benthic invertebrates that are in close contact to the sediments. These invertebrates then can be eaten by other wildlife and thus PCBs accumulate up the food chain. One of the most important aspects of the determination of the potential risk of PCBs to biota is the "food web" or "pathway analysis." Although exposure to important receptors can be postulated, the best method of assessing the potential for exposure is to measure the concentrations of PCBs in key dietary items. The relative merits of measuring and modeling exposures will be discussed later in this chapter.

The primary issue in the exposure assessment is the determination of the biologically available fraction of PCBs that are buried in sediments. PCBs might be buried deep enough in sediments to be below the biologically available zone. Furthermore, PCBs bound to sediments that are in the biologically available zone might be bound in such a manner that they are not biologically available. Specifically, some congeners might be less available because of the nature of their binding to sediment particles (Froese et al. 1998). The movement of PCBs out of sediments is a slow process. Otherwise, the concentrations of PCBs in contaminated sediments would dissipate to a point where they would no longer represent a toxicological risk to wildlife or humans. The slow movement of PCBs from the buried sediments also indicates that the available fraction is relatively small. The goal of the exposure assessment is to discern the fraction of PCBs that are available and the rate of release or movement into the food chain or the transport away from the source. In general, the goal of the exposure assessment is to determine the concentration of each congener that will be accumulated into various levels of a food chain of wildlife species in the vicinity of a location containing PCBs in the sediments.

The first step in the exposure assessment is to determine the species most likely to be exposed. As part of the problem-formulation phase, a conceptual model of exposure pathways is developed (see Box 6-2). This conceptual model can then be used to conduct a pathways analysis to determine the level of exposure expected for each trophic level or individual receptor (e.g., see Figure 6-1). These estimates of exposure can be either measured or predicted. In either case, the concentration of PCBs must be measured or predicted in

BOX 6-2 Conceptual Model Considerations

In preparing a conceptual model of the site, consideration should be given to the following (modified from EPA 1998):

- Sensitive human populations, including but not limited to the elderly, pregnant and nursing women, infants and children, and people suffering chronic illnesses.
- Circumstances in which a culturally or economically distinct population is exposed to PCBs.
- Sensitive and endangered wildlife species and critical habitats exposed to PCBs.
- Significant point and nonpoint sources of PCBs and any co-contaminants.
- Potential contaminant release mechanisms (e.g., volatilization, surface runoff and overland flow, leaching to groundwater, and tracking by humans and animals).
- Contaminant-transport pathways, such as surface-water flow, diffusion in surface water, and bioaccumulation and biomagnification in the food web.
- Cross-media transfer effects, such as volatilization to air and air phase transport.

either critical tissues of receptors or their diets. In general, to minimize the uncertainties in predictions, it is suggested to minimize the length of pathways along which predictions are to be made. Ultimately, it should be possible to link concentrations to top predators to concentrations in the sediments. That link is necessary to derive a proposed threshold concentration in sediments. The threshold concentration would be the cleanup criterion for a particular site. Uncertainties in the exposure assessment can be minimized by collecting measured values for certain key parts of the exposure pathways. For instance, measuring concentrations of PCBs in fish can serve as an integrated measure of the biologically available fraction of PCBs in sediments. The concentrations in fish can be used directly by comparing them to dietary toxicity reference values (TRV) or by using them to predict exposures to higher trophic levels. Similarly, the concentrations can be linked to concentrations in sediments with just a few links. The use of measured concentrations of PCBs in fish is suggested as the most relevant means of measuring exposure of receptors to PCBs in contaminated sediments.

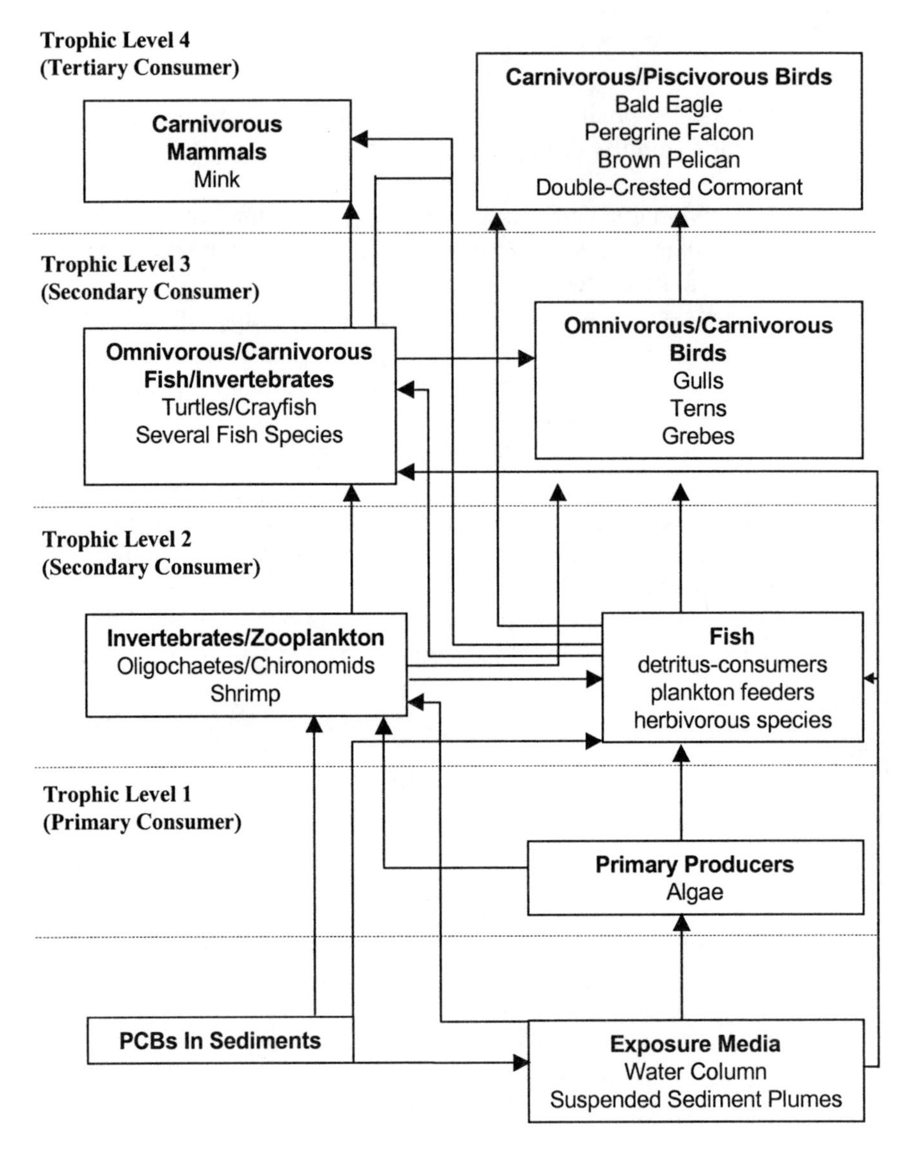

FIGURE 6-1 Food web with examples of representative species.

ANALYSIS PHASE

The analysis phase involves characterization of exposure to PCBs and other contaminants and development of quantitative relationships between contaminant concentrations and effects. Both activities involve the evaluation of available scientific data and an assessment of the relevance of the data to assessment endpoints and to exposure pathways. Exposure characterization describes sources of PCBs and other contaminants, their distribution in the environment, and their exposure to ecological and human populations. Characterization of the potential effects on humans and the environment involves the evaluation of PCB dose-response and other contaminant-response relationships or evidence that exposure to PCBs and other contaminants cause an observed response. Quantitative uncertainty analysis is typically performed in the analysis phase. The products of this phase are summary profiles that describe exposure and contaminant-response relationships.

For PCBs, the analysis of chemical exposure and contaminant-response relationships is complicated by the fact that PCBs are not a single compound, but rather a complex mixture of congeners whose composition in the environment can be drastically different from the original Aroclor mixtures. The changes in PCB-congener composition result from environmental processes, including differential volatilization, solubility, sorption, anaerobic dechlorination, and metabolism, and are referred to as "environmental weathering." Environmental weathering of PCBs is an important consideration in determining the fate and effects of PCBs, as presented in Boxes 6-3 and 6-4. Further details of the analysis phase, such as PCB exposure assessment, and human health and ecological effects are presented below.

PCB Exposure Assessment

The purpose of an exposure assessment is to determine the concentrations of PCBs in various environmental compartments, including sediment, water, benthic invertebrates, and fish, and to evaluate dietary exposures to PCBs of higher trophic level organisms, such as birds, aquatic mammals, and humans. The receptors of interest and the conceptual model for the site serve as the basis for the exposure-assessment studies.

The questions to be addressed in the exposure studies are as follows:

- What are the existing exposure levels of PCBs in the sediments?
- What are the expected exposure levels of PCBs for each potential risk-management option?

An assessment of present exposure is best addressed through direct measurement of PCBs in specific organisms or in their diet. Measurements of PCB effects in organisms may also be used (e.g., fish production or survival).

BOX 6-3 PCB Weathering in the Upper Hudson River

General Electric used PCB oils in the manufacture of electrical capacitors at plants in Hudson Falls and Fort Edwards, New York, from the late 1940s through 1977. Over this time, Aroclor 1242 was the primary PCB oil used in the plant sites. (Late in the production period, General Electric switched from Aroclor 1242 to Aroclor 1016, which is a reformulated version of Aroclor 1242 with a similar congener distribution.) The congener distribution for Aroclor 1242 is shown in panel A of Figure 6-2 and is largely composed of dichlorobiphenyls (BZ 4-15) and trichlorobiphenyls (BZ 16-39) (see Appendix H for list of BZ numbers).

During and subsequent to the production period, PCBs were released into the Hudson River, contaminating 200 miles of river from Hudson Falls to New York City. Sediment and fish samples collected in Thompson Island Pool (the first impoundment on the upper Hudson River, a few miles downstream of the General Electric plant sites) are shown in Figure 6-2 and provide a dramatic example of PCB weathering in the environment. The congener distribution for a surface- (0-2 cm) sediment sample in Thompson Island Pool (panel B) shows an enrichment of tetrachlorobipenyls (BZ 40-81) and pentachlorobiphenyls (BZ 82-127) compared with the original Aroclor 1242 mixture. This enrichment of the more-chlorinated-PCB congeners in surface sediments is attributed to the preferential binding of more-chlorinated PCB congeners to sediments.

The surface-sediment congener distribution also shows an enrichment in a few of the less-chlorinated congeners, such as BZ 1, 4, and 19, which are known dechlorination endproducts. This enrichment might be due to dechlorination in the surface sediments or particle mixing and diffusion of these congeners from deeper sediments where dechlorination is more pronounced (see panel C). Although the extent of dechlorination is extensive at this location in Thompson Island Pool, it does not appear to be as significant as that at other locations in the Hudson River and at other PCB sites with lower contamination levels.

The congener distribution for yellow perch from Thompson Island Pool is shown in panel D. These data indicate that as PCBs from the sediments or from continuing discharges from the plant sites are transferred through the food web, there is a clear shift in the distribution to more-chlorinated congeners. As discussed throughout this chapter, the changes in the congener distributions, which are collectively referred to as environmental weathering, have a profound effect on the transport, fate, bioaccumulation, and toxicity of PCBs and must be explicitly considered in the evaluation of risk.

BOX 6-4 Pattern Recognition

Principal components analysis was performed on the PCB_{total} -normalized concentrations of individual congeners (Figure 6-3) (Froese et al. 1998). A variance of 47% was explained by principal components 1 and 2 (PC1 and PC2). The results of the principal components analysis support the hypothesis that the pattern of relative concentrations of PCB congeners in sediments was different from that in tissues of organisms, including the benthic invertebrates in the sediments. Furthermore, the patterns of PCB congeners were significantly different in the tree swallows than in the benthic invertebrates. This difference indicates that the pattern of relative concentrations of PCBs changes because of such processes as weathering, bioaccumulation, and metabolic processes as the individual congeners move from one trophic level to the next.

Environmental weathering changes the relative concentrations of PCB congeners because of differential solubilities, volatilities, and sorption coefficients (Mackay et al. 1983). In addition, metabolism by microorganisms (Mavoungou et al. 1991) and animals (MacFarland and Clarke 1989) can cause relative proportions of some congeners to increase and others to decrease (Boon and Eijgenraam 1988; Borlakoglu and Walker 1989). Mean ratios of lipid-normalized mono- and non-ortho-substituted congeners to total concentrations of PCBs were not significantly different among trophic levels. Concentrations of PCB congeners 110, 81, and 77 were less in bird eggs than in invertebrates. That difference might be due to differential metabolism between birds and invertebrates (Boon et al. 1997). Alternatively, congeners 126, 157, and 156, perhaps due to their more fully occupied meta-positions, did not appear to be metabolized significantly in different biota (Boon et al. 1989).

Evaluation of future exposures under natural attenuation or other risk-management options are typically performed using simulation models. This approach would have to include other chemicals if significant co-contaminants are present at the site. Field monitoring and PCB exposure models are discussed below.

Field Monitoring

Present exposure levels of PCBs (and co-contaminants) are determined by measuring concentrations in relevant environmental media, such as sediments, water, benthic organisms, and fish, and determining dietary exposure rates,

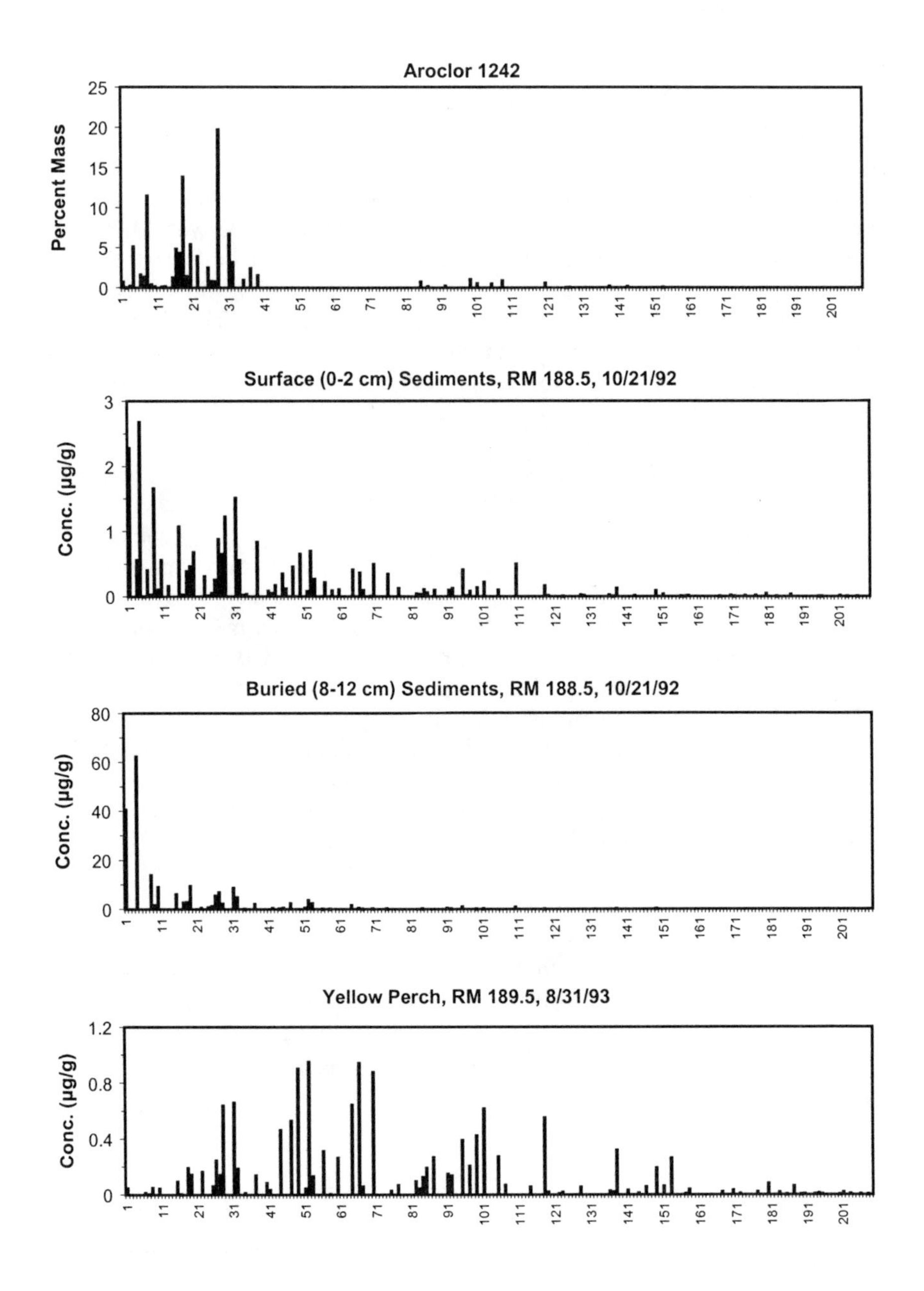

FIGURE 6-2 Congener distributions for PCB sources, sediments, and fish in the Upper Hudson River.

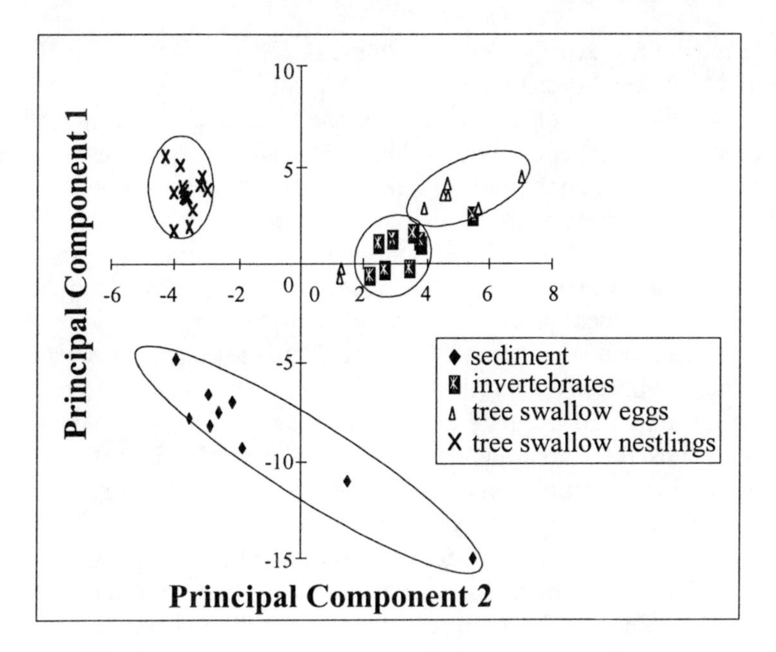

FIGURE 6-3 Principal components analysis of PCB_{total}-normalized concentrations of individual congeners. Source: Adapted from Froese et al. (1998).

particularly for higher trophic level organisms, such as birds, aquatic mammals, and humans. For this assessment, sample collection for sediments may include surface-sediment grab samples (representing the top 2-10 centimeters (cm) of sediment) or sliced-sediment core samples (representing the top 1-2 meters (m) of sediment with cores sliced in 2-20 cm intervals, depending on the specific sediment site). Water- column samples are collected and analyzed as either whole-water or filtered-water samples. Benthic organisms are analyzed as composite whole-organism samples. Fish are analyzed as individual or composite whole-fish samples for smaller fish and as fish fillets for larger, edible fish. Dietary exposure rates are determined from PCB concentrations in food items (e.g., fish) times food consumption rates. For birds and aquatic mammals, food consumption rates are estimated from observed feeding behavior, gut content studies, or bioenergetic calculations, which relate food consumption to organism respiration and growth. For humans, food consumption rates can be determined from individual diet surveys or food-market sale records. In calculating dietary exposure, PCB concentrations in contaminated prey are typically defined on a whole-body basis for birds and aquatic mammals, whereas PCB concentrations in fish fillet are used in calculating dietary exposure rates for humans. There might be specific situations in which calculations should take into account consumption of a specific portion of a harvested organism, such as the tomale of the lobster, which is considered to be a delicacy by some consumers, and can have PCB concentrations much higher than the claw or tail-muscle tissue (Farrington et al. 1986).

Because PCBs are a group of compounds and the absolute and relative concentrations of PCBs in sediments are changing as a function of space, time, and trophic level, the method used to quantify PCBs can have a great impact on the risk-assessment process. Therefore, a discussion of the various methods available is provided here, and a more detailed discussion is provided in Appendix F. There is a great deal of variation in the quality and quantity of information obtained by different methods, as well as in the costs of the analyses. Thus, there are tradeoffs between the type of information collected and the number of samples that can be studied. No single correct allocation of resources is appropriate for every site. Rather, a decision on allocation should be made in the problem-formulation stage of an assessment.

Analyses of PCBs in sediment, water, and fish samples are performed by gas chromatography. This requires extraction of PCBs from the sample and usually some cleanup of the PCB extract by one or more methods before analysis. In the analysis, a gas chromatograph (GC) is used to separate individual PCB congeners or combinations of congeners on the basis of physical and chemical properties, such as volatility or polarity. Over the past 20 years, open tubular capillary GC columns have replaced older packed GC columns for routine laboratory work. The use of open tubular capillary columns, which offer improved resolution, better selectivity, and increased sensitivity, is typically referred to as high-resolution gas chromatography (HRGC). Within the GC system, PCBs are detected by an electron capture detector (ECD), an electrolytic conductivity detector (ELCD), or mass spectrometry (MS). The chromatographic output of the GC allows identification of individual PCB congeners or combinations of congeners on the basis of the resolution of the capillary column and the detector.

Quantification of GC output is performed using Aroclor or congener-based methods. The Aroclor methods rely on comparison of select chromatographic peaks in the sample with the pattern of peaks in a series of pure Aroclor standards to estimate Aroclor concentrations. The method has been applied to chromatographs using both older packed GC columns and open tubular capillary columns and provides a useful comparison of present analyses to historical measurements. Comparison of chromatographic peaks in environmental samples with those found in commercial Aroclors, however, is of little value due to the alteration of PCB-congener distributions in the environment by physical-chemical weathering processes (e.g., differential volatilization, solubility, and sorption) and biochemical weathering reactions (e.g., anaerobic dechlorination and metabolism). Because congeners degrade at different rates depending on the environment, commercial Aroclor products are difficult to identify and difficult to quantify in the environment. The weathered multicomponent mixtures might have significant differences in peak patterns compared with Aroclor standards. The degree and position of chlorine substitution influences not only physical and chemical properties but also toxic effects. Thus, it is important to consider not only the total PCB concentration in a sample but also to characterize the distribution of individual PCB congeners in a sample. The changes in the absolute and relative concentrations of PCBs in the environment are one of the primary uncertainties in the risk assessment and one of the potential areas for disagreement among scientists and risk managers.

Congener-based methods provide a more accurate (albeit more costly) approach in quantifying total PCB concentrations in environmental samples. Analytical methods more commonly used at PCB-contaminated sites also provide direct quantification of 13 to 108 individual or coeluting congeners (see Appendix F) that can be used in more detailed exposure and toxicity evaluations.

Final measurement of PCB concentrations are reported in terms of Aroclor, total PCB, or concentrations of individual PCB congeners for sediment (as micrograms of PCB per gram of dry sediment), water (as nanograms of PCB per liter of water), and benthic organisms and fish (as micrograms of PCB per gram wet weight of organism). Organic carbon normalization of sediment data and lipid normalization of concentrations in benthic organisms and fish are often used to recognize the preferential sorption of PCBs into these phases. Measured PCB concentrations may be used directly to assess toxicity. For example, toxicity in fish and lower trophic organisms have been reported in terms of exposed water concentrations, exposed sediment concentrations, or tissue residue concentrations. For organisms at higher trophic levels, PCB concentrations in contaminated food are used to calculate dietary exposure rates that, for example, may be compared with allowable dosage rates (e.g., reference dose (RfD)), maximum allowable toxicant concentration (MATC), or toxicant reference value (TRV)).

In addition to assessing present exposure levels, nonchemical data should be collected at the site for use in developing PCB exposure models, assessing nonchemical impacts, and evaluating the applicability of various risk-management options (e.g., natural attenuation, source control, dredging, stabilization, ex situ treatment). Information may include access to the site, hydrodynamics and hydrology, climatology, time-series suspended material concentrations, and sediment bed properties (e.g., horizontal and vertical mapping of sediment grain size, mineralogy, water content and erodibility, presence of boulders and/or debris, and depth to bedrock or impermeable hardpan, i.e., hard-packed sediment). In addition, sampling can be performed to determine external solids loading and continuing PCB sources from upstream waters, contaminated industrial sites, wastewater treatment plants, combined sewer overflows, storm drains, and the atmosphere.

Accumulation of PCBs in Sediments

Data collected at a number of sites provide a general picture of PCB-contaminant behavior in sediments, benthic organisms, and fish.

At PCB-contaminated sites, sediments, particularly fine-grained, organic-rich sediments, serve as a long-term repository for PCBs. The horizontal and vertical distributions of PCBs in sediments are the result of temporal and spatial deposition patterns of fine- and coarse-grained sediment and the time history of PCB loadings to the system. The distribution of coarse- and fine-grained bottom sediments might exhibit a large degree of spatial variability; coarse-grained sediments are likely to be present in high-velocity regions and deposition of fine-grained sediment occur in more low-energy areas. The presence of cobbles, boulders, underlying bedrock or hard pan are also important considerations, particularly in evaluating dredging alternatives.

Since maximum production and use of PCBs in the United States occurred in the late 1960s and 1970s, the greatest concentrations of PCBs are often found at depth in sediment depositional areas. In addition, continuing sources of PCBs from industrial sites, wastewater discharges, storm-water discharges, and the atmosphere are still occurring (albeit at reduced rates) and might be adding to the inventory of PCBs in sediments. PCB input from contaminated ground water might add to PCB inventories in surface waters and sediments. At some sites, migration of PCBs from contaminated sediments may also be a concern. Although localized "hot spots" (i.e., areas with relatively high concentrations of PCBs) have been documented, the extent of PCB contamination is affected by water and sediment-transport processes and extends over larger areas of sediments.

As discussed previously, PCB-congener distributions in sediments are different from distributions of parent Aroclors because of physical, chemical, and biochemical weathering processes. These processes favor the preferential retention of more-chlorinated congeners in sediments and the transformation of certain more-chlorinated congeners to less-chlorinated congeners by dechlorination in anaerobic sediment layers. With the possible exception of a thin surficial layer, sediments that are likely to contain significant quantities of PCB contaminants (i.e., fine-grained, organic-laden sediments) are usually anaerobic. Under these conditions, reductive dechlorination, which results in the selective removal of chlorines from the PCB molecules, can occur.

Although PCB dechlorination does not have a dramatic effect on total mass of PCB concentrations, dechlorination might result in a significant reduction in some types of toxicity. For example, PCB congener 126 (3,3′,4,4′,5-pentachlorobiphenyl; IUPAC Number 126), which is one of the most potent PCB congeners exhibiting dioxin-like toxicity, has been reported to decrease by as much as 10- to 100-fold as a result of reductive dechlorination (Quensen et al. 1998).

Accumulation of PCBs in Benthic Organisms and Fish

PCB-contaminated sediments might serve as long-term sources of PCBs to sediment-dwelling organisms and to fish through various exposure pathways. These exposures are largely affected by PCBs in the top 5-10 cm of sediments and not by buried PCBs. This top layer, which is continuously reworked by sediment-dwelling organisms and remains in direct contact with the overlying water, is typically referred to as the "biologically active zone."

In some cases, desorption of PCBs from sediments in the biologically active zone can be kinetically inhibited and limit the bioavailability of PCBs to organisms. The degree to which PCB are bioavailable is a function of sediment properties, the time history of contamination, and PCB-congener properties. The resulting accumulation of PCBs in bottom-dwelling organisms is usually related to local contamination conditions. To account for preferential binding of PCBs to organic carbon in sediments and lipid content in the organisms, PCB concentrations in bottom-dwelling organisms and sediments are generally related using lipid-normalized accumulations of PCBs in organisms and organic carbon-normalized PCB concentrations in sediments. Because fish typically have a larger home range than bottom-dwelling organisms, aerial-weighted-average concentrations of PCBs in surface sediments, and not hot spot concentrations, are often considered to be a better indicator of potential PCB concentrations in fish.

PCB-congener distributions in water and fish differ from distributions in sediments because of the preferential release of less-chlorinated PCBs to the water, the preferential accumulation of more-chlorinated PCBs through the food web, and the potential metabolism of some PCB congeners. Transfer of PCBs through food webs typically results in biomagnification. Other contaminants, which might also be present in PCB-contaminated sediments, can accumulate in aquatic food webs. As discussed later in this chapter, most of the toxicity of PCBs in fish and higher trophic organisms is attributed to dioxin-like congeners and not total PCB concentrations. Thus, in most situations, the congeners causing the arylhydrocarbon receptor (AhR)-mediated effects are the critical contaminants for risk assessments and determine both the potential for risk of a particular mixture of PCB congeners and the allowable total concentration of PCBs.

Finally, PCBs buried in sediments below the biologically active zone should not be dismissed when evaluating PCB accumulation in bottom-dwelling organisms and fish. Erosion and reworking of sediments by large storms (e.g., the 100-year flood) or other catastrophic events (e.g., dam failure) have in some cases resulted in reintroduction of buried PCBs into the biologically active zone. The potential for such events should be characterized as part of the ERA.

PCB Exposure Models

Mechanistic mass-balance models have been used in conjunction with field-monitoring data at a number of sites (1) to quantify the relationships among external sources and PCB exposure concentrations in sediments, water, and biota, and (2) to project future contamination responses under various scenarios of natural recovery or other management options. The overall modeling approach used at PCB-contaminated sites is typically presented in terms of several components (or submodels) as outlined in Figure 6-4. The overall approach considers hydrodynamic and fluid transport, sediment transport, chemical fate and transport, and bioaccumulation models. Since PCBs have a strong affinity for organic carbon, organic carbon distributions have also been modeled explicitly for a few sites. Further details regarding the modeling components—hydrodynamic transport, sediment transport, organic carbon, PCB fate and transport, and bioaccumulation—are given below.

Hydrodynamic Transport

Hydrodynamic monitoring and model calculations serve as the basis for the overall assessment and are used to define the downstream transport of dissolved and particulate PCBs.

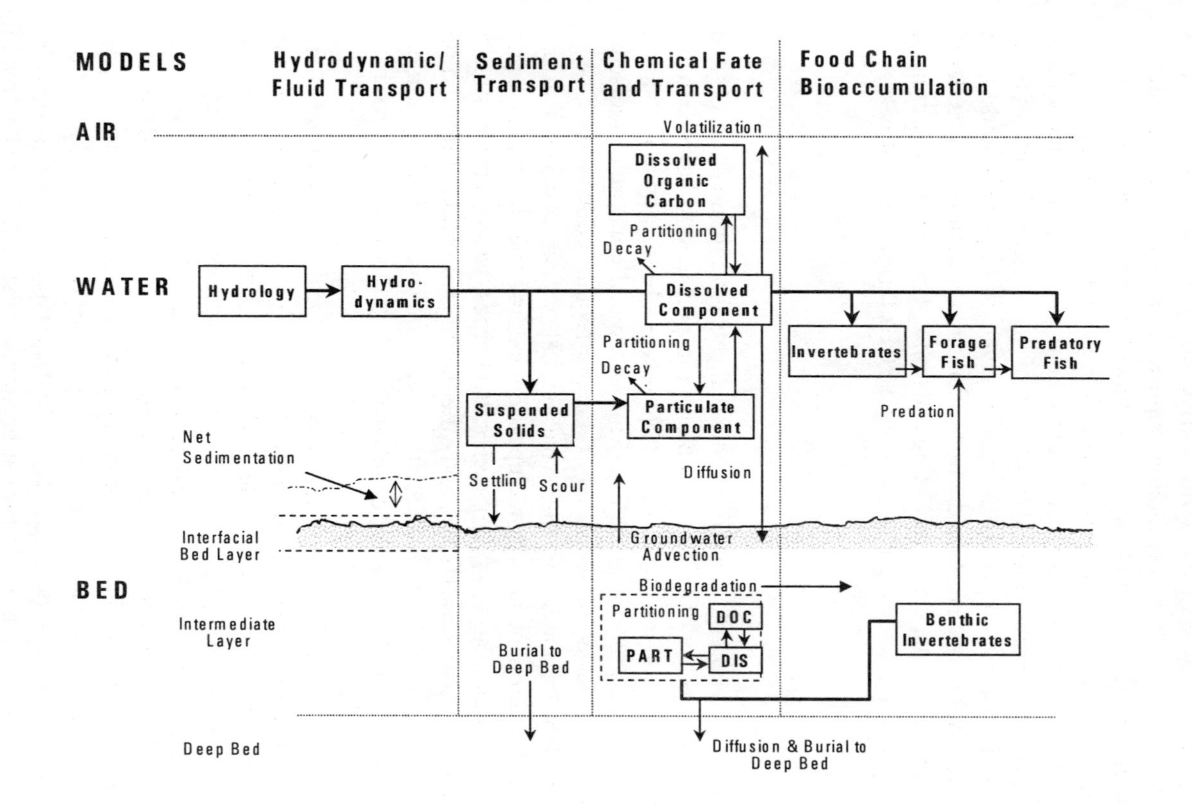

FIGURE 6-4 PCB exposure modeling framework. Abbreviations: DIS, dissolved; DOC, dissolved organic carbon; PART, particulate. Source: Adapted from QEA (1999a).

Hydrodynamic models have evolved over the past few decades from simple hydrodynamic calculations (e.g., for the James River Estuary (Farley et al. 1983)) to complex multidimensional hydrodynamic computer simulations (e.g., for the Hudson River (Blumberg et al. 1999; QEA 1999b; TAMS Consultants et al. 2000)). These models require site-specific data for calibration of bottom-roughness parameters. Adjustments in the geometry of numerical grid elements have also been used in model calibration. Such adjustments have been used to account for fine-scale features in the river channel that cannot be properly captured in the resolution of the numerical grid. In these evaluations, particular emphasis is usually given to high-flow and storm events, where the energy associated with higher water velocities is considered to have an important role in the reworking and remobilization of contaminated sediments and in transporting particle-bound PCBs further downstream.

Sediment Transport

Sediment transport plays a critical role in the water-column transport and ultimate bioavailability of PCBs. Sediment-transport models have been applied to rivers and estuaries with reasonable success, for example, in the Fox River (Gailani et al. 1991). Because of the site-specific nature of the sediment-transport process, these models typically require collection of large amounts of data for model parameterization, calibration, and verification. Required information includes flow-dependent loading rates for cohesive and noncohesive solids from upstream waters, tributaries, wastewater treatment plants, combined sewer overflows, and other sources of solids. Information is also necessary on channel bathymetry; stream flows and tidal elevations; and sediment maps of fine-grained material, coarse-grained material, and hard bottom along a channel.

For model calculations, settling velocities of cohesive sediments are typically defined by empirical correlation to particle size and concentration and water-column turbulence (Ziegler and Nisbet 1995). Sediment erosion potentials are also required and are determined empirically as a function of flow-dependent boundary shear stress (McNeil et al. 1996). Bottom shear stresses, which are a function of water velocities and are obtained directly from hydrodynamic modeling results, are used to drive the sediment-transport calculation. Calibration parameters include particle-size composition of the solids loading, median particle sizes, and active layer depth of the noncohesive sediments (Ziegler 1999a,b). In the calibration, model results are compared with observed suspended solids and spatial patterns of sediment accumulation (as determined from changes in channel bathymetry, from radionuclide dating of sediment cores, or from records of maintenance dredging for harbors). Model calibration is usually limited, however, by the availability of good quality data for high-energy storm events (e.g., the 10-year, 50-year, and 100-year storms). In many cases, available data are not sufficient to ensure a unique set of model calibration parameters. Additional emphasis should therefore be given to supplemental data-collection efforts and model-sensitivity studies to determine the best representation of sediment transport at the site.

Although there are inherent uncertainties associated with sediment-transport modeling, model results have greatly aided our understanding of sediment-transport dynamics at many PCB-contaminated sites. For example, in riverine systems, sediment erosion during low and moderate flows is negligible, and a large fraction of the total sediment transport typically occurs during a relatively small number of high-flow events. In this regard, sediment-transport modeling results provide critical information in assessing the depth of sediment scour during high-flow events and in identifying mixing of buried PCBs with the biologically active sediment layer, water column, or both.

Organic Carbon

Since PCBs have a strong affinity for particulate (POC) and dissolved organic carbon (DOC), organic carbon concentrations are needed to compute the partitioning of PCBs among freely dissolved, DOC-bound, and particulate phases. In many cases, measured POC and DOC concentrations are used directly in PCB partitioning calculations. Spatial and temporal variations in POC and DOC concentrations are observed at many sites and are associated with watershed runoff events, sewage-discharges locations, and seasonal phytoplankton productivity. In cases in which phytoplankton productivity has a dominant role in controlling spatial and temporal distributions of POC and DOC (e.g., Green Bay), POC and DOC concentrations may be specified from results of a dynamic eutrophication model calculation.

PCB Transport and Fate

Following evaluations for hydrodynamics, organic carbon, and sediment transport, chemical-fate calculations are performed using mass-balance numerical models to examine the transport and ultimate fate of PCBs in sediment and water-column segments. Because of large differences in physical-chemical behavior, biochemical reactions, and toxicity of PCBs (which will be discussed later), congener-specific or homologue-specific model calculations are considered to provide a better representation of PCB transport and fate than that provided by total PCB models. It is also important to note that model comparisons to a more detailed congener-specific or homologue-specific data set provide a more rigorous test of model calibration and hence a reduction in model uncertainty.

The partitioning of PCBs between the freely dissolved, DOC-bound, and particulate phases is essential in describing PCB transport, transfer, and transformation behavior. In modeling studies, partitioning reactions are usually assumed to be fast in comparison to other environmental processes and are typically modeled as instantaneous (or equilibrium) reactions. Partitioning between freely dissolved and particulate phases are either derived from field measurements or assumed to be a direct function of the octanol-water partitioning coefficient and the fraction of organic carbon on the suspended or sediment solids. To account for differences in PCB partitioning to lower-molecular-weight DOC compounds and octanol, partitioning of PCBs between the freely dissolved and the DOC-bound phases is typically assumed to be a fraction of the octanol-water partition coefficient.

PCBs bound to particles are subject to settling, resuspension, burial, and water-column suspended-solids transport processes, as defined by hydrodynamic and sediment-transport evaluations (Figure 6-4). In addition to sediment-transport processes, exchange of freely dissolved and DOC-bound PCBs can occur across the sediment-water interface and between the surface and the deeper sediments. Exchange between the sediment pore water and the overlying water column is dependent on the detailed hydrodynamic structure at the water-sediment interface (e.g., through hydrodynamic pumping of water through sediment bed forms) and can be greatly enhanced by biological activity. In the biologically active sediment layer, which can extend down to depths of approximately 5-10 cm, rates of diffusion and particle mixing are greatly enhanced by bioturbation. In the deeper sediment, particle mixing is virtually nonexistent, and exchange of freely dissolved and DOC-bound PCBs is largely a function of molecular diffusion and is generally considered to be slow.

Freely dissolved PCBs in the water column are also subject to volatilization across the air-water interface. The process preferentially affects lower-chlorinated PCB congeners because of the decreased hydrophobicity and increased likelihood that these congeners will be present in the water column as dissolved chemical. The overall loss of PCBs by volatilization can be substantial in water bodies with long hydraulic residence times.

Transformations of organic molecules can also occur by hydrolysis, photolysis, biodegradation, and reductive dechlorination reactions. Although PCBs were originally thought to be refractory, a number of studies have shown that certain PCB congeners can be degraded under aerobic conditions or microbially dechlorinated under anaerobic conditions in aquatic environments (Abramowicz 1990). For more details on aerobic and anaerobic transformation see Appendix E. Aerobic degradation of PCBs is therefore limited to the overlying water and possibly the surficial layer of oxic sediments, whereas anaerobic dechlorination of PCBs can occur deeper in the sediment column.

The major conclusion from aerobic degradation studies (Bedard et al. 1986; 1987a,b) is that biodegradation of PCBs can occur by the attack of a dioxygenase enzyme at vicinally unchlorinated (2,3 or 5,6) carbon atoms or at an unchlorinated (3,4 or 4,5) site. These attacks result in cleavage of the biphenyl ring and can be carried out by a variety of naturally occurring bacteria. Congeners with chlorines at both ortho (2,6) positions on either ring are generally not degraded as readily as congeners lacking this characteristic.

Under anaerobic conditions, organisms leave the biphenyl ring intact while removing chlorines from the ring, thereby producing less-chlorinated congeners. Although details of the dechlorination process are not fully understood, reductive dechlorination has been shown to proceed primarily through the selective removal of meta (3,5) and para (4) chlorines (Quensen et al. 1988; Abramowicz 1990; Abramowicz et al. 1993; Rhee et al. 1993a,b,c). Anaerobic dechlorination has been observed at a number of locations, including the upper Hudson River (see Figure 6-2). Anaerobic dechlorination is not thought to be important at sites with low concentrations of PCBs, possibly due to PCB concentrations being below a dechlorination threshold value.

Mathematical expressions for the various processes affecting the fate and transport of PCBs are fairly well established, and appropriate ranges for modeling coefficients have been determined from laboratory and field studies and previous modeling applications. A time sequence of PCB concentrations in the sediments and overlying water is typically used for model calibration. Because several modeling coefficients can be adjusted in the calibration procedure, a good comparison between PCB model results and field data alone does not guarantee the proper selection of all modeling coefficients. This difficulty in calibration is likely to be reduced but not eliminated in congener-specific models. Professional judgment therefore plays an important role in model calibration for PCB fate and transport. Uncertainties in model calibration usually stem from a lack of detailed knowledge of water-sediment exchange rates, sediment mixing rates, and depths of the biologically active sediment layer.

PCB Bioaccumulation

Accumulation of PCBs in organisms is typically viewed as a dynamic process that depends on direct uptake from water, dietary exposure, depuration (from back diffusion, excretion, and egestion), and metabolic transformation of PCBs within the organism. For phytoplankton and other plant species, direct uptake from water is described by diffusion of PCBs through cell membranes. For benthic invertebrates, direct uptake from water and ingestion of contaminated sediments might have important roles. For fish and higher trophic level organisms, PCB exposure from ingestion of contaminated prey will often dominate, and biomagnification of PCBs from one trophic level to the next is likely to occur (see Box 6-5).

Models of PCB bioaccumulation in organisms are available with several levels of detail, ranging from simple empirical models to complex food-web models. Simple empirical formulations, which use partition coefficients such as bioaccumulation factors (BAFs), biomagnification factors (BMFs) (Starodub et al. 1996), and biota-sediment accumulation factors (BSAFs),[1] describe distributions of PCBs in various environmental compartments (see Chapter 2 for definitions of BAFs and BMFs). BAF, BSAF, and BMF values are typically determined from field or laboratory data, and their application in bioaccumulation calculations is based on the assumption that partitioning of PCBs among water, sediment, and organisms will remain invariant over time.

Empirical BSAFs, which are not based on fugacity theory, are used to predict the accumulation of contaminants from sediments into higher trophic levels (Ankley et al. 1992a,b; Cook et al. 1993). These values implicitly consider the disequilibrium that exists, or can exist, between the sediment and pelagic species (Cook et al. 1993). When applied to aquatic species such as benthic invertebrates, BSAFs are defined as ratios of lipid-normalized concentrations of compounds to organic carbon-normalized concentrations of the same compounds in sediments (Ankley et al. 1992a). When predicting higher-order accumulations, for example, in birds that eat aquatic organisms, similar BMF ratios are used. Although the BSAF-BMF method can be empirical in nature (Cook et al. 1993), it also could be based on fugacity theory (Clark et al. 1988; Mackay and Paterson 1991; Ling et al. 1993) through the use of several assumptions, including that the system is at steady state. A further assumption under both theories is that the accumulation ratios (BSAFs and/or BMFs) are constants that can be applied from one location to another (Neely and Mackay 1982; Velleux and Endicott 1994). These descriptors are useful for static or slowly varying systems but are of little use in dynamic systems where the time responses for PCBs in organisms might be very different from PCB responses in water and sediments.

---

[1]BSAFs are used to define the PCB concentrations in organisms relative to concentrations in sediments.

BOX 6-5 Biota-Sediment Accumulation Factor (BSAF)/Biomagnification Factor (BMF) of PCBs Among Trophic Levels

A study was conducted on the changes in the absolute and relative concentrations and relative patterns of individual PCB congeners, as well as total PCB concentrations, among trophic compartments. It was determined whether toxic potentials of PCB mixtures change as a function of trophic level when accumulated from the sediments of Saginaw Bay. PCB concentrations were measured in sediments, emergent aquatic insects (primarily chironomidae), and eggs and nestlings of tree swallows (*Tachycineta bicolor*) from sites within the Saginaw River watershed (Nichols et al. 1995). In addition, the BSAF/BMF method was used to calculate sediment PCB concentrations theoretically protective of tree swallows on the basis of estimates of the toxicity of PCBs or 2,3,7,8-tetrachlorodibenzo-*p*-dioxin equivalents (TEQs) to other bird species (Giesy et al. 1994a; Ludwig et al. 1996).

Average lipid-normalized PCB_{total} concentrations were not different among the invertebrates, eggs, or nestlings. The average organic carbon-normalized PCB_{total} in sediments was about an order of magnitude less than tissue values. The fact that concentrations of PCBs were not different suggests that there is no net biomagnification of PCBs at these trophic levels. That is consistent with fugacity theory (Mackay et al. 1992) and suggests that BSAF-BMF methodologies might be appropriate for predicting the total mass of PCBs that would be accumulated at higher levels of the food chain in the Saginaw River. Furthermore, this observation indicates that the changes in relative concentrations of individual PCB congeners do not have a great influence on the total mass of PCBs predicted to occur in tissues of higher trophic levels. In addition, these results suggest that the concentrations of total PCBs in the tissues of the tree swallow eggs and nestlings were near steady state.

More detailed, dynamic models of PCB transfers through food webs have been developed over the past 10-15 years (Thomann and Connolly 1984, Thomann et al. 1992a,b; Gobas 1993, Gobas et al 1993). In this approach, accumulation of PCBs by individual organisms is described by uptake of PCBs from water, sediments, contaminated prey, depuration, and metabolism. PCB uptake by higher trophic species is explicitly linked to accumulations in lower trophic levels by the food-web feeding structure. Overall, the food-web bioaccumulation models are similar in their construct and reflect a cross-fertilization of ideas among investigators (see comparison of Thomann and Gobas models in Burkhard 1998). For slowly varying systems (such as large lakes), steady-state bioaccumulation model calculations are often sufficient. For dynamic systems (such as contaminated rivers or estuaries and bays with migratory fish), time-variable model calculations are usually required.

The formulation of dynamic bioaccumulation models is based on fish bioenergetic relationships (e.g., the relationship of fish weight to growth, respiration, food consumption, and gill transfer rates) and laboratory-derived kinetic relationships based on information for PCB uptake and depuration processes. The models require site-specific information on food-web feeding structure, whole-body lipid content, and in certain cases, fish migration patterns. A time sequence of measured PCB tissue residue concentrations in exposed organisms is used in calibrating PCB-uptake efficiencies and depuration rate coefficients. Model calibration, however, is often limited by available data. The greatest uncertainties in model calibration usually stem from a lack of information on food-web feeding preferences and inadequate data sets for PCB tissue residues. The latter limitation is exacerbated by variations in PCB tissue residues that are often observed for a given species (i.e., intraspecies variability). Multiple samples of fish are usually required to address this issue of intraspecies variability and to properly define the statistics of exposed populations; this data need is particularly problematic at sites where there are endangered or threatened species such as the Pacific salmon in Puget Sound. Such analyses are also confounded by the need to correlate tissue levels to site of exposure, and such correlations might be difficult to determine for migrating species of fish, birds, and mammals. Following model calibration, simulations are performed to determine future PCB tissue residue concentrations in food-web species on the basis of projected exposure concentrations in sediment and overlying water that are determined from model simulations for PCB transport and fate. For example, projected responses for PCB concentrations in fish from Thompson Island Pool on the upper Hudson River are discussed in Box 6-6. The relative importance of downstream transport, volatilization, burial, dechlorination, and food-chain transfer can also be examined using model simulation results.

Summary of Exposure Modeling Results

Modeling the fate and bioaccumulation of PCBs at contaminated-sediment sites requires a clear understanding of hydrodynamics, sediment transport, and organic carbon behavior.

BOX 6-6 The Use of Models in Evaluating Management Options

As part of its Hudson River PCB Reassessment, EPA Region II has developed mathematical models describing the transport, fate, and bioaccumulation of PCBs in the upper Hudson River (TAMS 2000) and has applied the models to examine the effectiveness of various management options. Model projections for PCB concentrations in Thompson Island Pool fish are shown in Figure 6-5 for five possible remediation scenarios:

- No action (no upstream source control).
- Monitored natural attenuation (MNA) with upstream source control.
- Capping, with removal to accommodate the cap and upstream source control, followed by MNA.
- Select removal (of approximately 2.65 million cubic yards of contaminated sediment) and upstream source control, followed by MNA.
- Removal (of approximately 3.82 million cubic yards of contaminated sediment) and upstream source control, followed by MNA.

Similar modeling studies were performed by the General Electric Company (QEA 1999b). Model projections for PCB concentrations in largemouth bass in Thompson Island Pool are shown in Figure 6-6. Both the EPA and General Electric models gave comparable results for Thompson Island Pool. Differences in the model projections in Figures 6-5 and 6-6 appeared to be largely attributed to different assumptions used in describing the remedial actions (i.e., implementation and effectiveness of source controls, dredging volumes and areas, commencement and duration of dredging, and capping of dredged areas). EPA and General Electric model projections for sections of the river below Thompson Island Dam show less agreement.

This example demonstrates how models can be used to project the relative effectiveness of various management options and how models developed by independent investigators can be used to test the consistency of model formulation and calibration of transport, fate, and bioaccumulation of PCBs at specific sites.

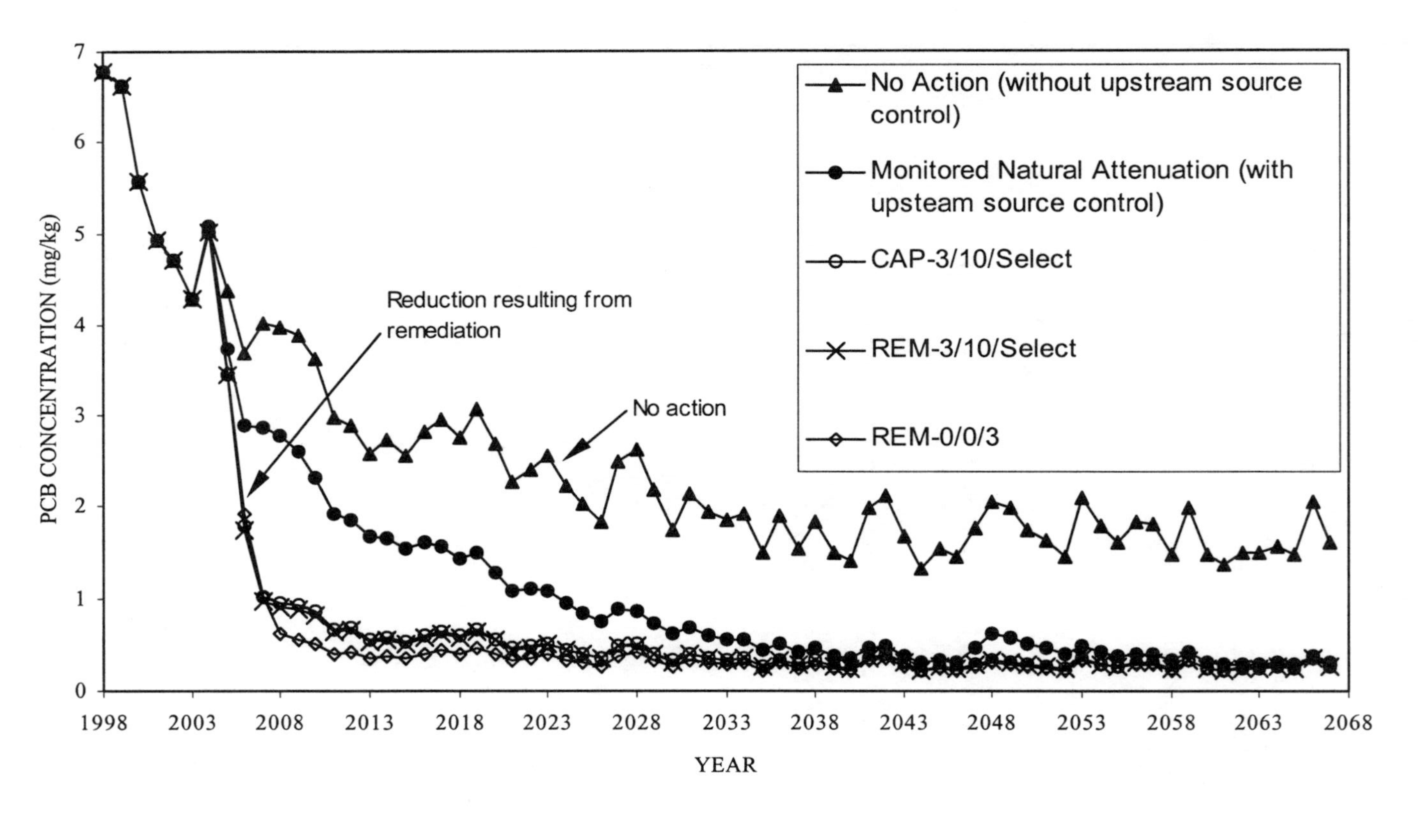

FIGURE 6-5 Model projections for PCB_{3+} concentrations in Thompson Island fish from 1998 to 2068 for various remedial alternatives as outlined by EPA, Region II.

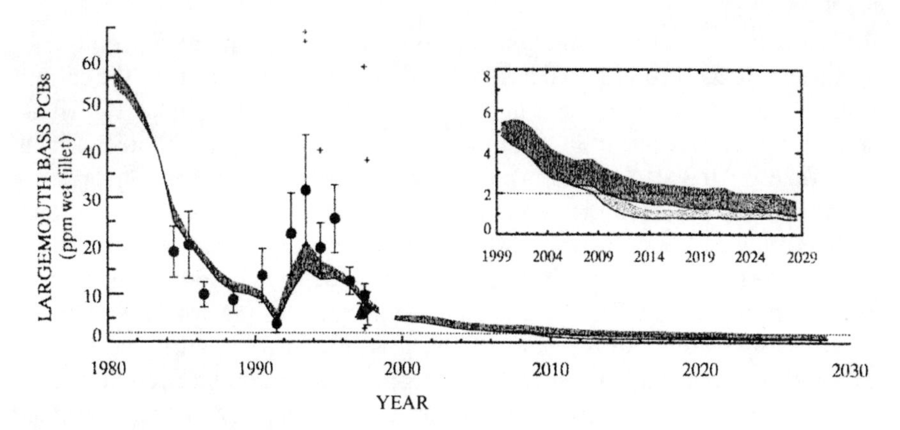

FIGURE 6-6 Predicted PCB_{3+} concentrations in Thompson Island Pool largemouth bass over time period from 1980 to 2029 for both natural recovery (upper line represents the uncertainty in the projection) and dredging of Reacher 8, 7 and cohesive sediments between 2005 and 2016 (lower line) to a residual concentration of zero. Dotted line at 2 ppm represents Food and Drug Administration tolerance limit; circles represents New York State Department of Environmental Conservation; triangles represents exponent; diamonds represents General Electric; crosses indicate values excluded from the annual averages. Source: Adapted from QEA (1999b).

Hydrodynamic models are rather advanced and can be used to determine a description of temporal and spatial distributions in flow. Sediment-transport models are still in the development stage and require a large collection of data for model calibration. The greatest uncertainty in sediment-transport model predictions is associated with high-flow events (e.g., 10-year, 50-year, or 100-year flooding events) when the largest amount of sediment is expected to be mobilized. Organic carbon distributions in the sediment and overlying water are usually determined by field sampling and are assumed to be constant from year to year. For certain studies, field measurements have been complemented by eutrophication model results.

Formulations of models to predict PCB fate and bioaccumulation are well-developed. In applying these models to PCB-contaminated-sediment sites, care should be taken to include reliable projections of external PCB loads, and to critically examine the model calibration procedures. Once pathways are established and models are calibrated and validated, they can be used to predict future trends and the results of various remedial actions. However, because several modeling coefficients can be adjusted in the calibration procedure, a good comparison of PCB model results to field data alone does not guarantee the proper selection of sources, pathways, and coefficients. This difficulty in calibration is likely to be reduced but not eliminated in congener-specific models. Professional judgment, therefore, is important in the calibration of models for both fate and bioaccumulation. The most significant sources of uncertainty in PCB models of the transport and fate of PCBs are specification of water-sediment exchange rates, of sediment mixing rates, and of the depth of the biologically active sediment layer. Sources of uncertainty in PCB bioaccumulation models stem from a lack of information on food-web feeding preferences and inadequate data sets to define the statistical and temporal variations in PCB tissue residues. Because those factors are critical in determining future exposures to PCBs, it is often beneficial to have model calibrations performed by more than one group and to conduct detailed model comparisons. Review of model results by expert panels is also recommended. Publication of model results in peer-review literature, although desirable, should not be considered a sufficient criterion for model acceptance. Even with these uncertainties in calibration, PCB exposure models have provided and continue to provide important information for the decision-making process.

One of the major issues in conducting exposure assessments is predicting the movement of PCB congeners from the sediments to biological receptors. Due to the fact that the absolute and relative concentrations of PCB congeners can change over time and that different congeners with different chemical-physical characteristics result in different partitioning characteristics, the exposure profile can change over time and depending on trophic levels.

PCB Effects Assessment

The purpose of effects assessment is to determine PCB dose-response and other stressor-response relationships and show evidence that exposure to PCBs or other stressors causes an observed human or ecological response. In this context, the emphasis in the assessment of the human response is on incremental effects in the individual, and the assessment of the ecological response typically measures an adverse effect in a given population or component of the food web (e.g., the effects of PCBs on survival or reproduction of a given species). Commercial PCB mixtures are known to elicit a broad spectrum of toxic responses that are dependent on several factors, including chlorine content, purity, dose, species and strain, age and sex of animal, and route and duration of exposure (Giesy and Kannan 1998). Metabolism must be considered when assessing toxicity, because the persistence of a congener can affect its toxicity. Immunotoxicity, carcinogenicity, and developmental toxicity as well as biochemical effects of commercial PCB mixtures have been extensively investigated in various laboratory animals, fish, and wildlife species.

In this section, the congener-specific approaches for assessing exposure and effects—TCDD-toxicity equivalents (TEQs) and BSAFs—are first described, followed by a discussion of ecological and human health effects associated with PCBs.

Congener-Specific Exposure and Effects

In the risk-characterization process, estimates of exposure and hazard are required as discussed earlier in this chapter. When assessing the potential for PCBs in sediments to result in exposure and or effects in biota, simple thermodynamic relationships are often applied. These relationships, sometimes referred to as transfer coefficients—BCF, BAF, and BSAF (discussed in the section PCB Bioaccumulation earlier in this chapter)—are simple ratios of the concentrations of PCBs in various environmental compartments. Exposure and effects estimates are complicated by the fact that PCBs are a mixture, and the relative concentrations of individual congeners in the mixture are subject to change. For this reason, the simple ratios do not apply to the total concentrations of PCBs but rather to individual congeners and are a function of their chemical properties and those of the environment in which they occur. Thus, it is generally inappropriate to apply such simple relationships to total PCB concentrations. To account for the changes in the relative concentrations and thus toxic potency, the TEQ approach has been developed. Because the concentration of TEQs in an environmental or biological sample is a function of both the concentrations of individual congeners and their relative toxic potencies, it is inappropriate to apply simple thermodynamic ratios to concentrations of TEQs. Since both TEQ and BSAF techniques are applied in risk assessments of PCB-contaminated sediments, both of these concepts and their utility are discussed below.

Toxicity Equivalent Values

When considering exposure to and toxicity of PCBs, it is important to take into account the concentrations of individual PCB congeners, in addition to the total PCB concentrations, because the various congeners elicit different types and magnitudes of effects. Coplanar PCB congeners can bind to and cause pleiotrophic effects through the AhR. The toxic effects of noncoplanar PCBs are not mediated through the AhR. Some of the neurotoxic effects of PCBs appear to result from the noncoplanar PCBs. (See Appendix G for a further discussion of the neurotoxic effects of PCBs.) For the effects mediated by the AhR, the critical toxicants to which wildlife are exposed are not total PCBs but rather concentrations of the congeners that are structurally similar to 2,3,7,8-tetrachlorodibenzo-*p*-dioxin (TCDD) and bind to the AhR (Ludwig et al. 1996; Giesy et al. 1994a,b). Concentrations of TEQs represent the total potential of the dioxin-like PCB congeners to cause TCDD-like toxicity. The values represent a weighted average of the concentrations of individual non-ortho and mono-ortho-substituted PCB congeners corrected for their relative potencies to cause AhR-mediated effects. However, it is important to note that the TEQ approach does not account for all effects, such as some neurotoxicity, that can be mediated by the noncoplanar PCBs through mechanisms not involving the AhR.

The ratio between the TEQs in sediments and that in invertebrates is a function of bioavailability. The TEQ concentration, based on organic carbon-normalized and lipid-normalized concentrations, decreased by a factor of about 3.3 from the sediments to the invertebrates (see Table 6-1). This factor probably represents the measure of bioavailability of the non- and mono-ortho-substituted congeners due to factors not accounted for by organic carbon normalization. The organic carbon in sediments is not exactly the same as invertebrate lipids (Karickhoff et al. 1979), and it has been suggested that the correction factor from sediment organic carbon to tissue lipids is approximately 1.7, with lipids dissolving more solute than sediments (Sabljic et al. 1995). If the ratio of normalized concentrations from sediment to invertebrate is corrected for this factor, a ratio of 1.9 is calculated. Theoretically, based on fugacity theory, if the organic carbon in the sediments is equivalent to that in biota, this value should be 1.0. Thus, only about half of the mass of PCB in

TABLE 6-1 Average TEQs Calculated from the Average Sum of Non- and Mono-Substituted PCBs and the Resulting TEQs Ratios Across Trophic Levels

Sample	Sum Non- and Mono-PCB		TEQs		TEQs
Matrix	ng/g Lipid	SD	ng/g Lipid	SD	Trophic Ratios
Sediment[1]	2.2×10^3	4.2×10^2	1.1×10^{-1}	2.2×10^{-2}	—
Invertebrate	8.3×10^2	6.8×10^2	3.0×10^{-2}	2.0×10^{-2}	Invertebrate/ sediment, 0.3
Egg	1.4×10^3	7.8×10^2	8.8×10^{-2}	4.1×10^{-2}	Egg/invertebrate, 3.0
Nestling	2.4×10^3	3.5×10^3	1.1×10^{-1}	1.6×10^{-1}	Nestling/egg, 1.3

[1]Sediment values are ng/g organic carbon.
Abbreviations: ng/g, nanograms per gram; SD, standard deviation.
Source: Adapted from Froese et al. (1998).

the sediment, normalized to organic carbon, is available for equilibrium partitioning—that is, the biologically available fraction is approximately 0.5. The estimate of bioavailability is complicated somewhat by the fact that invertebrates might have sediment in their guts (Ingersoll et al. 1995). Thus, the ratio between the two compartments may be biased either up or down, depending on the effect that the gut sediment has on the fugacity of the system.

Comparison of calculated TEQ values (Table 6-2) with PCB_{total} indicates that the average lipid-normalized PCB_{total} concentrations did not change across trophic levels, but the distributions of the congeners changed significantly. Potency is defined as the ratio of TEQs normalized to a PCB congener, PCB 153, that contributes a significant proportion of the total PCB mass and does not selectively weather or degrade (Jones et al. 1993). The potency ratio used in the study of Froese et al. (1998) was defined as the quotient of TEQ concentrations of adjacent trophic levels. For the samples analyzed by Froese et al., most of the calculated concentration of TEQ (87%; SD = 3%) was contributed by PCB 126.

No change in the relative potential from the egg to the nestling (potency ratio (nestling/egg) = 1.0) was observed by Froese et al. (1998) (Table 6-2). However, the potency changed by nearly a factor of 3 from the sediment-dwelling benthic invertebrates to the tree swallow eggs (potency ratio (egg/invertebrate) = 2.6). A potency ratio of 1.0 would be expected for the nestling/egg ratio, because there was little opportunity for significant metabolism or accumulation of PCBs from the egg to the nestling. For the initial trophic transfer, the adult bird accumulates, metabolizes, and transfers the organochlorines into the developing egg. The nestlings retain the egg burden of compounds, which are diluted by growth, but also begin further accumulation through feeding (Jones et al. 1994). There was a sharp decrease in the toxic potential from the sediment to the invertebrates (ratio = 0.01). As discussed above, this decrease is probably due in part to the differences in bioavailability of the lipophilic compounds in the organic carbon on the sediment versus in the lipid of the invertebrates. The relative proportion of those congeners that contribute the most to the toxic potential have been shown to increase with trophic level from sediments to birds (Jones et al. 1993, 1994). The relative potency factors observed here are similar to those that can be calculated for tree swallow eggs and nestlings from Green Bay, Wisconsin (Ankley et al. 1993). Using those data, the relative potencies were calculated based on total PCBs of 1.6×10^{-5}, 1.8×10^{-5} and 1.1×10^{-5} for tree swallow eggs, newly hatched chicks, and 16-day old nestlings, respectively. The average of these potencies (1.5×10^{-5}) is 2.2 times greater than that of the Saginaw Bay tree swallow eggs and nestlings (6.6×10^{-6}).

PCB153 is often used for normalizing PCB data, because it is a major component of several technical PCB mixtures (Clark et al. 1988). PCB 153

TABLE 6-2 Relative Concentration Ratios of Congeners 153 + 132 to PCB_{total}, Relative Potencies of TEQ Normalized to Either PCB 153 or PCB_{total}, and the Corresponding Potency Ratios

		Potency Based on PCB 153		Potency Based on PCB_{total}		Potency Based on $PCB_{coplanar}$		
Sample Matrix	PCB 153/ PCB_{total}	Potency	Potency Ratio[a]	Potency	Potency Ratio	Potency	Potency Ration	Enrichment[a]
Sediment	0.014	4.8×10^{-3}		6.6×10^{-5}		5.1×10^{-5}		120
Invertebrate	0.032	4.8×10^{-5}	0.01	1.5×10^{-5}	0.02	3.6×10^{-5}	0.7	1.5
Tree swallow egg	0.048	1.3×10^{-4}	2.6	7.1×10^{-6}	3.9	4.8×10^{-5}	1.7	8.0
Tree swallow nestling	0.055	1.3×10^{-4}	1.0	6×10^{-6}	1.2	6.2×10^{-5}	0.8	4.6

[a]Potency Ratios and Enrichment are calculated as described below in Equations 1 and 2, respectively. The potency ratio is always the potency of the higher trophic level matrix divided by the potency of the matrix that is one trophic level lower.
Source: Adapted from Froese et al. (1998).

is a di-ortho-substituted hexachlorinated congener that does not have vicinal unsubstituted carbon atoms and thus is not significantly metabolized by most organisms. On a DB-5 chromatographic column, PCB 153 co-elutes with PCB 132. The relative ratios of these two congeners are variable and depend on a number of factors, including metabolic processes across trophic levels or specific species (Niimi et al. 1996). PCB 132 can be metabolized, whereas PCB 153 cannot. Additionally, other PCB congeners are metabolized, ultimately resulting in the relative ratio of PCB 153 increasing with trophic level (Table 6-2). The effect of this increase is that the use of the PCB 153 + 132 peak for normalization might underestimate the potency, particularly if there are significant metabolism differences between tr1 to tr2 (Equation 1).

$$\text{Potency Ratio} = \frac{TEQ_{tr2}}{PCB\ 153_{tr2}} \bigg/ \frac{TEQ_{tr1}}{PCB\ 153_{tr1}}, \tag{1}$$

where $PCB\ 153_{tr}$ is the concentration of PCB congener 153 in trophic level; TEQ_{tr} is the equivalent concentration of 2,3,7,8-TCDD in trophic level; and tr1 and tr2 is the trophic levels 1 and 2, respectively.

However, the error caused by this effect is small, because the ratio of the PCB 153 + 132 peak to PCB_{total} does not change drastically among trophic levels (Table 6-2). When the relative potencies normalized to either the mass of 153 + 132 or that of PCB_{total} congeners, the potential bias of each normalization method must be considered, particularly when comparing data from different sources. Ratios of the relative potencies among trophic levels demonstrated a decrease in the relative toxic potency from sediment to invertebrates but an increase in the relative toxic potency between invertebrates and tree swallow eggs (Table 6-2).

The relative enrichment of TEQs with respect to the original Aroclor mixture in the same trophic level can also be calculated (Table 6-2). Enrichment was defined as the PCB 153-normalized ratios of TEQ concentrations in samples to TEQ concentrations of the Aroclor. This ratio provides a measure of the extent of change of the relative toxic potency of the PCB mixture in the environment, measured as TEQ increases from the original combination of technical Aroclor mixtures that were released into the environment (Equation 2).

$$\text{Enrichment} = \frac{TEQ_{sample}}{PCB\ 153_{sample}} \bigg/ \frac{TEQ_{Aroclors}}{PCB\ 153_{Aroclors}}, \tag{2}$$

where TEQ_{sample} is 2,3,7,8-TCDD TEQs contributed by PCB in the sample; $TEQ_{Aroclors}$ is 2,3,7,8-TCDD TEQs from Comstar-weighted original Aroclor contributions; PCB 153_{sample} is the concentration of PCB 153 in the sample; and PCB $153_{Aroclors}$ is the concentration of PCB 153 from Comstar-weighted original Aroclor contributions.

Enrichment of TEQs for invertebrates (Table 6-2) indicates that little change occurred in the PCB composition due to weathering or metabolic processes. However, enrichment occurred for tree swallow eggs and nestlings. That was likely due to metabolic processes that selectively dechlorinate and/or excrete some congeners, while allowing others, primarily the more AhR-active non- and mono-ortho-substituted congeners, to accumulate (Boon and Eijgenraam 1988; Tanabe et al. 1987). The relative enrichment of TEQs for sediments is 120, which is 15 to 80 times greater than the apparent relative enrichment for biota in the trophic levels above the sediment. It is unclear why a relatively greater enrichment of TEQs was observed in sediment.

Biota-Sediment-Accumulation-Factor Values

The BSAF has been proposed as an empirical relationship to predict the lipid-normalized concentrations of residues from the total organic carbon (TOC)-normalized concentration of the same residue in sediments (Ankley et al. 1992a,b; Tracey and Hansen 1996). The BSAF approach has been proposed for use as a regulatory tool in risk-assessment methodologies involving contaminated sediments (Parkerton et al. 1993).

BSAF values measured by Froese et al. (1998) varied, depending on whether they were calculated on the basis of the PCB_{total}, the sum of non- and mono-ortho-substituted PCBs, or TEQs (see Table 6-3). BSAF calculations based on total PCBs were between 8 and 11, and those based on non- and

TABLE 6-3 BSAF Values[a] for Each Matrix Based on Total PCBs, the Sum of the Non- and Mono-Ortho-Substituted PCB Congeners, and TEQs

Matrix	PCB_{total}	Sum of Non- and Mono-PCBs	TEQs
Invertebrates	11	0.4	0.3
Tree swallow eggs	8.8	0.6	0.8
Tree swallow nestlings	9.3	1.1	1.0

[a]Each value represents the ratio of the lipid-normalized concentration in tissue divided by the organic carbon-normalized concentration in sediments.
Source: Adapted from Froese et al. (1998).

mono-ortho-substituted congeners ranged from 0.4 to 1.1. The BSAF is operationally defined and can be site- and species-specific (Lake et al. 1990). The actual BSAF values measured are dependent on the chemical-physical properties of both the residue and sediments. Although TOC may be similar among sediments, there are qualitative, in addition to quantitative, differences that are not accounted for in the BSAF. In addition, the relative importance of TOC, inorganic properties, and size of sediment particles can have an influence on the BSAF values. Also, duration of the residence in the sediment and exposure of biota to the sediments can influence the BSAF value particularly for superhydrophobic compounds that take a long time to reach steady-state with the sediment and biota (Hawker and Connell 1985). Finally, composite samples of emergent invertebrates integrate the overall exposure of organisms to an area in which actual concentrations of residues (such as PCBs), type of sediment, and TOC content might be quite patchy. Thus, in calculating a BSAF, relatively great variations in values can be observed if a small number of samples of TOC-normalized sediment-residue concentrations are used to estimate the denominator of the BSAF. That was the case in the study by Froese et al. (1998) in which the average TOC-normalized total PCB concentration in sediments was 1.7 ng of PCB/g of TOC (standard deviation = 2.6 μg of PCB/g of TOC), a range of more than 34-fold between the least and greatest values. If the greatest value was used instead of the average value, the BSAF would be approximately 1.0. Thus, the anomalously high value of the BSAF for total PCBs is probably due to such variation. Based on this type of sensitivity analysis, a range of as much as 35-fold would be expected in BSAF values calculated in this manner. The use of BSAF values in risk assessments assumes that these values do not vary among locations or that an overall average value can be calculated for a region. Although that is not necessarily an invalid assumption, the range of BSAF values should be considered in addition to the average values in interpreting the results of risk assessments.

Fugacity theory predicts that if the organic carbon in the sediment and the lipid in the animal tissues is equivalent as a solvent for the contaminant of interest, the BSAF should be 1.0 in systems at steady state (Hoke et al. 1994). That theoretical value is generally not observed in data collected from the field, because the octanol-equivalent fat fraction for sediment dry-weight organic matter is about 0.3 (Karickhoff et al. 1979; Sabljic et al. 1995). Thus, the BSAF is approximately 1.7 if it is calculated from organic carbon-normalized concentrations in the sediment and lipid-normalized concentrations in organisms.

BSAF values for total PCBs are often greater than would be expected if based on fugacity theory. The high values might be related to the contribution of PCBs in sediments within the guts of the invertebrates. Similarly, anomalously large BSAF values have been observed for accumulation of other compounds from sediments by invertebrates (Eadie et al. 1985; Landrum 1989; Landrum et al. 1992). However, BSAF values of 1-2 have also been reported for PCB_{total} (Ankley et al. 1992a,b). When BSAF values were based on the sum of non- and mono-ortho-substituted PCB congeners or TEQ, BSAF values were less than or equal to 1.0 and increased with increasing trophic level (Table 6-3).

BSAF values have been reported to range from 0.1 to 10 for benthic invertebrates (DiToro et al. 1991). The BSAF values observed for the distribution of total PCBs between sediments and infaunal invertebrates can range by up to 2 orders of magnitude, but a global average value of 1.7 has been suggested for use in risk assessments when BSAF values have not been determined for a particular site (Landrum and Poore 1988). For instance, BSAFs for accumulation of PCBs from marine sediments by infaunal organisms, such as mollusks (*Mercinaria mercinaria)* and polychaetes (*Neghtys incisa),* ranged from 1.7 to 4.6, depending on the PCB congener (Lake et al. 1990), and BSAFs for accumulation of PCB 153 by the mayfly (*Hexagenia limbata*) ranged from 4.5 to 15.5 (Drouillard et al. 1996). Other BSAF values for mayfly accumulation of PCB congeners from sediments have been reported to range from 4.8 to 6.5 (Boese et al. 1995). BSAFs for accumulation of individual PCB congeners by the mussel (*Malacoma nasta*) ranged from 0.19 for PCB 209 to 4.74 for PCB 118 (Landrum and Poore 1988). Those authors found a mean BSAF of 2.84 for all of the PCB congeners that they examined. In a compilation of reported values (Tracey and Hansen 1996), BSAFs ranged from 0.80 for the mayfly and 0.93 for the blue mussel (*Mytilus edulis*) to 5.11 for the bivalve mollusc (*Yoldia limatula*) and 5.30 for the midge (*Chironimid* spp.). The mean of median BSAF values for various species was 2.1 (Tracey and Hansen 1996).

The results of the assessment of the patterns of relative concentrations of PCB congeners as a function of trophic level indicate that when BSAF values are based on the sum of non- and mono-ortho-substituted PCB congeners or TEQs, the observed BSAF values are similar to fugacity theory predictions. Thus, the concentrations of TEQs measured in tissues of tree swallows might be used to predict an integrated measure of concentrations of TEQ in sediments, and vice versa, by the application of a fairly simple partitioning model.

Effects of PCBs on Wildlife

In the sections below, the committee summarizes the information on adverse effects associated with exposure to PCBs on wildlife and humans. The types and thresholds for effects of PCBs is presented for lower trophic organisms—that is, plankton and invertebrates, fish, birds, and nonhuman mammals.

(More detailed information on the toxic effects of PCBs on organisms may also be found in Appendix G.)

Generally, PCBs and other dioxin-like chemicals are not particularly toxic to lower trophic-level biota, including algae, zooplankton, and invertebrates, because these species lack the AhR that mediates the critical mechanism of toxicity. Numerous studies have been conducted with both individual congeners and Aroclor mixtures to assess the toxicity of PCBs to invertebrates. The EPA ECOTOX: Ecotoxicology Database contains 1,687 individual toxicity data citations for PCBs in aquatic organisms, including algae, zooplankton, invertebrates, and fish.

Because PCBs usually are associated with, or bound to, sediments, attempts have been made to relate PCB concentrations in sediment to effects concentrations. These effects concentrations, as well as the national ambient water quality criteria for total PCBs, are presented in Appendix G. It should be noted, however, that these so-called sediment effects concentrations have several limitations. Specifically, there is a lack of causal association due to the presence of co-contaminants and a poor correlation to observed effects. In addition, the bioavailability and toxicity of PCBs in sediments can vary, depending on the sediment organic carbon concentration. In general, the primary uses of invertebrate PCB data for ecological risk assessments are to provide information on site-specific bioavailability and for input to dietary models for biota that consume invertebrates. Because of the poor correlation between concentrations of PCBs and effects in benthic invertebrates and the inherent limitations of the apparent-effects-threshold (AET)-type approaches, such as effects range medium (ERM), due to the co-occurrence of other contaminants, the values developed by these approaches are underestimates of the actual threshold concentration for adverse effects. Therefore, the use of these methods is not recommended.

A more appropriate approach would be to apply the equilibrium partitioning (EqP) theory to calculate sediment quality criteria (Giesy and Hoke 1990). In this method, the site-specific organic carbon content of the sediment is used in conjunction with the organic carbon partitioning coefficient for the compound (Koc) to predict the concentration of PCBs in sediment pore water. This predicted concentration is then compared with the water-quality criterion for fresh or saltwater, as appropriate to calculate a hazard quotient (HQ). By assuming a HQ value of 1.0, the critical concentration in bulk sediments can be calculated. There are Aroclor data available for the calculation, but as noted above, they appear to be of limited value because of the relatively high concentrations required to elicit effects. Furthermore, coplanar PCBs are not expected to be substantially more toxic to invertebrates than other PCBs because of the absence of the cellular receptor required to mediate toxicity.

Assessments of ecological risks for receptors of concern (such as fish, birds, and mammals) are made by comparing observed or projected chemical concentrations (or dosage rates) to toxicity reference values (TRVs). In this approach, TRVs represent concentrations of chemicals in water, food, or the tissue of the receptor (or dietary dosage rates) that will not cause toxicological effects. Ideally, TRVs are derived from chronic toxicity studies in which an ecologically relevant endpoint was assessed in the species of concern, or a closely related species. Although TRVs can be based on no-observed-adverse-effects levels (NOAELs), they also can be calculated from the lowest-observed-adverse-effects levels (LOAELs). With robust data sets, TRVs calculated from LOAELs might incorporate less uncertainty than TRVs derived from NOAELs. Alternatively, TRVs can be expressed as the geometric mean of the NOAEL and LOAEL to provide a conservative estimate of a threshold of effect (Tillitt et al. 1996).

The development of TRVs for site assessment is based on information on chronic toxicity effects in laboratory and field-based studies of target, sentinel, or surrogate species. Toxic effects of PCBs include developmental effects, reproductive failure, liver damage, cancer, wasting syndrome, and death (Metcalfe and Haffner 1995). These effects are primarily attributed to the non-ortho and mono-ortho-substituted or coplanar PCBs (Safe 1990; Giesy et al. 1994a; Metcalfe and Haffner 1995). The toxicity of the coplanar PCBs, TCDD, and other polychlorinated dibenzohydrocarbons (PCDHs) is expressed through their ability to bind to and activate the AhR (Poland and Knutson 1982; Gasiewicz 1997; Blankenship et al. 2000). Although most of the toxicity of PCBs is attributed to dioxin-like congeners acting through an AhR-mediated mechanism of action (Blankenship et al. 2000), PCB congeners with more than two ortho-substituted chlorine atoms have also been reported to cause chronic (primarily neurobehavioral) toxicity through non-AhR-mediated effects (Giesy and Kannan 1998). These effects, which have been observed, only occur at much greater concentrations than AhR-mediated effects. Although PCB congeners with more than two ortho-substituted chlorine atoms constitute the greatest mass of total PCBs in the environment, toxicity associated with the coplanar, dioxin-like congeners will almost invariably be dominant in higher organisms (i.e., fish, birds, and mammals) that possess the AhR.

Causal links between chronic toxicity effects and Aroclor or total PCB concentrations have been reported in a number of laboratory studies. These results, however, might not be directly applicable to field conditions, because they were largely performed using parent Aroclor mixtures and do not account for possible reduction or enrichment in toxic potency by environmental weathering (Williams et al. 1992; Corsolini et al. 1995a,b; Quensen et al. 1998). For example, microbially mediated anaerobic reductive dechlorination is common in sediments and can eliminate the most toxic coplanar congeners (Zwiernik et al. 1998). In contrast, certain aquatic animals selectively bioaccumulate the coplanar dioxin-like congeners. Such accumulation can result in a greater relative proportion of these toxic congeners in tissues than in Aroclor mixtures (Williams et al. 1992; Corsolini et al. 1995a,b).

Based on the dominance of AhR-mediated effects, an alternative approach is to correlate observed toxicity to the concentrations (or equivalent concentrations) of dioxin-like congeners. This approach has been shown to work quite well for birds and mammals (Giesy et al. 1994a; Leonards et al. 1995). In determining equivalent dioxin-like concentrations, each PCB, dioxin, and other PCDH congener is considered to bind with different affinity to AhR and therefore to exhibit a different potency for eliciting biological effects (Safe 1990; Ahlborg et al. 1994; Van den Berg et al. 1998). Although the toxic potency of each of these chemicals varies, the potencies can be normalized to that of TCDD, the most potent of the PCDHs, using a toxicity equivalency factor (TEF) (Van den Berg et al. 1998). The concentration of TCDD toxicity equivalents (TEQs) is then determined by multiplying the concentration of a chemical and its TEF as follows (Equation 3):

$$\text{ng of TEQ / kg of sample} = \sum_{i=1}^{n}\left[\text{PCDH}_i \times \text{TEF}_i\right], \qquad (3)$$

where PCDH_i represents the concentration of a specific PCDH, and n is the total number of PCDH congeners contributing to dioxin-like toxicity.

Risk assessments are meant to be accurate representations of the potential for risk but are designed to be protective rather than predictive. Even though PCBs are a mixture of compounds with different relative potencies, the concentration of the entire mixture to which an organism is exposed is determined by the congeners or class of congeners with the maximum toxic potency. Because the potential for adverse effects is a function of the exposure concentration and sensitivity of the organism of interest, the best way to determine the critical toxicant is by comparing hazard quotients (Equation 4).

$$\text{HQ} = \frac{\text{measured or modeled concentration}}{\text{TRV}}, \qquad (4)$$

The compound or compounds with the greatest HQ values are the critical toxicant—that is, the toxicants that would have the greatest overlap between exposure and TRV values would determine the maximum allowable exposure to the entire complex mixture.

When this analysis has been conducted for a variety of species at a variety of locations, it was found that when the allowable concentration of PCB mixtures is based on the HQ for TEQs for dioxin-like effects, concentrations of non-AhR-active congeners are less than the threshold for effects caused by the nondioxin-like congeners. In other words, if the risk assessments are based on TEQs, wildlife will be protected from the effects of nondioxin-like compounds as well.

A large number of laboratory and field toxicity studies have been conducted over the years using Aroclor, total PCB, individual PCB congeners, TCDD, and TEQ concentrations as a measure of PCB exposure. Results for lower trophic aquatic organisms, fish, birds, and mammals are summarized below. However, because organisms are exposed to environmentally weathered mixtures that contain not only PCB congeners but also polychlorinated dibenzo-*p*-dioxins (PCDDs), polychlorinated dibenzofurans (PCDFs), polychlorinated naphthalenes (PCNs), and possibly other AhR-active compounds; the potential effects of PCBs should be interpreted in the context of these other compounds (Giesy et al. 1994b). However, at most PCB-contaminated sites, PCB coplanar congeners will likely be the primary contributor to AhR-mediated toxicity.

Short discussions of the thresholds for toxic effects for lower trophic level organisms, fish, birds, and nonhuman mammals are given in the following sections. (A detailed analysis of mechanisms of analysis and TRVs is given in Appendix G.)

Toxicity Results for Lower Trophic Organisms

PCBs and other dioxin-like chemicals are not particularly toxic to lower trophic organisms, such as phytoplankton, zooplankton, and invertebrates, because these species lack the cellular AhR that is required to mediate dioxin-like toxicity. PCBs do exhibit a minimum level of toxicity due to nonspecific or "narcotic" effects in these organisms. This level of toxicity is related to the critical body concentration (McCarty and Secord 1999).

Toxicity data indicate that waterborne or tissue PCB concentrations that might cause adverse effects for lower trophic level organisms are much greater than PCB concentrations reported for contaminated environments (Niimi et al. 1996). For example, Niimi et al. (1996) reported adverse effects in phytoplankton and zooplankton at PCB concentrations in water greater than 0.5-1.0 μg/L, which is roughly 100-1,000 times greater than PCB concentrations reported for contaminated ecosystems. Effects thresholds for PCB-contaminated sediments have been suggested (e.g., effects range low (ERL) and effects range median (ERM) values of 0.0227 and 0.18 mg of PCB/kg of dry weight) (Long and Morgan 1991). These threshold values, however, are suspect because of the lack of a causal relationship between PCB-sediment concentrations and effects. Confounding factors affecting the relationship between sediment concentrations and effects can include differences in PCB bioavailability as a function of sediment organic carbon and the presence of co-contaminants. Therefore, ERM and ERL values are deemed to be unreliable and should not be used for ERAs.

Toxicity Results for Fish

Adult fish are exposed to PCBs and related compounds via water, sediment, and food. Eggs and embryos can accumulate these lipophilic chemicals from the female during vitellogenesis. Bioaccumulation of PCBs by fish is dependent on the physical and chemical characteristics of individual congeners and on the biotransformation and elimination rates of congeners by fish. Early-life stages are generally the most sensitive to chemical contaminants. Thus, accumulation of these persistent chemicals during this life stage is critical to the characterization of risks from PCB exposures.

There are considerable field and laboratory data (from water, dietary, intraperitoneal-injection, and egg-injection exposure) available in which concentrations of PCB congeners, TCDD, and TEQs in eggs or other fish tissue have been measured and related to particular adverse effects. Exposure to TEQs results in effects, such as mortality during early development and thyroid hyperplasia in adult fish (Walker and Peterson 1991). Other toxic effects and histopathological lesions seen during early development are characterized by cardiovascular and circulatory changes, edema, hemorrhages, and mortality (Walker and Peterson 1991). Results of risk assessments demonstrate that maximum allowable tissue concentrations in fish tissue are more dependent on the potential for effects in piscivorous wildlife than on direct effects in the fish themselves.

The majority of fish toxicity data for PCBs is from laboratory experiments. For example, 96-hour LC_{50} values (the dose that is lethal to 50% of a test population) for fathead minnows are reported to be in the range of 8-15 μg/L for different PCB mixtures (Nebecker et al. 1974). The major route of exposure to fish in the wild, however, is not likely to be through the water but rather via the food chain. However, tissue-based effects thresholds are independent of exposure pathway, because both effects and measured concentrations in the eggs of rainbow trout were the same whether exposure occurred by maternal transfer, water uptake, or injection (Walker et al. 1996). Concentrations of PCBs in tissue, therefore, provide a better means of assessing toxicity, because they bypass the issue of water-versus-food-web exposure.

Life stages of marine fish species were evaluated for PCB concentrations in tissues and for critical effects (Jarvinen and Ankley1999). Niimi et al. (1996) presented a similar table with greater tissue residue-based effects threshold values. Effects are observed in the high microgram per kilogram to low milligram per kilogram range.

If tissue residue concentrations are not available, residue concentrations can be estimated from measured water or sediment concentrations by using equilibrium partitioning relationships, such as the bioconcentration factors (BCF), bioaccumulation factors (BAF), BSAFs, or more-complex trophic-transfer models.

Toxicity Results for Birds

There are sufficient toxicological data available for birds to evaluate both dietary and tissue residue-based effects levels for PCBs. However, data are not available for all of the pertinent species at particular locations. Thus, uncertainty factors are still applied to determine TRVs. As with other organisms, birds demonstrate considerable differences in species sensitivities to PCBs and related dioxin-like compounds. In particular, chickens, which are the most frequently used species for PCB exposures, are among the most sensitive of avian species. Therefore, it is important to use species-specific toxicity data whenever possible.

Toxic effects of PCBs and related compounds in birds have been studied by in vivo exposure, in ovo exposure by injection, and in vitro exposure with cultured avian hepatocytes. Due to better control of exposure, dose, and timing, egg-injection studies are often of more use in deriving TRVs. In most cases, embryotoxic and teratogenic effects of PCBs seem to be the most sensitive and ecologically relevant endpoints in birds (Hoffman et al. 1998). Several in ovo studies of dioxin-like compounds are adequate for determining tissue-based-toxicity LOAEL, NOAEL, and EC_{50} (the concentration that causes an effect in 50% of a test population) values that can be used in developing TRVs for certain PCB congeners (Powell et al. 1996a,b; Henshel et al. 1997; Hoffman et al. 1998).

Toxicological studies (Ankley et al. 1993; Murk et al. 1994; Tillitt et al. 1996; Boon et al. 1997) have been found that correlate effects from PCB dietary exposure to Aroclors and that correlate tissue residue effects levels in eggs to Aroclors, total PCBs, PCB congeners, and TCDD and TCDD-equivalents (see Box 6-7). These studies show good consistency. Aroclor results

BOX 6-7 Special Consideration of PCB 77

PCB 77, one of the more potent non-ortho PCB congeners, contributes one of the greatest mole fractions to the TEQ content of most technical PCB mixtures. Due to its greater susceptibility to metabolism and degradation, it requires special consideration. Although originally considered resistant to degradation, and readily biomagnified because of its lack of ortho substitution (Tanabe et al. 1987), PCB 77 behaves differently from other non-ortho PCBs and differently from other members of the tetrachlorinated homologue group (Willman et al. 1997; Froese et al. 1998). Therefore, it is *not* appropriate to assume that PCB 77 or other individual congeners have similar properties within a homologue group.

Based on biomagnification factors derived from field studies, it is unlikely that PCB 77 would biomagnify from fish to bird eggs to the same extent as congeners 126 and 169. In a study of bird eggs and hatchlings from the lower Fox River and Green Bay, Wisconsin, it was observed that PCB 77 decreased when hatchlings reached an age of approximately 25 days old, whereas all other congeners continued to accumulate (Ankley et al. 1993). Those authors suggested that the decrease was due to metabolism of PCB 77. This observation is in agreement with other studies that have observed metabolic degradation and depletion of 77 in bird microsomes (Murk et al. 1994), marine mammals (Boon et al. 1997), and mink (Tillitt et al. 1996) relative to other congeners. Similarly, Bright et al. (1995) reported that PCB 77 was diminished rather than enriched with increasing trophic status (going from sediment to sea urchins to four-horn sculpins). In a study of Green Bay, Willman et al. (1997) found that PCB 77 decreased in relative abundance both within total PCBs and within its homologue group with increasing trophic level. Due to the predominance of PCB 77's contribution to total TEQs (using avian TEFs) in PCB-contaminated sediments, PCB 77 is a critical congener to model when characterizing risk to avian species. Thus, modeling of congener 77 must take the above factors into account.

are of limited use, because the studies were conducted using unweathered Aroclor mixtures, and very few of the studies were conducted on wildlife species. Total PCB results, particularly results for the last 10 years, have largely been performed using "weathered" PCBs. For risk assessments, PCB concentrations in bird eggs can be estimated using biomagnification factors.

Toxicity Results for Nonhuman Mammals

Evaluations on the effects of environmental contamination on the health of aquatic mammals have presented a considerable challenge. Although the results of semi-field studies have been confounded by the occurrence of co-contaminants in the diet and by the lack of dose-response relationships, studies in such animals as seals have suggested that PCBs were the main cause of the observed toxic effects (de Swart et al. 1994; Ross et al. 1995). Furthermore, the exposures mimicked observed field situations, and, therefore, these values were considered suitable for deriving a toxicity benchmark to assess risks of PCBs to aquatic mammals.

A few studies are available for assessing toxicity in seals (de Swart et al. 1994; Ross et al. 1995), otter (Leonards et al. 1995), and mink (Tillitt et al. 1996). However, logistic considerations make the sampling of a large number of wild animals difficult, and ethical concerns discourage in vivo studies on captive animals. Thus, much of the mammalian toxicology is based on study with rodents. However, mink are among the most sensitive animals that have been studied to date. Risk assessments based on mink often result in the least allowable concentrations in sediments, and if mink are used as a surrogate in risk assessment, most other species, including humans, should be protected.

Lipid-normalized NOAEL values of 11 microgram per gram (μg/g) in seal blood and of 9 μg/g in otter liver were reported by Kannan et al. (2000) for PCBs. A lipid-normalized LOAEL value of 210 picogram/gram (pg/g) for TEQs in seal blubber was reported by Ross et al. (1995). That value is less than the reported range of 400-2,000 pg/g for otters and mink, implying that seals are more sensitive to TEQs.

Summary of Effects in Wildlife

PCBs have been demonstrated to cause a wide range of effects in animal models. Because of the co-occurrence of other stressors and accessory factors, it is more difficult to demonstrate PCB-specific effects in wildlife and almost impossible to demonstrate in human populations. The best evidence of the effects in populations outside the laboratory are those effects in wildlife, especially birds. Effects in mammals and fish are less obvious, in part because it is more difficult to follow individual organisms throughout their development. In PCB-contaminated environments, there is good evidence of their effects in some wild populations of birds. The effects attributed to exposure to PBS include alteration of enzyme activities, suppression of the immune system, embryo-lethality, and developmental abnormalities. The effects observed in the field have been substantiated in the same wildlife species and surrogate animal models under laboratory conditions. Depending on the degree of contamination, the effects in birds can be minor to severe. In some locations, the weight of evidence is strong that PCBs have caused adverse effects in individuals and that this exposure has resulted in population-level effects. In other situations, the potential for effects must be inferred.

Human Health Effects from PCBs

Much of the information on the human health effects of PCBs have been gleaned from three major types of studies: occupational studies (Bertazzi et al. 1987; Brown 1987; Sinks et al. 1992 Kimbrough et al. 1999); retrospective and prospective studies of populations exposed to relatively high concentrations of PCBs through accidental exposure (e.g., Yusho and Yucheng incidents; Kuratsune et al. 1972; Masuda 1985; Rogan et al. 1988); and longitudinal prospective cohort studies of selected populations, such as the Michigan PCB study (Humphrey 1983), the Michigan aging sport fishermen study (Schantz 1996), and the Michigan angler study (Courval et al. 1999). In addition to those studies of populations exposed postnatally, scientists are studying several cohorts of children exposed to low-to-moderate (Rogan et al. 1988; Huisman et al. 1995; Patandin et al. 1999) or moderate-to-high (Lonky et al. 1996; Mergler et al. 1998) levels of PCBs in utero. In the studies on developmental effects, the more highly exposed cohorts had parents who ate PCB-contaminated fish (generally caught from the North American Great Lakes waters), used contaminated cooking oil (e.g., Yusho and Yucheng incidences), or for the Michigan cohort, were exposed to PCBs in food (mostly meat and dairy) from PCB-treated silos on farms. The more moderately exposed infants were exposed to PCBs from their parents diet, which presumably reflects the diet of the "typical" Dutch or American citizen, since the mothers were recruited from the general population.

These developmental cohort studies have produced similar results, demonstrating that the more highly exposed infants exhibited neurodevelopmental deficits or changes in behavior. For example, the infants in the 90th percentile exposure group in the study by Rogan et al. (1986) exhibited hypotonicity and hyporeflexia based on the Brazelton Neonatal Behavioral Assessment (Rogan et al. 1986). Interestingly, most of the neurobehavioral delays or deficits observed have been correlated more with measures of prenatal exposure to PCBs than with measures of lactational exposure to PCBs, even when the neurobehavioral observations (including learning) were made years later (up to 11 years later for the study by Jacobson and Jacobson 1996) (Jacobson et al. 1985, 1990; Gladen et al. 1988). Similarly, alterations in immunological parameters were most strongly correlated with prenatal rather than postnatal measures of PCB exposure, although the effects from postnatal exposure disappear more quickly (primarily within 2-3 years following exposures) (Weisglas-Kuperus et al. 1995).

These observations in humans are supported by longitudinal studies of rhesus monkeys exposed to PCBs in utero (while the mother was being fed PCBs) or in breast milk (as long as 3 years after maternal exposure to PCBs). In one study, mothers were fed Aroclor 1016 at 0, 0.007, or 0.028 milligram per kilogram (mg/kg) of body weight per day throughout the gestational and lactational periods. Hyperpigmentation in the infants was observed in both dose groups. However, decreased birth weight and impaired performances of spatial discrimination, spatial learning, and memory were observed only in the high-dose group (Levin et al 1988; Schantz et al. 1991). An additional set of studies followed the offspring of females dosed with Aroclor 1248 at 0, 0.5, 1, and 2.5 parts per million (ppm), 3 days per week throughout the prenatal and the 4-month lactational period. Offspring born up to 3 years after exposure to the mothers receiving the highest dose were also followed and tested. The infants of the mothers in the highest dose group had reduced body weight and deficits in discrimination-reversal learning, hyperactivity when young, and hypoactivity as adolescents (Bowman et al. 1978, 1981). Adolescent and young adult monkeys, born up to 3 years after the mothers were no longer fed the PCBs, still exhibited neurobehavioral deficits on a spatial alteration task.

Because of the relatively great sensitivity of the developing fetus and the correspondence between human and animal data, many risk assessors now choose to use the developmental neurotoxicity data as the basis for setting reference doses (RfDs) for human health risk assessment. A review of RfDs derived by various agencies in the first half of the 1990s (Rice 1995) shows that even the highest RfD with the least protective assumptions (0.1 μg/kg/day for the less-chlorinated PCBs) (Aroclors 1016 and 1242), and 0.05 μg/kg/day for the more-chlorinated PCBs) results in an acceptable level in fish tissue of approximately 0.05 mg/kg. However, this value used only the RfD for the more-chlorinated PCBs, because bioaccumulation and biotic metabolic processes favor accumulation of the more-chlorinated congener over the more readily (but still slowly) metabolized, less-chlorinated PCB congeners.

Therefore, even though potentially higher levels of exposure might result from postnatal exposure in breast milk, the more biologically sensitive and significant exposure period is prenatal. To address that, most states have set advisories for fish consumption at lower PCB concentrations for women of child-bearing age than for men.

Cancer

Cancer, including stomach, squamous-cell carcinoma (Svensson et al. 1995), and non-Hodgekin's lymphoma (Rothman et al. 1997), has been linked to PCB exposure. However, the link between exposures to PCBs and cancer incidence has been somewhat weaker for some cancers (Sinks et al. 1992; Brown 1987; Bertazzi et al. 1997). Unlike the effects of prenatal and perinatal exposure to radiation, onset of cancer does not appear to be an effect of developmental exposure to PCBs. Rather, PCBs have been shown to be promoters of tumorigenesis after initiation by a genotoxic agent that, unlike PCBs, can form DNA adducts with subsequent DNA damage. Thus, it is unlikely that PCBs are complete carcinogens (Delzell et al. 1994).

Limitations of Human Health Risk Assessments for PCBs

The classic problem with determining relative risk due to mixtures of compounds, like PCBs, is posed by the following question, "How do we account for the toxicity of the individual congeners when potencies differ between congeners and between endpoints (effects) for a given congener?" One limitation of the epidemiological studies of the effects of PCBs is the problem of measuring rare effects induced in plastic organ systems or for endpoints where there is a strong cellular repair mechanism. Cancer is a rare effect, and the correlation between exposure to a given chemical, such as PCBs, and the increased incidence of cancer is difficult, if not impossible, to make. First, the manifestation of the effect (cancer or a tumor) is delayed, and estimates of prior exposure, which could have been in utero or several decades earlier, are almost impossible to accurately assess retrospectively. All of the epidemiological studies of occupational exposures are weak in this respect (Brown 1987; Sinks 1992; Kimbrough et al. 1999). Longitudinal prospective studies can reduce this limitation, but it is difficult to follow a large enough cohort over the time period required to assess the association between exposure and future cancer incidence.

When assessing biomarkers that might be linked to later incidents (and incidence) of disease, changes in these biomarkers, such as induced P4501A activity, are also linked to chemicals that are often co-contaminants of PCBs.

In addition, the animal body is so plastic, and has so many compensatory and repair mechanisms that even when biomarkers, such as indicators of immune function (e.g., numbers of T or B lymphocytes, antibody titers, or evaluation of neutrophil function or T-cell activation), are compromised and exposure is known (e.g., the Dutch infants prospective study by Weisglas-Kuperus et al. 1995), it is difficult, if not impossible, to observe a statistically significant increased incidence of infection (Tryphonas 1998).

An additional limitation of the assessment of potential effects of PCBs in humans relates to multiple sources. In laboratory studies, all exposures can be theoretically controlled. Thus, the experimenter can choose the route and absolute exposure level. For wildlife studies, exposure can be assessed by taking tissue biopsies throughout the exposure period or by sacrificing animals or eggs to assess population or subpopulation exposure levels. For human studies, biopsies are a possibility but are not always taken and are rarely taken multiple times throughout the exposure period. Biopsies assess net exposure from all sources, but it can be prohibitively expensive to run full congener analyses on all biopsies taken. Few studies, even those for which almost full congener analyses are available, have analyzed the correlation between each individual congener and the endpoint of interest. Therefore, some congeners, albeit present in relatively high concentrations in some environmental samples (e.g., PCBs 47, 48, or 52), are routinely ignored. To address that, data sets for which there are more complete data, should be fully evaluated to determine the relative congener potencies for the less-studied congeners and their endpoints.

RISK CHARACTERIZATION

Risk characterization provides a formalized method for combining information from exposure to PCBs with dose-response relationships to determine the likelihood that adverse effects are occurring or will occur. Because information on exposure and effects is incomplete, this characterization should include an evaluation of the inherent uncertainties underlying the risk calculations. Risk-management decisions might result in competing risks; therefore, it is not appropriate to select extreme safety factors that result in gross overestimation of human and ecological risks from PCB-contaminated sediments. If no alternative risks exist, it might be more appropriate to take a very conservative approach in estimating the risks of PCBs in sediments. However, that is not the case, and the most accurate estimate of risk is required for rational decision-making. It is possible to overestimate the potential risks by 100- to 1,000-fold with the use of safety factors. Ecological and human risk characterizations are described below, followed by a discussion of uncertainty. This discussion is largely based on standard approaches that are used by EPA to characterize risk at PCB-contaminated sites.

Ecological Risk from Chemical Exposure

The potential risks of PCB exposure to ecological receptors are evaluated by comparing the results of exposure assessment (e.g., measured or modeled concentrations of chemicals in receptors of concern) with TRVs. In this approach, a hazard quotient (HQ) is calculated as in Equation 2. HQ values exceeding a value of 1 are typically considered to indicate potential risk to the average individual within the local population (e.g., reduced or impaired reproduction). If an HQ indicates that risks are present for the average individual, then risks are typically considered to be present for the local population. If an HQ suggests that effects are not expected to occur for the average individual, then they are considered likely to be insignificant for the local population. Recent advancements in modeling and geographic information systems have made it more feasible to consider ranges of variability in both measured and predicted exposures and responses. Application of probabilistic approaches, when appropriate, is encouraged. EPA has developed guidelines for the application of such approaches such as those of the EPA Workgroup on Probabilistic Risk Assessment for Pesticides (ECOFRAM) (EPA 1999).

Human Risk from Chemical Exposure

Risk characterizations for human health effects are developed on an individual, and not population-wide, basis. As a consequence, risk characterizations need to be performed for various subgroups within the population to account for differences in sensitivity and exposure rates (e.g., subsistence fishers). For each identified subgroup, chemical dose, or chronic daily intake (CDI), is determined from concentrations of PCBs in food and water consumption or air inhaled. For example, the CDI for ingestion of PCB-contaminated fish is given (Equation 5):

$$\mathrm{CDI}(\mathrm{mg\ \ PCB/kg \bullet day}) = \frac{\mathrm{C \times CR \times EF \times ED}}{\mathrm{BW \times AT}}, \tag{5}$$

where C is the exposure concentration (e.g., mg of PCB/kg of fish); CR is the contact rate (e.g., g of fish ingested/day); EF is the exposure frequency; ED is the exposure duration that, for example, may be determined from U.S. Census data based on population mobility; BW is the average body weight of the individual over the exposure period; and AT is the time over which exposure is averaged.

Variations in exposure are expected within an identified subpopulation because of individual differences in fish consumption rates, body weight, and PCB concentrations in consumed fish. Therefore, EPA guidance calls for an evaluation of both a central estimate of risk and an estimate of risk to an individual for a reasonable maximum exposure (RME). According to EPA guidance (1998a), central-tendency estimates are intended to reflect central estimates of exposure or dose, and RME estimates are intended to reflect persons at the upper end (i.e., above about the 90th percentile) of the distribution. Since the RME estimate is typically based on estimates of likely high-end exposure factors, the actual percentile associated with the RME is difficult to determine.

A preferred approach for estimating the RME uses probabilistic methods, such as a Monte Carlo analysis (EPA 1997a). The RME is expressed as probabilities of overlap between the distribution of the exposure and the effects (Solomon et al. 2000). The advantage of the Monte Carlo method is that an explicit estimate of the likelihood, or probability, of exposure can be obtained given sufficient data on parameter distributions (e.g., a population's fish ingestion rate, body weight, exposure period, and PCB concentrations in fish). In the Monte Carlo method, the range and relative likelihood of exposure is calculated by randomly selecting parameter values from their probability distributions and calculating a CDI. CDI calculations are repeated many times using this approach to obtain a distribution of chemical exposures. A set of selected CDI percentiles (e.g., the 90th, 95th and 98th percentiles) are then determined for subsequent risk calculations. Risk calculations for human health are performed separately for hazards associated with noncancer effects, such as reproductive impairment, developmental disorders, disruption of specific organ functions, learning problems, and for incremental risk of developing cancer due to PCB exposure.

For noncancer hazards, risk is expressed by the ratio of the CDI to an RfD using the HQ in Equation 6:

$$\text{Hazard Quotient } \left(\text{HQ}\right) = \frac{\text{CDI}}{\text{RfD}}. \qquad (6)$$

HQs are then summed over all chemicals of potential concern and all applicable exposure routes to determine the total hazard index (HI). If an HI is greater than 1, unacceptable exposures might be occurring, and there might be a concern for potential noncancer effects. The relative value of the HI, however, should not be translated into an estimate of the severity of the hazard or a prediction for a specific disease. This index is simply an index to guide decision-making and further analysis.

For cancer, risk is expressed as a probability and is based on the cancer potency of the chemical, known as the cancer slope factor (CSF, Equation 7). The CSF is the plausible upper-bound estimate of carcinogenic potency used to calculate risk from exposure to carcinogens by relating estimates of lifetime CDI to the incremental risk of an individual developing a cancer over a lifetime. This calculation is based on total concentrations of PCBs, and the CSF is based on technical Aroclor mixtures.

$$\text{Cancer Risk} = \text{CDI} \times \text{CSF}. \qquad (7)$$

To account for certain PCB congeners exhibiting dioxin-like toxicity, supplemental calculations for dioxin cancer risks should be performed using measured, modeled, or estimated concentrations of TCDD-toxicity equivalents (TEQs) in estimating CDI values. The primary, although not exclusive, source of information for CSFs and RfDs used in Superfund risk assessments is the EPA Integrated Risk Information System (IRIS) database, which is updated periodically to incorporate new data. Although the committee notes that EPA's cancer classification for dioxins remains controversial as of publication of this report, the values are used as examples. For example, acceptable upper-bound risks for cancer are in the range of 1×10^{-4} to 1×10^{-6} (or an increased probability of developing cancer of up to 1 in 10,000 to 1 in 1,000,000).

Uncertainty

When potential risks posed by PCBs have been assessed for cancer, reproductive impairment, and neurobehavioral effects, the allowable daily consumption of PCBs in the diet (primarily from fish) were similar (Williams et al. 1992). Due to the conservative nature of the linear extrapolation applied in developing the RfD based on cancer risk, the allowable exposure for cancer risk would be approximately 2-fold less than that for noncancer endpoints. This level of uncertainty is relatively small, compared with the other uncertainties encountered in risk assessments. Furthermore, when risk assessments have been based on AhR-mediated effects (dioxin-like effects on reproduction) and non-AhR-mediated effects (nondioxin-like effects of di-ortho-substituted PCBs on neurobehavioral endpoints), the allowable exposures were similar (Giesy and Kannan 1998), AhR-mediated endpoints resulting in approximately 2-fold greater HQ values. Thus, basing risk assessments on AhR-mediated effects should be protective of non-AhR-mediated effects. However, multiple lines of evidence based on both mechanisms of action should be considered to correct for site-specific differences in PCB congener patterns.

At each step of the risk-assessment process, there are sources of uncertainty that can ultimately affect the decision-making process. These include the following:

- *Sampling error and representativeness*: Chemical exposure estimates may include systematic error and random error associated with sample location (e.g., river mile), spatial heterogenities (e.g., coarse-grained versus fine-grained sediment, local vegetation), intraspecies variability, and temporal variation (e.g., seasonal variations associated with hydrological events and biological cycles).
- *Sample analysis and quantification*: Chemical measurements might not represent true concentrations in the environment because of possible inaccuracies in extraction and laboratory analysis of PCBs and co-contaminants. Measurement error might also be associated with dry weight and organic carbon content of sediments and wet weight and lipid content of biota. Measurement of PCB congeners at or near the detection limit is also a possible source of error for the dioxin-like congeners (e.g., PCB 126, concentrations of which are generally small but can be an important contributor to concentrations of TEQs).
- *Conceptual model*: Conceptual models are constructed to provide a realistic yet tractable description of chemical exposure pathways. If a certain contaminant pathway of either in-place or external sources of PCBs is not included or not properly characterized in the conceptual model, errors will be propagated throughout the analysis.
- *Natural variation and parameter error*: Natural variations in measurement of PCB concentrations, lipid content, age, body weight, feeding rate, and duration of exposure for an endpoint species might result in errors in specifying parameters for exposure assessment.
- *Model error*: Inappropriate selection or aggregation of variables (e.g., use of total PCBs), incorrect functional forms, and incorrect boundary conditions can lead to errors in exposure estimates.
- *Toxicological uncertainties*: A large number of toxicological studies were conducted in the laboratory using unweathered Aroclor mixtures. Uncertainties in toxicological endpoints may therefore be associated with laboratory stress of organisms and differences in laboratory and field exposures (e.g., weathering, presence of sediments, effects of dissolved organic carbon, effects of co-contaminants, and bioaccumulation effects). Intraspecies variability, interspecies variability, less-than-lifetime exposures, and acute-versus-chronic

toxicity testing are additional sources of uncertainty in defining dose-response relationships.

The exposure potential for PCBs via inhalation is based on the chemical-physical properties of the congeners. Natural-fate processes, such as anaerobic dechlorination of PCBs in sediments, result in the transformation of more-chlorinated, less-volatile PCBs (such as those found in commercial PCB mixtures) to less-chlorinated, more-volatile PCBs. As a result, the inhalation exposure potential from excavated sediments might be greater than the potential expected based on volatility data for commercially available Aroclors (Chiarenzelli et al. 1998). Furthermore, increased ambient-air PCB concentrations can occur during remediation operations because of increased physical disturbance of the sediments. These increased concentrations have been reported during wet excavation at several PCB-contaminated sites (e.g., Bloomington, Indiana) (Bloomington Herald Times 2000).

Summary of Risk Characterization

Currently applied methods for characterizing ecological and human health risks appear appropriate for PCB-contaminated sediments. Risk and the associated uncertainties of risk should be determined on a site-specific basis. For the overall analysis of risk from chemical exposure, the general sense is that there is typically a moderate degree of uncertainty associated with exposure assessment; an order of magnitude of uncertainty associated with ecological dose-response; and greater than an order of magnitude of uncertainty associated with the quantification of human health effects. As a result, the continued use of uncertainty factors in risk calculations is recommended to ensure to the extent possible that risk estimates are protective. At the same time, if this application of safety factors is not minimized and used in a transparent manner, the potential risks associated with PCBs in sediment will be overstated. Overstated risks can unduly alarm the public and force an active management solution that will not reduce risk and thus will be inappropriate. Furthermore, if some risks are overstated while others are not considered, such as those associated with active remediation, a biased decision will be reached, and the environment might be damaged with a reduction of risk, or in some cases, risks might be increased due to inappropriate or unnecessary remedial actions.

ECONOMIC, SOCIAL, AND CULTURAL IMPACTS

In addition to the potential effects of PCBs on human and ecological health, PCB contamination might result in various economic, social, and cultural impacts on the affected parties. For the purposes of assessment, these impacts fall into two general classes: direct (and indirect) human use effects and non-use values (Marine Policy Center 2000). For the first class, human uses that are curtailed or terminated because of PCB contamination and that may result in economic damages are typically considered. These uses include the following:

- *Lost benefits from commercial fishing*: Direct economic impacts are associated with losses in sale of the catch and increased unemployment of fishers in the region.
- *Lost benefits to recreational fishers (recreational experience)*: Economic impacts are felt directly through a decrease in tourism and marina activity. Reopening of river reaches as a "catch-and-release" fishery or with consumption advisories still results in lost benefits from personal consumption of recreationally harvested fish, which is an important component of the fishing "experience."
- *Reduction in property values*: Highly desirable waterfront property might be reduced in value because of the presence of PCB-contaminated sediments, particularly near PCB hot spots. This loss might occur even if there is no apparent danger to human health.
- *Increased costs of port, harbor, and navigation channel dredged-material disposal*: PCB contamination in dredging zones might be felt directly through the increased costs in disposal of contaminated dredged material or indirectly through curtailment in navigational dredging and associated loss in economic benefits from port activities and marina operations.
- *Increased health effects due to changes in diet from a fish-based diet to a less healthy diet*: For example, Ken Jock, of the St. Regis Mohawk Tribe (NRC Public Session, Albany, New York, November 8-9, 1999), cited increases in diabetes and thyroid diseases that are believed to be associated with changing from a fish-based diet.
- *Loss of benefits to recreational swimmers and boaters*: Although scientific information suggests that human health risk from incidental contact with river water and sediments is low, some individuals might avoid swimming and boating in contaminated areas because of a fear of being exposed to PCBs.
- *Increased costs of drinking water*: Although dissolved PCB concentrations in waters overlying PCB-contaminated sediments are typically below the maximum contaminant level goal for PCBs in drinking water (0.5 μg/L), individuals might incur costs, such as the purchase of bottled water because of uncertainty about PCB contamination.
- *Miscellaneous costs.* For example, Bill Acker, a concerned citizen (NRC Public Session, Green Bay, Wisconsin, September 27-29,1999), cited

a personal hardship associated with unexpected costs of disposing of PCB-contaminated zebra mussels harvested from his beachfront on Green Bay.

The second general class of impacts are the so-called non-use (or passive) values that are intangible and arise from the satisfaction that individuals experience from a particular environment, ecosystem, or river "culture," in the absence of any physical use. Many of these impacts are difficult to quantify but ultimately are related to one's willingness to sacrifice other benefits for the preservation of the environment for present and future generations. Examples of non-use include the following:

- *Protected species and their habitats*: Protecting the viability and promoting the recovery of species in a watershed might be an environmental good that has value to humans even if they do not use the species directly (e.g., the reemergence of bald eagles on the Hudson River).
- *Ecosystem services*: Non-use values might arise from the knowledge that the ecosystem, or even critical components of the ecosystem, is functioning properly.
- *Human culture*: The existence of a distinct community or culture that is strongly linked to the ecosystem might be threatened by contamination. For example, Ken Jock, of the St. Regis Mohawk Tribe (NRC Public Session, Albany, New York, November 8-9, 1999), cited how the loss of fishing and hunting grounds along the St. Lawrence River has caused many of the men to leave the reservation to find work, resulting in a breakdown of family and a loss of tribal traditions.

COMPARATIVE RISK ASSESSMENT

Determining an overall assessment of risk from PCB-contaminated sediments is a challenging task that requires sound scientific information and consensus-building among affected parties. Acceptable risk levels for human health, ecological, economic, social, and cultural impacts should therefore be established at the onset of the process. The relative importance of various risk factors should be discussed continually by the affected parties throughout the risk-assessment process. If the nature and extent of all the risks are not considered by the affected parties, a risk that affects only a small population or one that affects a large population might be given the same emphasis. In comparing one risk against the same type of risk, establishing the extent of each risk is extremely important. Approaches for comparing risk are discussed in Chapter 8.

Because the notion of what is an acceptable risk can vary greatly among the affected parties, rectifying differences among the parties can be an extremely difficult task. This problem is only exacerbated by the large spatial extent of many PCB-contaminated sites. For example, civic group leaders from areas along the upper Hudson (i.e., above the Federal Dam at Troy, New York) were very accepting of fishing bans and fish consumption advisories as a long-term solution for PCB contamination (see Box 6-8). However, civic group leaders from areas below the Federal Dam placed greater emphasis on human health effects and a safe Hudson River fishery (NRC Public Session, Albany, New York, November 8-9, 1999). These perceptions stem from the fact that the downstream communities see only benefits from active remediation. Upstream communities, however, which will be disrupted by dredging operations and will be exposed to greater risks during the implementation of any management options, recognize that there are conflicting risks and a balance is needed between the risks and the benefits of such activities. The challenge in resolving such a disagreement is balancing the risk to both affected parties, which will be a negotiated settlement. However, this need not always be the case as is evidenced by the experiences of communities near the Duwamish Waterway in Washington (see Box 6-9).

Examples of questions that need to be considered in comparative risk assessment include the following:

- Should human health be the primary driver in assessing overall risk (even though there might be a great deal of uncertainty associated with human health risk estimates)?
- How should economic, social, and cultural impacts be factored into risk-based decisions? For example, should jobs and the local economy take precedence over ecological concerns?
- How should short-term and long-term risks be compared? For example, should short-term increases in PCB exposure levels during remedial activity be acceptable in achieving long-term reduction of risk?

To foster an open discussion of risks, summary information should be tabulated and periodically disseminated to all affected parties. For example, a summary table, such as Table 6-4, can be constructed to provide a concise comparison of the primary risks that are driving the decision process. For each risk, information should be provided on the populations or subgroups affected, a quantitative or qualitative measure of impact, an estimate of uncertainty (or in the case of disagreement, the range of conflicting risk estimates), and appropriate references for each risk estimate. Information in the table can be expanded or revised throughout the decision process (e.g., to account for additional concerns of affected parties or new estimates of existing or potential risks).

BOX 6-8 Risk Perception Along the Hudson River

Research into what influences the public's perception of risk (Hance et al. 1988) has separated risk into the more easily measured "hazard" and the more subjective and emotion-laden "outrage." Some of the factors that affect "outrage" were garnered by the committee during their visit to the Hudson River, and discussion with affected parties include the following:

- Many of the public groups from along the lower Hudson River expressed a preference for dredging whatever PCB-contaminated sediments could be removed. In contrast, the communities along the upper Hudson had mixed feelings on whether to dredge and were divided on the issue.
- For communities along the lower Hudson, exposure to PCBs is controlled by others upriver and is involuntary, because they simply live downstream from the source and derive no direct benefit (historically or currently) from the source (i.e., the General Electric plant). Although they live downstream from the potential dredging area, they might be exposed to higher concentrations of PCBs flowing downriver or carried in the air. However, the most obvious visible impacts of the dredging would not adversely affect them.
- Along the upper Hudson, the use of the river for boating and catch-and-release fishing is at least as important as the desire to eat the fish. Although the presence of PCBs in the sediments does not directly affect their ability to boat on the river, active remediation would directly and negatively affect the communities' ability to use the upper Hudson River for boating and fishing.

To some people on the upper Hudson River, the process of dredging is perceived as "coercion" and is seen as being controlled by "others." Thus, to these people, it is the process of dredging itself that is perceived as the risk, a risk to their enjoyment of the river and a risk to their economic livelihoods if they have marinas along the river.

Clearly, people's perceptions of the most appropriate course of action relative to PCBs in the sediments of the Hudson River are divergent and, in some cases, diametrically opposed. These opinions are based on many factors, including perceptions of who gains and who loses, that are beyond the scope of a scientific review of the risks involved in the management of PCB-contaminated sediments.

BOX 6-9 Responsive Process versus Unresponsive Process—Examples from the Duwamish Waterway and the Hudson River

In Puget Sound, PCBs contaminated the Duwamish Estuary and Commencement Bay, among other places. The regional EPA personnel and local regulatory officials coordinating the remediation planning and completion processes involved many members of the community from the beginning of the process. Community members helped choose remediation solutions and participated in many of the less hazardous remediation and restoration activities. In this example, the remediation activities were, in part, individually (or community) controlled and were perceived of as "voluntary" (chosen by the community). In addition, the community perceived the remediation and restoration process to be responsive to their desires.

By contrast, the communities along the Hudson River are not a part of any active decision-making processes, despite an elaborate community involvement structure that has been established (see www.epa.gov/hudson/public-participation.htm). By all accounts, the community is allowed to provide input or feedback to the planning processes at selective times; times chosen by the EPA project managers, not the community. The Hudson River community involvement process used by EPA does not appear to allow community involvement in any decision-making or even in problem-formulation phases and does not appear to be responsive to community needs and frustrations. This lack of community involvement has contributed to the impass relative to the appropriate course of action to be taken on the Hudson River.

CONCLUSIONS AND RECOMMENDATIONS

Risk management of PCB-contaminated-sediment sites should comprehensively evaluate the broad range of risks posed by PCB-contaminated sediments and associated management actions. These risks should include societal, cultural, and economic impacts as well as human health and ecological risks.

Risk assessments and risk-management decisions should be conducted on a site-specific basis and should incorporate all available scientific information.

Current management options can reduce risks but cannot completely eliminate PCBs and PCB exposure from contaminated-sediment sites. Because

TABLE 6-4 Summary of Comparative Risks

Category	Examples of Types of Risk	Populations or Sub-groups Affected	Quantitative or Qualitative Measure of Impacts	Estimates of Uncer-tainty	Refer-ence
Human health effects	Cancer and noncancer risks associated with PCB exposure				
Ecological effects	PCB toxicity to organisms, loss of habitat during and after risk-management activities				
Societal impacts	Changes in a community social structure due to loss of traditional work patterns; reduced maternal-offspring contact due to avoidance of breast-feeding				
Cultural impacts	Loss of American Indian fishing rites				
Economic impacts	Loss of commercial and recreational fisheries, decline in property values, increased costs of dredged material disposal				

all options will leave some residual PCBs, the short- and long-term risks they pose should be considered when evaluating management strategies.

Research should focus on an improved understanding of the risks associated with exposures to PCB-contaminated sediments. Because the composition of PCB mixtures in the environment changes over time, risk characterizations should be based on the specific congeners present at a site. New studies on the toxicity and fate of PCBs in the environment are being conducted, and the results from these studies should be used to inform risk assessments at contaminated sites.

PCBs are often found at a site in conjunction with other chemicals of concern (e.g., pesticides, polycyclic aromatic hydrocarbons, dioxins, furans, and metals). Research is needed to determine the interactions among these chemicals and the impact that these interactions can have on site-specific risk assessments and subsequent risk-management efforts.

Models used to describe all relevant PCB exposure pathways—from the contaminated sediments, through the aquatic food web, and to specific receptors—must consider exposures to sensitive populations, including but not limited to the elderly, pregnant women, infants, and children; culturally or economically unique populations; and sensitive and endangered wildlife and their habitats. These models, which have inherent uncertainty, require calibration that should be conducted by multiple researchers. It is important that the model results are peer reviewed.

References

Abramowicz, D.A. 1990. Aerobic and anaerobic biodegradation of PCBs: a review. Crit. Rev. Biotechnol. 10(3):241-251.

Abramowicz, D.A., M.J. Brennan, H.M. van Dort, and E.L. Gallagher. 1993. Factors influencing the rate of polychlorinated biphenyl dechlorination in Hudson River sediments. Environ. Sci. Technol. 27(6):1125-1131.

Ahlborg, U.G., G.C. Becking, L.S. Birnbaum, A. Brouwer, H.J.G.M. Derks, M. Feeley, G. Golor, A. Hanberg, J.C. Larsen, A.K.D. Liem, S.H. Safe, C. Schlatter, F. Wærn, M. Younes, and E. Yrjänheikki. 1994. Toxic equivalency factors for dioxin-like PCBs. Chemosphere 28(6):1049-1067.

Ankley, G.T., P.M. Cook, A.R. Carlson, D.J. Call, J.A. Swenson, H.F. Corcoran, and R.A. Hoke. 1992a. Bioaccumulation of PCBs from sediments by oligochaetes and fishes: comparison of laboratory and field studies. Can. J. Fish. Aquat. Sci. 49(10):2080-2085.

Ankley, G.T., K. Lodge, D.J. Call, M.D Balcer, L.T. Brooke, P.M. Cook, R.G. Kreis Jr., A.R. Carlson, R.D. Johnson, G.J. Niemi, R.A. Hoke, C.W. West, J.P. Giesy, P.D. Jones, and Z.C. Fuyin. 1992b. Integrated assessment of contaminated sediments in the lower Fox River and Green Bay, Wisconsin. Ecotoxicol. Environ. Saf. 23(1):46-63.

Ankley, G.T., G.T. Niemi, K.B. Lodge, H.J. Harris, D.L. Beaver, D.E. Tillitt, T.R. Schwartz, J.P. Giesy, P.D. Jones, and C. Hagley. 1993. Uptake of planar polychlorinated biphenyls and 2,3,7,8-substituted polychlorinated dibenzofurans and dibenzo-*p*-dioxins by birds nesting in the Lower Fox River/Green Bay, Wisconsin. Arch. Environ. Contam. Toxicol. 24(3):332-344.

Bedard, D.L., M.L. Haberl, R.J. May, and M.J. Brennan. 1987a. Evidence for novel mechanisms of polychlorinated biphenyl metabolism in Alcaligenes eutrophus H850. Appl. Environ. Microbiol. 53(5):1103-1112.

Bedard, D.L., R. Unterman, L.H. Bopp, M.J. Brennan, M.L. Haberl, and C. Johnson.

1986. Rapid assay for screening and characterizing microorganisms for the ability to degrade polychlorinated biphenyls. Appl. Environ. Microbiol. 51(4):761-768.

Bedard, D.L., R.E. Wagner, M.J. Brennan, M.L. Haberl, and J.F. Brown, Jr. 1987b. Extensive degradation of Aroclors and environmentally transformed polychlorinated biphenyls by Alcaligenes eutrophus H850. Appl. Environ. Microbiol. 53(5):1094-1102.

Bertazzi, P.A., L. Riboldi, A. Pesatori, L. Radice, and C. Zocchetti. 1987. Cancer mortality of capacitor manufacturing workers. Am. J. Ind. Med. 11(2):165-176.

Bertazzi, P.A., C. Zocchetti, S. Guercilena, D. Consonni, A. Tironi, M.T. Landi, and A.C. Pesatori. 1997. Dioxin exposure and cancer risk: a 15-year mortality study after the "Seveso accident". Epidemiology 8(6):646-652.

Blankenship, A.L., K. Kannan, S.A. Villalobos, J. Falandysz, and J.P. Giesy. 2000. Relative potencies of individual polychlorinated naphthalenes and halowax mixtures to induce Ah receptor-mediated responses. Environ. Sci. Tehnol. 34(15):3153-3158.

Bloomington Herald Times. 2000. EPA offers to relocate some neighbors of site: PCB levels in air high enough to delay cleanup at Lemon Lane Landfill. Pp. A1. July 21, 2000.

Blumberg, A.F., L.A. Khan, and J.P. St. John. 1999. Three-dimensional hydrodynamic model of New York Harbor Region. J. Hydraul. Eng. 125(8):799-816.

Boese, B.L., M. Winsor, H. Lee, S. Echols, J. Pelletier, and R. Randall. 1995. PCB congeners and hexachlorobenzene biota sediment accumulation factors for Macoma nasuta exposed to sediments with different total organic carbon contents. Environ. Toxicol. Chem. 14(2):303-310.

Boon, J.P., and F. Eijgenraam. 1988. The possible role of metabolism in determining patterns of PCB congeners in species from the Dutch Wadden Sea. Mar. Environ. Res. 24(1-4):3-8.

Boon, .J.P., F. Eijgenraam, J.M. Everaarts, and J. Duinker. 1989. A structure-activity relationship (SAR) approach towards metabolism of PCBs in marine animals from different trophic levels. Mar. Environ. Res. 27(3-4):159-176.

Boon, J.P., J. van der Meer, C.R. Allchin, R.J. Law, J. Klungsoyr, P.E.G. Leonards, H. Spliid, E. Storr-Hansen, C. Mckenzie, and D.E. Wells. 1997. Concentration, dependent changes of PCB patterns in fish-eating mammals: structure evidence for induction of cytochrome P450. Arch. Environ. Contam. Toxicol. 33(3):298-311.

Borlakoglu, J.T., and C.H. Walker. 1989. Comparative aspects of congener specific PCB metabolism. Eur. J. Drug Metab. Pharmacokinet. 14(2):127-131.

Bowman, R.E., M.P. Heironimus, and J.R. Allen. 1978. Correlation of PCB body burden with behavioral toxicology in monkeys. Pharmacol. Biochem. Behav. 9(1):49-56.

Bowman, R.E., M.P. Heironimus, and D.A. Barsotti. 1981. Locomotor hyperactivity in PCB-exposed rhesus monkeys. Neurotoxicology 2(2):251-268.

Bright, D.A., W.T. Dushenko, S.L. Grundy, and K.J. Reimer. 1995. Effects of local and distant contaminant sources: polychlorinated biphenyls and other organochlorines in bottom-dwelling animals from an Arctic estuary. Sci. Total. Environ. 160-161:265-283.

Brown, D.P. 1987. Mortality of workers exposed to polychlorinated biphenyls—An update. Arch. Environ. Health. 42(6):333-339.

Burkhard, L.P. 1998. Comparison of two models for predicting bioaccumulation of hydrophobic organic chemicals in a Great Lakes food web. Environ. Toxicol. Chem. 17(3):383-393.

Chiarenzelli, J., R. Scrudato, B. Bush, D. Carpenter, and S. Bushart. 1998. Do large-scale remedial and dredging events have the potential to release significant amounts of semivolatile compounds to the atmosphere? Environ Health Perspect 106(2):47-49.

Clark, T., K. Clark, S. Paterson, D. Mackay, and R.J. Norstrom. 1988. Wildlife monitoring modeling and fugacity. Environ. Sci. Technol. 22(2):120-127.

Cook, P.M., R.J. Erickson, R.L. Spehar, S.P. Bradbury, and G.T. Ankley. 1993. Interim Report on Data and Methods for Assessment of 2,3,7,8-Tetrachlorodibenzo-*p*-dioxin Risk to Aquatic Life and Associated Wildlife. U.S. Environmental Protection Agency, Washington, D.C. NTIS/PB93-202828.

Corsolini, S., S. Focardi, K. Kannan, S. Tanabe, A.Borrell, and R. Tatsukawa. 1995a. Congener profile and toxicity assessment of polychlorinated biphenyls in dolphins, sharks and tuna collected from Italian coastal waters. Mar. Environ. Res. 40(1):33-54.

Corsolini, S., S. Focardi, K. Kannan, S. Tanabe, and R. Tatsukawa. 1995b. Isomer-specific analysis of polychlorinated biphenyls and 2,3,7,8-tetrachlorodibenzo-*p*-dioxin equivalents (TEQs) in red fox and human adipose tissue from central Italy. Arch. Environ. Contam. Toxicol. 29(1):61-68.

Courval, J.M., J.V. DeHoog, A.D. Stein, E.M. Tay, J. He, H.E. Humphrey, and N. Paneth. 1999. Sport-caught fish consumption and conception delay in licensed Michigan anglers. Environ. Res. 80(2 Pt 2):S183-S188.

Delzell, E., J. Doull, J.P. Giesy, D. Mackay, I.C. Monro, and G.M. Williams. 1994. Interpretive review of the potential adverse effects of chlorinated organic chemicals on human health and the environment: polychlorinated PCBs. Regul. Toxicol. Pharmacol. 20(1 Part 2):S1-S1056.

de Swart, R.L., P.S. Ross, L.J. Vedder, H.H. Timmerman, S. Heisterkamp, H. Van Loveren, J. Vos, P.J.H. Reijnders, and A.D.M.E. Osterhaus. 1994. Impairment of immune function in harbor seals (Phoca vitulina) feeding on fish from polluted waters. Ambio 23(2):155-159.

DiToro, D.M., C.S. Zarba, D.J. Hansen, W.J. Berry, R.C. Swartz, C.E. Cowan, S.P. Pavlou, H.E. Allen, N.A. Thomas, and P.R. Paquin. 1991. Technical basis for establishing sediment quality criteria for nonionic organic chemicals using equilibrium partitioning. Environ. Toxicol. Chem. 10:1541-1583.

Drouillard, K.G., J.J.H. Ciborowski, R. Lazar, and D.D. Haffner. 1996. Estimation of the uptake of organochlorines by the mayfly Hexagenia limbata (Ephemeroptera; Ephemeridae). J. Great Lakes Res. 22(1):26-35.

Eadie, B.J., W.R. Faust, N.R. Morehead, and P.F. Landrum. 1985. Factors affecting bioconcentration of PAH (polynuclear aromatic hydrocarbon) by the dominant benthic organisms of the Great Lakes. Pp. 363-377 in Polynuclear Aromatic Hydrocarbons: Mechanisms, Methods and Metabolism, M. Cooke and A.J. Dennis, eds. Columbus, OH: Battelle Press.

EPA (U.S. Environmental Protection Agency). 1997a. Uncertainty Factor Protocol for Ecological Risk Assessment. Toxicological Extrapolations to Wildlife Receptors. Ecosystems Protection and Remediation Division. U.S. Environmental Protection Agency, Region VIII, Denver, CO. February. [Online]. Available: http://www.epa.gov/Region8/superfund/risksf/trv-ucfs.pdf .

EPA (U.S. Environmental Protection Agency). 1997b. Ecological Risk Assessment Guidance for Superfund: Process for Designing and Conducting Ecological Risk Assessments. EPA 540-R-97-006. Office of Emergency and Remedial Response, U.S. Environmental Protection Agency, Washington, DC. June. NTIS PB97-963211.

EPA (U.S. Environmental Protection Agency). 1998. Risk Assessment Guidance for Superfund: Volume1. Human Health Evaluation Manual (Part D, Standardized Planning Reporting and Review of Superfund Risk Assessments). Interim. EPA/540/R-97/033. OSWER-9285.7-01D. Office of Emergency and Remedial Response, U.S. Environmental Protection Agency, Washington, DC. NTIS PB97-963305.

EPA (U.S. Environmental Protection Agency). 1998a. Exposure Factors Handbook. Vol. I, II, III. National Center for Environmental Assessment, Office of Research and Development, U.S. Environmental Protection Agency. [Online]. Available: http://www.epa.gov/ncea/exposfac.htm. [September 01, 2000].

EPA (U.S. Environmental Protection Agency). 1999. ECOFRAM Update: Probabilistic Risk Assessment Tools for Pesticides. Office of Pesticide Programs, U.S. Environmental Protection Agency. 2 pp. October 21, 1999. [Online]. Available: http://www.epa.gov/oppefed1/ecorisk/update.html .

Farley, K.J., J.A. Mueller, and D.J. O'Connor. 1983. Distribution of kepone in the James River Estuary. J. Environ. Engr. 109(2):396-413.

Farrington, J.W., A.C. Davis, B.J. Brownawell, B.W. Tripp, C.H. Clifford, and J.B. Livramento. 1986. The biogeochemistry of polychlorinated biphenyl in the Acushnet River Estuary, Massachusetts. Pp. 174-197 in Organic Marine Geochemistry, M.L. Sohn, ed. ACS Symposium Series 305. Washington, DC: American Chemical Society.

Froese, K.L., G.T. Ankley, D.A. Verbrugge, G.J. Niemi, C.P. Larsen, and J.P. Giesy. 1998. Bioaccumulation of polychlorinated biphenyls from sediments to aquatic insects and tree swallow eggs and nestlings in Saginaw Bay, Michigan, USA. Environ. Toxicol. Chem. 17(3):484-492.

Gailani, J., C.K. Ziegler, and W. Lick. 1991. Transport of suspended solids in the lower Fox River. J. Great Lakes Res. 17(4):479-494.

Gasiewicz, T.A. 1997. Dioxins and the Ah receptor: probes to uncover processes in neuroendocrine development. Neurotoxicology 18(2):393-413.

Giesy, J.P., and R.A. Hoke. 1990. Freshwater sediment quality criteria: toxicity bioassessment. Pp. 265-348 in Sediments: Chemistry and Toxicology of In-Place Pollutants, R. Baudo, J.P. Giesy, and H. Muntau, eds. Ann Arbor: Lewis.

Giesy, J.P., and K. Kannan. 1998. Dioxin-like and non-dioxin-like toxic effects of polychlorinated biphenyls (PCBs): implications for risk assessment. Crit. Rev. Toxicol. 28(6):511-569.

Giesy, J.P., J.P. Ludwig, and D.E. Tillitt. 1994a. Dioxins, dibenzofurans, PCBs, and colonial fish-eating water birds. Pp. 249-307 in Dioxins and Health, A. Schecter, ed.. New York: Plenum Press.

Giesy, J.P., J.P. Ludwig, and D.E. Tillitt. 1994b. Deformities in birds of the Great Lakes region: assigning causality. Environ. Sci. Technol. 28(3):128A-135A.

Gladen, B.C., W.J. Rogan, P. Hardy, J. Thullen, J. Tingelstad, and M. Tully. 1988. Development after exposure to polychlorinated biphenyls and dichlorodiphenyl dichloroethene transplacentally and through human milk. J. Pediatr. 113(6):991-995.

Gobas, F.A.P.C. 1993. A model for predicting the bioaccumulation of hydrophobic organic chemicals in aquatic food webs: application to Lake Ontario. Ecol. Model. 69:1-17.

Gobas, F.A.P.C., J.R. McCorquodale, and G.D. Haffner. 1993. Intestinal absorption and biomagnification of organochlorines. Environ. Toxicol. Chem. 12(3):567-576.

Hance, B.J., C. Chess, and P.M. Sandman. 1988. Improving Dialogue with Communities: A Risk Communication Manual for Government, Environmental Communication Research Program, New Jersey Agricultural Experimental Station. Trenton, NJ: Division of Science and Research.

Hawker, D.W., and D.W. Connell. 1985. Relationships between partition coefficient, uptake rate constant, clearance rate constant and time to equilibrium for bioaccumulation. Chemosphere 14(9):1205-1219.

Henshel, D.S., B. Hehn, R. Wagey, M. Vo, and J.D. Steeves. 1997. The relative sensitivity of chicken embryos to yolk- or aircell-injected 2,3,7,8-tetrachlorodibenzo-*p*-dioxin. Environ. Toxicol Chem. 16(4):725-732.

Hoffman, D.J., M.J. Melancon, P.N. Klein, J.D. Eisemann, and J.W. Spann. 1998. Comparative developmental toxicity of planar polychlorinated biphenyl congeners in chicken, american kestrels, and common terms. Environ. Toxicol. Chem. 17(4):747-757.

Hoke, R.A. G.T. Ankley, A.M. Cotter, T. Goldenstein, P.A. Kosian, G.L. Phipps, and F.M. VanderMeiden. 1994. Evaluation of equilibrium partitioning theory for predicting acute toxicity of field-collected sediments contaminated with DDT, DDE and DDD to the amphipod Hyalella azteca. Environ. Toxicol. Chem. 13(1):157-166.

Huisman, M., C. Koopman-Esseboom, V. Fidler, M. Hadders-Algra, C.G. van der Paauw, L.G. Tuinstra, N. Weisglas-Kuperus, P.J. Sauer, B.C. Touwen, and E.R. Boersma. 1995. Perinatal exposure to polychlorinated biphenyls and dioxins and its effect on neonatal neurological development. Early Hum. Dev. 41(2):111-127.

Humphrey, H.E.B. 1983. Population studies of PCBs in Michigan residents. Pp. 299-310 in: PCBs: Human and Environmental Hazards, F.M. D'Itri, and M.A. Kamrin, eds. Boston, MA: Butterworth.

Ingersoll, C.G., G.T. Ankley, D.A. Benoit, E.L. Brunson, G.A. Burton, F.J. Dwyer, R.A. Hoke, P.F. Landrum, T.J. Norberg-King, and P.A. Winger. 1995. Toxicity and bioaccumulation of sediment-associated contaminants using freshwater invertebrates: a review of methods and applications. Environ. Toxicol. Chem. 14(11):1885-1894.

Jacobson, J.L., and S.W. Jacobson. 1996. Intellectual impairment in children exposed to polychlorinated biphenyls in utero. N. Engl. J. Med. 335(11):783-789.

Jacobson, S.W., G.G. Fein, J.L. Jacobson, P.M. Schwartz, and J.K. Dowler. 1985. The effect of intrauterine PCB exposure on visual recognition memory. Child. Dev. 56(4):853-860.

Jacobson, J.L., S.W. Jacobson, and H.E. Humphrey. 1990. Effects of exposure to PCBs and related compounds on growth and activity in children. Neurotoxicol. Teratol. 12(4):319-326.

Jarvinen, A.W., and G.T. Ankley. 1999. Linkage of Effects to Tissue Residues: Development of a Comprehensive Database for Aquatic Organisms Exposed to Inorganic and Organic Chemicals. Pensacola, FL: Society of Environmental Toxicology and Chemistry (SETAC). 364 pp.

Jones, P.D., J.P. Giesy, J.L. Newsted, D.A. Verbrugge, D.L. Beaver, G.T. Ankley, D.E. Tillitt, and K.B. Lodge. 1993. 2,3,7,8-tetrachlorodibenzo-*p*-dioxin equivalents in tissues of birds at Green Bay, Wisconsin, USA. Arch. Environ. Contam. Toxicol. 24(3):345-354.

Jones, P.D., J.P. Giesy, J.L. Newsted, D.A. Verbrugge, J.P. Ludwig, M.J. Ludwig, H. Auman, T.J. Kubiak, and D. Best. 1994. Accumulation of 2,3,7,8-tetrachlorodibenzo-*p*-dioxin equivalents by double crested (Phalacrocorax auritus, Pelicaniformes) chicks in the North American Great Lakes. Ecotoxicol. Environ. Saf. 27(2):192-209.

Kannan, K., A.L. Blankenship, P.D. Jones, and J.P. Giesy. 2000. Toxicity reference values for the toxic effects of polychlorinated biphenyls to aquatic mammals. Hum. Ecol. Risk Assess. 6(1):181-201.

Karickhoff, S.W., D.S. Brown, and T.A. Scott. 1979. Sorption of hydrophobic pollutants on natural sediments. Water Res. 13(3):241-248.

Kimbrough, R.D., M.L. Doemland, and M.E. LeVois. 1999. Mortality of male and female capacitor workers exposed to polychlorinated biphenyls. J. Occup. Environ. Med. 41(3):161-171.

Kuratsune, M., T. Yoshimura, J. Matsuzaka, and A. Yamaguchi. 1972. Epidemiological study on Yusho, a poisoning caused by ingestion of rice oil contaminated with a commercial brand of polychlorinated biphenyls. Environ. Health Perspect. 1:119-128.

Lake, J.L., N.I. Rubinstein, H.I.I. Lee, C.A. Lake, J. Heltshe, and S. Pavignano. 1990. Equilibrium partitioning and bioaccumulation of sediment-associated contaminants by infaunal organisms. Environ. Toxicol. Chem. 9(8):1095-1106.

Landrum, P.F. 1989. Bioavailability and toxicokinetics of polycyclic aromatic hydrocarbons sorbed to sediments for the amphipod Pontoporeia hoyi. Environ. Sci. Toxicol. 23(5):588-595.

Landrum, P.F., and R. Poore. 1988. Toxicokinetics of selected xenobiotics in Hexagenia limbata. J. Great Lakes Res. 14(4):427-437.

Landrum, P.F., W.A. Frez, and M.S. Simmons. 1992. The effect of food consumption of the toxicokinetics of benzo(a)pyrene and 2,2',4,4',5,5'-hexachlorobiphenyl in Mysis relicta. Chemosphere 25(3):397-415.

Leonards, P.E.G., T.H. de Vries, W. Minnaard, S. Stuijfzand, P. de Voogt, W.P. Cofino, N.M. van Straalen, and B. van Hattum. 1995. Assessment of experimental data on PCB-induced reproduction inhibition in mink, based on an isomer- and congener-specific approach using 2,3,7,8-tetrachlorodibenzo-*p*-dioxin toxic equivalency. Environ. Toxicol. Chem. 14(4):639-652.

Levin, E.D., S.L. Schantz, and R.E. Bowman. 1988. Delayed spatial alternation deficits resulting from perinatal PCB exposure in monkeys. Arch. Toxicol. 62(4):267-273.

Ling, H., M. Diamond, and D. Mackay. 1993. Application of the QWASI fugacity/ equivalence model to assessing sources and fate of contaminants in Hamilton Harbor. J. Great Lakes Res. 19(3):582-602.

Long, E.R., and L.G. Morgan. 1991. The Potential for Biological Effects of Sediment Sorbed Contaminants Tested in the National Status and Trends Program. Office of Oceanography and Marine Assessment, National Ocean Service, Rock ville, MD. GRA&I 14. NTIS PB91-172288.

Lonky, E., J. Reihman, T. Darvill, J. Mather Sr., and H. Daly. 1996. Neonatal behavioral assessment scale performance in humans influenced by maternal consumption of environmentally contaminated Lake Ontario fish. J. Great Lakes Res. 22(2):198-212.

Ludwig, J.P., H. Kurita-Matsuba, H.J. Auman, M.E. Ludwig, C.L. Summer, J.P. Giesy, D.E. Tillitt, and P.D. Jones. 1996. Deformities, PCBs and TCDD-equivalents in double-crested cormorants (Phalacrocorax auritus) and Caspian terns (Hydroprogne caspia) of the upper Great Lakes 1986-1991: testing the hypotheses of cause-effect relationships. J. Great. Lakes Res. 22(2):172-197.

MacFarland, V.A., and J.U. Clarke. 1989. Environmental occurrence, abundance, and potential toxicity of polychlorinated biphenyl congeners: considerations for a congener-specific analysis. Environ. Health Perspect. 81:225-239.

Mackay, D., and S. Paterson. 1991. Evaluating the multimedia fate of organic chemicals: a level III fugacity model. Environ. Sci. Technol. 25(3):427-436.

Mackay, D., W.Y. Shiu, J. Billington, and G.L. Huang. 1983. Physical chemical properties of polychlorinated biphenyls. Pp. 59-70 in: Physical Behavior of PCBs in the Great Lakes, D. Mackay, S. Patterson, S.J. Eisenreich, and M.S. Simmons. eds. Ann Arbor, MI: Ann Arbor Science.

Mackay, D., S. Paterson, and W.Y. Shiu. 1992. Generic models for evaluating the regional fate of chemicals. Chemosphere 24(6):695-717.

Marine Policy Center. 2000. Cost-Effective Reduction of PCB-contamination in the Hudson River and Estuary. Final Report to the Hudson River Foundation for Science and Environmental Research, New York, NY. Grant No. 008/991. Research Project R/B-147-PD. Marine Policy Center, Woods Hole Oceanographic Institution, Woods Hole, MA.

Masuda, Y. 1985. Health status of Japanese and Taiwanese after exposure to contaminated rice oil. Environ Health Perspect. 60:321-325.

Mavoungou, R., R. Masse, and E. Sylvestre. 1991. Microbial dehalogenation of 4 4' dichlorobiphenyl under anaerobic conditions. Sci. Total. Environ. 101(3):263-268.

McCarty, J.P., and A.L. Secord. 1999. Nest-building behavior in PCB-contaminated tree swallows. AUK 116(1):55-63.

McNeil, J., C. Taylor, and W. Lick. 1996. Measurements of erosion of undisturbed bottom sediments with depth. J. Hydraul. Eng. 122(6):316-324.

Mergler, D., S. Belanger, F. Larribe, M. Panisset, R. Bowler, M. Baldwin, J. Lebel, and K. Hudnell. 1998. Preliminary evidence of neurotoxicity associated with eating fish from the Upper St. Lawrence River Lakes. Neurotoxicity 19(4-5):691-702.

Metcalfe, C.D., and G.D. Haffner. 1995. The ecotoxicology of coplanar polychlorinated biphenyls. Environ. Rev. 3(2):171-190.

Murk, A., D. Morse, J. Boon, and A. Brouwer. 1994. In vitro metabolism of 3,3',4,4'-tetrachlorobiphenyl in relation to ethoxyresorufin-o-deethylase activity in liver microsomes of some wildlife species and rat. Eur. J. Pharmacol. 3(2-3):253-61.

Nebecker, A.V., F.A. Puglisi, and D.L. Defoe. 1974. Effect of polychlorinated biphenyl compounds on survival and reproduction of the fathead minnow and flagfish. Trans. Am. Fish. Soc. 103(3):562-568.

Neely, W.B., and D. Mackay. 1982. Evaluative model for estimating environmental fate. Pp. 127-143 in: Modeling the Fate of Chemicals in the Aquatic Environment, K.L. Dickson, A.W. Maki, and J. Cairns Jr., eds. Ann Arbor, MI: Ann Arbor Science.

Nichols, J.W., C.P. Larsen, M.E. McDonald, G.J. Niemi and G.T. Ankley. 1995. Bioenergetics-based model for accumulation of PCBs by nesting tree swallows, Tachycineta bicolor. Environ. Sci. Technol. 29(3):604-612.

Niimi, A.J., H.B. Lee, and D.C.G. Muir. 1996. Environmental assessment and ecotoxicological implications of the co-elution of PCB congeners 132 and 153. Chemosphere 32(4):627-638.

Parkerton, T.F., J.P. Connolly, R.V. Thoman, and C.G. Uchrin. 1993. Do aquatic effects of human health end points govern the development of sediment-quality criteria for nonionic organic chemicals? Environ. Toxicol. Chem. 12(3):507-523.

Patandin, S., C.I. Langting, P.G. Mulder, E.R. Boersma, P.J. Sauer, and N. Weisglas-Kuperus. 1999. Effects of environmental exposure to polychlorinated biphenyls and dioxins on cognitive abilities in Dutch children at 42 months of age. J. Pediatr. 134(1):33-41.

Poland, A., and J.C. Knutson. 1982. 2,3,7,8-tetrachlorodibenzo-*p*-dioxin and related halogenated aromatic hydrocarbons: examination of the mechanism of toxicity. Ann. Rev. Pharmacol. Toxicol. 22:517-554.

Powell, D.C., R.J. Aulerich, J.C. Meadows, D.E. Tillitt, J.P. Giesy, K.L. Stromberg, and S.J. Bursian. 1996a. Effects of 3,3',4,4',5-pentachlorobiphenyl (PCB 126) and 2,3,7,8-tetrachlorodibenzo-*p*-dioxin (TCDD) injected into the yolks of chicken (Gallus domesticus) eggs prior to incubation. Arch. Environ. Contam. Toxicol 31(3):404-409.

Powell, D.C., R.J. Aulerich, K.L. Stromborg, and S.J. Bursian. 1996b. Effects of 3,3',4,4'-tetrachlorobiphenyl, 2,3,3',4,4'- pentachlorobiphenyl, and 3,3',4,4',5-pentachlorobiphenyl on the developing chicken embryo when injected prior to incubation. J. Toxicol. Environ. Health 49(3):319-338.

PCCRARM (Presidential/Congressional Commission on Risk Assessment and Risk Management). 1997. Framework for Environmental Health Risk Management: Final Report. Washington, DC: The Commission.

QEA (Quantitative Environmental Analysis). 1999a. PCBs in the Upper Hudson

River. Vol. 2. A Model of PCB Fate, Transport, Bioaccumulation. Prepared for General Electric, Albany, New York. May.

QEA (Quantitative Environmental Analysis). 1999b. PCBs in the Upper Hudson River. Errata. Prepared for General Electric, Albany, New York. July.

Quensen, J.F., J.M. Tiedje, and S.A. Boyd. 1988. Reductive dechlorination of polychlorinated biphenyls by anaerobic microorganisms from sediments. Science 242(4879):752-754.

Quensen, J.F, 3rd., M.A. Mousa, S.A. Boyd, J.T. Sanderson, K.L. Froese, and J.P. Giesy. 1998. Reduction of aryl hydrocarbon receptor-mediated activity of polychlorinated biphenyl mixtures due to anaerobic microbial dechlorination. Environ. Toxicol. Chem. 17(5):806-813.

Rhee, G.-Y., B. Bush, C.M. Bethoney, A. DeNucci, H.-M. Oh, and R.C. Sokol. 1993a. Anaerobic dechlorination of aroclor 1242 as affected by some environmental conditions. Environ. Toxicol. Chem. 12(6):1033-1039.

Rhee, G.-Y., B. Bush, C.M. Bethoney, A. DeNucci, H.-M. Oh, and R.C. Sokol. 1993b. Reductive dechlorination of aroclor 1242 in anaerobic sediments: pattern, rate and concentration dependence. Environ. Toxicol Chem. 12(6):1025-1032.

Rhee, G.-Y., R.C. Sokol, B. Bush, and C.M. Bethoney. 1993c. Long-term study of the anaerobic dechlorination of aroclor 1254 and without biphenyl enrichment. Environ. Sci. Technol. 27(4):714-719.

Rice, D.C. 1995. Neurotoxicity of lead, methylmercury, and PCBs in relation to the Great Lakes. Environ. Health Perspect. 103(suppl. 9):71-87.

Rogan, W.J., B.C. Gladen, K.L. Hung, S.L. Koong, L.Y. Shih, J.S. Taylor, Y.C. Wu, D. Yang, N.B. Ragan, and C.C. Hsu. 1988. Congenital poisoning by polychlorinated biphenyls and their contaminants in Taiwan. Science 241(4863):334-336.

Rogan, W.J., B.C. Gladen, J.D. McKinney, N. Carreras, P. Hardy, J. Thullen, J. Tinglestad, and M. Tully. 1986. Neonatal effects of transplacental exposure to PCBs and DDE. J. Pediatr. 109(2):335-341.

Ross, P.S., R.L. de Swart, P.J.H. Reijnders, H. van Loveren, J.G. Vos, and A.D.M.E. Osterhaus. 1995. Contaminant-related suppression of delayed-type hypersensitivity and antibody responses in harbor seals fed herring from the Baltic Sea. Environ. Health Perspect. 103(2):162-167.

Rothman, N., K.P. Cantor, A. Blair, D. Bush, J.W. Brock, K. Helzlsouer, S.H. Zahm, L.L. Needham, G.R. Pearson, R.N. Hoover, G.W. Comstock, and P.T. Strickland. 1997. A nested case-control study of non-Hodgkin lymphoma and serum organochlorine residues. Lancet 350(9073):240-244.

Sabljic, A., H. Gutsen, H. Verhaar, and J. Hermens. 1995. QSAR modeling of soil sorption: improvements and systematics of log Koc vs. log Kow corrections. Chemosphere 31(11-12):4489-4514.

Safe, S. 1990. Polychlorinated biphenyls(PCBs), dibenzo-*p*-dioxins(PCDDs), dibenzofurans (PCDFs), and related compounds: environmental and mechanistic considerations which support the development of toxic equivalency factors (TEFs). Crit. Rev. Toxicol. 21(1):51-88.

Schantz, S.L. 1996. Developmental neurotoxicity of PCBs in humans: what do we know and where do we go from here? Neurotoxicol. Teratol. 18(3):217-227.

Schantz, S.L., E.D. Levin, and R.E. Bowman. 1991. Long-term neurobehavioral effects of perinatal polychlorinated biphenyl (PCB) exposure in monkeys. Environ. Toxicol. Chem. 10(6):747-756.

Sinks, T., G. Steele, A.B. Smith, K. Watkins, and R.A. Shults. 1992. Mortality among workers exposed to polychlorinated biphenyls. Am. J. Epidemiol. 136(4):389-398.

Solomon, K., J. Giesy, and P. Jones. 2000. Probabilistic risk assessment of agrochemicals in the environment. Crop Protect. 19(8-10):649-655.

Starodub, M.E., P.A. Miller, G.M. Ferguson, J.P. Giesy, and R.F. Willis. 1996. A protocol to develop acceptable concentrations of bioaccumulative organic chemicals in sediments for the protection of piscivorous wildlife. Toxicol. Environ. Chem. 54(1-4):243-259.

Svensson, B.G., Z. Mikoczy, U. Stromberg, and L. Hagmar. 1995. Mortality and cancer incidence among Swedish fishermen with a high dietary intake of persistent organochlorine compounds. Scand. J. Work Environ. Health 21(2):106-115.

TAMS Consultants, Inc., Limno-Tech, Inc., Menzie-Cura & Associates, Inc., and Tetra Tech, Inc. 2000. Phase 2 Report- Review Copy, Further Site Characterization and Analysis, Vol. 2D - Revised Baseline Modeling Report Hudson River PCBs Reassessment RI/FS. Prepared for U.S. Environmental Protection Agency, Region 2 and U.S. Army Corps of Engineers, Kansas City District. January. [Online]. Available: http://www.epa.gov/hudson/rbmr-bk3&4.pdf [December19, 2000]

Tanabe, S., N. Kannan, A. Subramanian, S. Watanabe, and R. Tatsukawa. 1987. Highly toxic coplanar PCBs occurrence, source, persistency and toxic implications to wildlife and humans. Environ. Pollut. 47(2):147-163.

Thomann, R.V., and J.P. Connolly. 1984. Model of PCB in the Lake Michigan trout food chain. Environ. Sci. Technol. 18(2):65-71.

Thomann, R.V., J.P. Connolly, and T.F. Parkerton. 1992a. An equilibrium model of organic chemical accumulation in aquatic food webs with sediment interaction. Environ. Toxicol. Chem. 11(5):615-629.

Thomann, R.V., J.P. Connolly, and T.F. Parkerton. 1992b. Modeling accumulation of organic chemicals in aquatic food webs. Pp.153-186 in: Chemical Dynamics in Fresh Water Ecosystems, F.A.P.C. Gobas and J.A. McCorquodale, eds. Chelsea, MI: Lewis .

Tillitt, D.E., R.W. Gale, J.C. Meadows, J.L. Zajicek, P.H. Peterman, S.N. Heaton, P.D. Jones, S.J. Bursian, T.J. Kubiak, J.P. Giesy, and R.J. Aulerich. 1996. Dietary exposure of mink to carp from Saginaw Bay. 3. Characterization of dietary exposure to planar halogenated hydrocarbons, dioxin equivalents, and biomagnification, Environ. Sci. Technol. 30(1):283-291.

Tracey, G.A., and D.J. Hansen. 1996. Use of biota-sediment accumulation factors to assess similarity of nonionic organic chemical exposure to benthically-coupled organisms of differing trophic mode. Arch. Environ. Contam. Toxicol. 30(4):467-475.

Tryphonas, H. 1998. The impact of PCBs and dioxins on children's health: immunological considerations. Can. J. Public Health 89(Suppl.1):49-52, 54-57.

Van den Berg, M., L. Birnbaum, A.T.C. Bosveld, B. Brunström, P. Cook, M. Feeley, J.P. Giesy, A. Hanberg, R. Hasegawa, S.W. Kennedy, S.W., T. Kubiak, J.C. Larsen, F.X. van Leeuwen, A.K. Liem, C. Nolt, R.E. Peterson, L. Poellinger, S. Safe, D. Schrenk, D. Tillitt, M. Tysklind, M. Younes, F. Waern, and T. Zacharewski. 1998. Toxic equivalency factors (TEFs) for PCBs, PCDDs, PCDFs for humans and wildlife. Environ. Health Perspect. 106(12):775-792.

Velleux, M., and D. Endicott. 1994. Development of a mass balance model for estimating PCB export from the Lower Fox River to Green Bay. J. Great Lakes Res. 20(2):416-434.

Walker, M.K., and R.E. Peterson. 1991. Potencies of polychlorinated dibenzo-*p*-dioxin, dibenzofuran and biphenyl congeners, relative to 2,3,7,8-tetrachloro-dibenzo-*p*-dioxin for producing early life stage mortality in rainbow trout (Oncorhynchus mykiss). Aquatic Toxicol. 21:219-238.

Walker, M.K. P.M. Cook, B.C. Butterworth, E.W. Zabel, and R.E. Peterson. 1996. Potency of a complex mixture of polychlorinated dibenzo-*p*-dioxin, dibenzofuran, and biphenyl congeners compared to 2,3,7,8,-tetrachlorodibenzo-*p*-dioxin in causing early life stage mortality. Fundam. Appl. Toxicol. 30(2):178-186.

Weisglas-Kuperus, N., T.C. Sas, C. Koopman-Esseboom, C.W. van der Zwan, M.A. De Ridder, A. Beishuizen, H. Hooijkaas, and P.J. Sauer. 1995. Immunologic effects of background prenatal and postnatal exposure to dioxins and polychlorinated biphenyls in Dutch infants. Pediatr. Res. 38(3):404-410.

Williams, L.L., J.P. Giesy, N. DeGalan, D.A. Verbrugge, D.E. Tillitt, G.T. Ankley, and R.A. Welch. 1992. Prediction of concentrations of 2,3,7,8-TCDD equivalents (TCDD-EQ) from total concentrations of PCBs in fish fillets. Environ. Sci. Technol. 26(6):1151-1159.

Willman, E.J., I.B. Manchester-Neesvig, and D.E. Armstrong. 1997. Influence of ortho-substitution on patterns of PCB accumulation in sediment, plankton and fish in a freshwater estuary. Environ. Sci. Technol. 31(12):3712-3718.

Ziegler, C.K. 1999a. Sediment Stability at contaminated sediment sites. Pp.26-27 in: Contaminated Sediment Management Technical Papers, Sediment Management Work Group, Detroit, MI. Fall 1999. [Online]. Available: http://www.smwg.org/ [January 02, 20001].

Ziegler, C.K. 1999b. Effective decision-making models for evaluating sediment management options. Pp.9-11 in: Contaminated Sediment Management Technical Papers, Sediment Management Work Group, Detroit, MI. Fall 1999. [Online]. Available: http://www.smwg.org/ [January 02, 20001].

Ziegler, C.K., and B.S. Nisbet. 1995. Long-term simulation of fine-grained sediment transport in large reservoir. J. Hydraul. Eng. 121(11):773-781.

Zwiernik, M.J., J.F. Quenson, 3rd, and S.A. Boyd. 1998. FeSO4 amendments stimulate extensive anaerobic PCB dechlorination. Environ. Sci. Technol. 32(21): 3360-3365.

7

Assessing Management Options

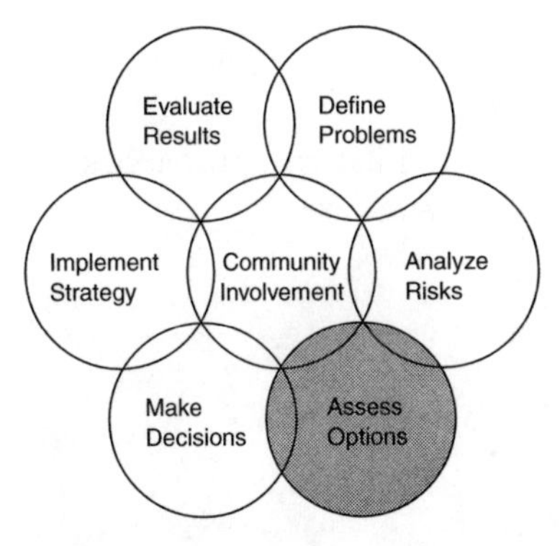

The preceding stages in the framework for environmental risk management lead to the formulation of risk management goals, the identification of the chemicals, sources, affected media, and potential risks of contaminated sediments. Only after the risk-management goals have been defined and the risks assessed should management options be examined. A range of technologies are applicable to the management of contaminated sediments, and these technologies must be identified and analyzed, and their benefits, effectiveness, costs, feasibility, and adverse consequences compared. All technologies might be appropriate under some site and contaminant conditions, but no technology exists that is generally applicable or preferred for the management of all contaminated sediments. Even at a single site, the application of multiple technologies is likely to be required to achieve risk-management goals. The lack of a single generally applicable and effective option means that site-specific analyses are needed to determine which combination of technologies best meet the identified goals (as determined in stage 1) at the least cost and with the most benefit to the environment and the community.

The identification and selection of a risk-management strategy and its component management options should be based on a range of considerations, including effectiveness, permanence, implementability, risks associated with implementation, costs, and state and community acceptance. These factors must be evaluated relative to the identified goals and based on site-specific conditions. The affected parties should be included in the identification, selection, and evaluation processes. Without a clear statement of goals, it is not possible to evaluate and compare the effectiveness of any management option at meeting those goals. Without adequate consideration of site-specific conditions, the comparison is incomplete. The primary purpose of this chapter is to outline the process of identifying and evaluating contaminated-sediment management options and identify the characteristics of specific technologies that influence their selection at a particular site.

Assessing Options

- Identifying Options
- Analyzing Options for
  - Benefits
  - Effectiveness
  - Costs
  - Feasibility
  - Adverse Consequences

IDENTIFYING OPTIONS

Among the many different regulatory and nonregulatory approaches to reducing and managing risks of contaminated sediments are

- Socioeconomic options.
  - Institutional controls.
  - Offsets.
- Source control.
- Natural attenuation and recovery.
  - Biodegradation.
  - Sedimentation.
- In situ treatment.
  - Enhanced natural attenuation.
  - Capping.
- Multicomponent removal and ex situ treatment.
  - Dredging technologies.
  - Pretreatment technologies.
  - Ex situ treatment, storage, and disposal technologies.
  - Technologies for management of residual contaminants.

Often the identification of potential management options occurs simultaneously with the identification and analysis of risks and potential risks. Input from all affected parties, including the public, industry, local community-based organizations, and government, is critical to the identification of an appropriate range of possible options. In the identification of options, the initial emphasis should be on completeness. Possible options should not be arbitrarily eliminated prior to a systematic and thoughtful evaluation by all participants. The participation of concerned groups at this stage can also help to identify potential beneficial outcomes of some options that might not be readily apparent to others. Options that incorporate beneficial outcomes beyond that achieved simply by reducing the risks of contaminated sediments generally have a greater chance of acceptance and success, because they may be seen as win-win options. Examples of such beneficial outcomes include creation of green space, habitat, and wetlands or the use of dredged material for fill.

Many of the potential technologies for management of contaminated sediments were initially developed to manage contaminated soils. Unfortunately, many of these technologies are difficult to apply to contaminated sediments, and they might impose potentially unacceptable risks. Sites with contaminated sediments are often poorly controlled, dynamic systems containing large volumes of moderately contaminated material. An analysis of the Superfund Record of Decisions from 1982 to1997[1] showed that the average contaminated-soil site considered for ex situ treatment contained 38,000 cubic yards of contaminated material and that the average site considered for in situ treatment contained approximately 105,000 cubic yards of contaminated material. Sites at which contaminated sediments occur, however, often contain in excess of 1,000,000 cubic yards of contaminated material and generally are not directly accessible. Soils can be removed in a relatively dry state for further processing, and sediments are removed as slurries with a high proportion of water that must be processed. The control of releases of contaminants is much more difficult during removal of submerged sediments than during removal of soil for ex situ treatment. The difficulty is due to limited control of the subaqueous environment as well as the sometimes profound chemical and physical changes that the sediment undergoes during removal (e.g., anaerobic to aerobic and wet to dry).

Identifying management options for contaminated sediment is also complicated by the multiple technologies often involved. The application of ex situ treatment or disposal of sediments, for example, typically introduces a complete train of technologies, including removal by dredging, temporary storage or pretreatment to reduce water content or volume, final dredged material treatment or disposal, and management of any residually contaminated materi-

---

[1]Treatment Technologies for Site Cleanup: Annual Status Report, 9th Ed. U.S. Environmental Protection Agency 542-R99-001, April 1999.

als. In situ treatments, such as capping, are normally coupled with source control and often require long-term monitoring and maintenance. Natural-attenuation processes (i.e., biodegradation and sedimentation) are also a component of all contaminated-sediment management options, because these processes are expected to have an impact on any residuals left after application of other management approaches. Generally, large contaminated-sediment sites also require the application of different options at different portions of the site, each containing multiple technologies. The identification of sediment management options must recognize the entire train of technologies that constitute each option so that a fair evaluation and comparison of these options can be accomplished.

EVALUATING MANAGEMENT OPTIONS

The evaluation of management options requires definition of

- The goals and objectives of the management actions (see Chapter 5).
- A valid conceptual model of the sediment system to be managed (see Chapter 6).

Without definition of the management goals and objectives, it is not possible to measure the effectiveness of a management option. Without a valid conceptual model of the site, it is not possible to define how a management option can successfully meet the risk-reduction goals and objectives. The conceptual model must be based on site-specific conditions, and experience at other sites must be used to aid our understanding and not simply be presumed to represent default conditions at the site. Although a valid conceptual model of a site is a minimal requirement for success, it is recognized that many sites require much more. Large, complicated sites posing substantial risks and potentially large cleanup costs generally require development of an extensive database and sophisticated prognostic models to compare management options adequately and evaluate the risk reduction they can achieve.

The primary purpose of evaluating options at this stage is to develop the information and data necessary to compare and select a viable alternative from among the options. The evaluation is not designed to make that selection. The evaluation of options is often conducted by technical personnel who might not fully recognize or consider such factors as acceptability to the community. For this reason, the process of evaluating options is separated from making decisions, the next stage in the decision-making framework (see Chapter 8). Although options may be eliminated from further consideration during evaluation due to technical infeasibility or other factors, elimination must be done cautiously to avoid prematurely eliminating options that might ultimately prove effective. For this reason, affected parties should participate in the development of the evaluation criteria and have input into the evaluation process.

The options for managing contaminated sediments vary greatly with respect to the site-specific factors that must be evaluated:

- Feasibility.
- Expected benefits and effectiveness.
- Expected costs.
- Potential adverse consequences.
- Distribution of benefits, risks, and costs among various potentially affected groups.

All technologies have advantages and disadvantages when applied at a particular site, and it is critical to the evaluation that these be identified as completely as possible. For example, reducing risks from contaminated sediment in the aqueous environment might create other risks in the terrestrial environment. These risks might be to the same community affected by the in situ sediments or to other communities faced with transportation, treatment, or disposal of contaminated dredged material. The evaluation of sediment management options must take into account the entire train of technologies that constitute an option and the costs and potential risks throughout the life cycle of the options. Evaluation, screening, and ultimately, selection of an option depends on recognition of the full environmental, human, social, and economic costs of implementing that option. Incomplete evaluation, particularly neglect of required components of a management option or their environmental and economic costs, biases the evaluation.

Even if all components of a management option are considered, costs are difficult to estimate and compare. Management costs that are routinely available often use different bases and assumptions and can be influenced dramatically by site-specific conditions, such as the debris or the physical characteristics of the sediments. The existing database of costs is inadequate to provide anything more than a crude estimate of the costs of the application of a particular management option at a particular site. (See Table 5-1a,b for costs associated with management options, primarily dredging, at several PCB-contaminated-sediment sites.) Standardization of costing approaches has been advocated by such groups as the American Society of Testing and Materials; a proposed standard is under development by that organization. Such standardization, when combined with a detailed evaluation of site-specific conditions that might affect costs, should improve the ability to predict management costs accurately and allow comparison of costs of various management options and remediation technologies.

The evaluation of options also requires assessment of their effectiveness. The primary goal of contaminated-sediment management is the protection of resources at risk, such as human or ecological health, commercial or recreational fishing stocks, or a particular endangered species. Ultimately, it would be preferable to choose management options that best protect affected resources or lead to their recovery. That takes time, however, and interim indicators of effectiveness, such as contaminant loss or exposure and risk to sensitive populations during and immediately after implementation of the option, are generally necessary. Assessing or predicting even short-term human health and ecological risks directly for all options at this stage can be exceedingly difficult.

An alternative for the evaluation and comparison of management options is to use contaminant mass flows as a surrogate measure of exposure and, ultimately, risk—that is, a technology may generally be assumed to result in reduced exposure and risk if it leaves less residual contamination in the sediment and puts fewer contaminants into the air and water than an alternative technology. Evaluation of contaminant mass flows for each management option can be most useful in the comparative rather than the absolute evaluation of potentially applicable management options by providing a systematic screening tool. A comparative analysis of mass flows can also help identify those components of an overall management strategy that largely control the overall exposure or risk and, therefore, should receive the most resources and effort for detailed evaluation.

Although contaminant mass flows can be useful, exposure and risk to human and ecological health ultimately drives the need for risk management and the success or failure of any management option. In particular, absolute rather than comparative analyses are needed to balance short-term acute risks, for example, from removal options with long-term risks, such as from in situ containment options. The actual magnitude of the resulting exposure and risk must be assessed for decision-makers to make these tradeoffs in the next stage of the framework.

Mass release or flow analysis can also be used to estimate concentrations in surficial sediment or overlying water. Ultimately, it is these quantities that are most closely related to risk. The long-term behavior of the concentration can then be used to define exposure and risk as a function of time. For example, if dredging and onshore treatment or disposal are used, the water concentrations increase during the dredging action but, if effective, decrease rapidly beyond that expected of natural-attenuation processes. Finally, if in situ capping is used, the concentrations in water and the risk might be reduced dramatically over the short-term, but there is a possibility of cap failure at some later time, resulting in increased exposure and risk unless maintenance and repair are performed.

The relative importance of the exposure and risk is dependent upon the magnitude of the concentrations and the type of adverse effects caused by the contaminant. Long-term cancer risk in humans, for example, is usually assumed to be related to average exposure or dose, whereas noncarcinogenic risk is usually assumed to be significant only if a certain threshold is exceeded. Thus, the cancer risks associated with various management options can be compared by evaluating the cumulative exposure over time, and the noncarcinogenic risks can be compared by evaluating the maximum exposure or the period exceeding the threshold. Other more complicated risk factors must be evaluated by more sophisticated analyses.

There are at least two reasons why current risk might be of more concern than deferral of risks into the future. First, it is common practice to discount future gains and losses in recognition of rates of time preference and productivity of capital assets (e.g., Cropper and Portney 1990; Portney and Weyant 1999). So a given risk today is commonly viewed as more of a concern than an equivalent risk at some future date. Second, changing technology might make PCB-related risks less of an issue in the future. For example, if PCB-contaminated sediments are capped beneath clean sediments and if that cap is expected to last for a long enough time (say one century), it is possible that new technologies can be developed in the mean time to alleviate the risk. On the other hand, some might question whether it is morally acceptable to take actions that benefit society today, if those actions push risks off onto future generations (Kadak 2000).

Component Technologies of Sediment Management Options

A wide variety of management options exist for addressing contaminated-sediment problems. Socioeconomic options might provide offsets that improve human or ecological health or quality of life. Institutional controls attempt to reduce exposure to contaminants associated with sediments, for example, by restricting fishery usage or by providing alternative sources of fish.

Source control, often by treatment, removal, or containment, seeks to eliminate the causes of the contamination to ensure the permanence of management actions. Natural-attenuation processes are likely to affect the risk of the contaminants in the sediments. These processes combined with institutional controls are expected to limit exposure to the contaminated sediments. The primary natural process likely to reduce risk to PCBs at contaminated-sediment sites is stable burial by deposition of clean sediment, a process that physically separates contaminants from organisms in the biologically active region near the sediment-water interface. A variety of more active in situ and ex situ options also exist for contaminated sediments. The application of ex situ treatment or disposal typically introduces a complete train of component technologies, including removal by dredging, temporary storage or pretreatment to reduce water content or volume, final dredged material treatment or disposal, and management of any residually contaminated materials, including air and water waste streams. All of these options pose varying degrees of effectiveness for any particular PCB-contaminated site. As indicated previously, the identification and comparative evaluation of these options as to effectiveness, cost, and contaminant losses and residuals must take into account all components of the individual options.

These general approaches to sediment management are addressed below. Excellent reviews of the remediation technologies including a description of their design, application, and effectiveness are already available (see Averett et al. 1990; EPA 1994a; NRC 1997). As a result, only outlines of the various approaches and the potential advantages, disadvantages, limitations, and applications are discussed here. The primary goal is to indicate the risks associated with the various options or their components and the conditions under which those risks might be problematic. An important consideration in the selection of a management option is balancing effectiveness with costs. Unfortunately, costs are very site-specific and highly variable and are often estimated on different bases, making generalizations on costs difficult. As a result, the costs of various management options are not discussed here except in general terms. To be useful, project-costing information must be reported much more completely with identification of major cost factors at a particular site.

Socioeconomic Options

Options for reducing exposure to PCBs include approaches that do not directly address the contaminated sediments in which they reside. Two types of approaches in this area include the use of institutional or administrative controls to limit exposure and access by groups at risk and the use of offsets to meet environmental goals by reducing risks from sources other than the contaminated sediments.

Institutional Controls

In the parlance of a previous report by the NRC (1997), institutional controls are "interim controls" implemented to minimize exposure to contami-

BOX 7-1 Administrative or Institutional Controls

Are nonengineered instruments, such as administrative and legal controls, that minimize the potential for exposure to contamination by limiting land or resource use.

- Are generally to be used in conjunction with other risk-management options, such as treatment or containment.
- Can be used during all stages of the cleanup process to accomplish various cleanup-related objectives.
- Should be "layered" (i.e., use of multiple mechanisms) or implemented in a series to provide overlapping assurances of protection from contamination.

nants and reduce risk to humans and the environment until long-term management reduces risks to acceptable levels (see Box 7-1).

There are several general categories of institutional controls: government controls; informational devices; proprietary controls; and enforcement tools with institutional control components. Institutional controls involving PCB-contaminated sediments often take the form of government controls, such as fishing bans or fishing catch-and-release requirements, or informational devices, such as fishing advisories. These controls can also include physical measures, such as fences and signs, to limit activities that might result in exposure to contamination at a site. For example, in an effort to prevent or minimize exposure to fish with PCB contamination above a target risk level, signs warning against eating fish have been posted at most public boat-launch areas and recreational areas at Lake Hartwell in South Carolina since 1987 (Hahnenberg 1995, as cited in NRC 1997). As noted in the 1997 NRC report, in some cases fishing restrictions have been in place for so long that they have become de facto permanent solutions. For instance, PCB-contaminated fish and sediments were found in the upper Hudson River in the early 1970s. Health advisories against fish consumption from the lower river and a complete ban on fishing in the upper river have been in effect since the mid-1970s (Harkness et al. 1993).

The effectiveness of fishing controls is an open question. The committee responsible for the 1997 NRC report was unable to find enough information to document or analyze the risk reduction of either fishing bans or advisories. The committee summarized a study by Belton et al. (1985) that illustrates the compliance problems involved. This study addressed a potential 60-fold increase in the risk of human cancer associated with the lifetime consumption of PCB-contaminated fish from the Hudson-Raritan estuary area. The effectiveness of public-health advisories as risk-reduction measures was evaluated by a careful, multidisciplinary study of recreational fishers. Approximately 59% of those surveyed fished for the purpose of catching food. More than 50% of the respondents were aware of the warnings, and those who did not consume the fish were generally persuaded by a perception of unacceptable risks. But 31% of those who ate their catch did so despite believing it was contaminated. The researchers concluded that the broad-scale rejection of the health advisories was due to a combination of factors: the way the media were used, the nature and delivery of the health advisory, and personal predispositions that tended to reduce the credibility or usefulness of the communication.

In a more recent study (NYSDOH 2000), approximately half of the Hudson River anglers surveyed remained unaware of fish advisories, and approximately one-third ate the fish from the river. Of the anglers surveyed, however, only 6-7% indicated that their primary reason for fishing was for food, and 90% indicated that they were fishing primarily for recreation. Those that did eat the fish often shared the fish with women and children despite NYSDOH recommendations that these portions of the population eat no fish from the Hudson River.

Fishing controls might have more impact if fish advisories are available in multiple languages, are handed out with fish licenses, or are readily accessible on the Internet. For example, the state of Indiana maintains an updated fish consumption advisory on its official website. Also, local health or environmental management personnel should be encouraged to visit local fishing holes and advise people about the risks of eating the fish they catch (i.e., enforce catch-and-release laws associated with a fish advisory).

Institutional controls involving PCB-contaminated sediments may also take the form of legal restrictions on the use of aquatic or adjacent upland property (proprietary controls). These legal restrictions are often described in a restrictive covenant on the property that runs with the land executed by the property owner and recorded with the register of deeds for the county in which the site is located. The restrictive covenant may contain measures prohibiting uses (e.g., residential) or activities (e.g., recreational fishing) more likely to result in exposure to residual contamination on the property. The restrictive covenant may also provide for continued access to the site by appropriate agencies. The restrictive covenant may be removed, upon notice to and approval by the appropriate federal or state agency, if and when residual contamination on the site finally achieves acceptable risk levels.

Like fishing bans, proprietary controls are only as effective as the willingness of local, state, or federal authorities to monitor and enforce such restrictions. Given that restrictive covenants may allow for residual contamination on the site, questions remain whether such measures truly reduce risks from exposure over the long term (see Chapter 4 for more discussion of community concerns regarding institutional controls).

In summary, institutional controls "may be advisable when sediment contamination poses an imminent danger and immediate risk reduction is required" (NRC 1997). The committee finds that institutional controls might also be advisable as a supplement to long-term controls and technologies when management actions that would reduce contaminant risks to acceptable levels are impracticable and long-term solutions necessarily involve natural-attenuation processes. Although few data are available concerning the effectiveness of institutional controls, the measures identified above appear to be practical and are likely to reduce risk to some (albeit unknown) degree. Other advantages include low costs and ease of implementation.

When institutional controls form part of the management strategy, they should be negotiated at the same time as the remainder of the strategy rather than after the strategy has proved to be ineffective. Negotiating institutional controls as part of the management strategy will allow maximal flexibility in achieving a solution acceptable to directly affected parties. To be effective, however, the controls must be understandable, monitored, and enforced.

Offsets

Offsets can be an effective way to attain environmental goals in a technically feasible and cost-effective manner and can facilitate negotiations among affected parties. Offsets trade off an increase in environmental harm by certain actions against an equivalent or greater reduction in harm by other actions. For example, suppose there is a goal to reduce aggregate pollution emissions from all sources by 50%. An offset might reduce overall emissions by 50% or more by reducing emissions by more than 50% at sites that are inexpensive to control and simultaneously allowing current levels of emission at sources where pollution control is very expensive. The goal is to think creatively to identify and pursue win-win solutions that are preferred by all parties.

Typically, offsets focus on controlling pollution emissions. However, such control is not likely to be effective for a pollutant like PCBs, where zero emission is the environmental goal. Instead, offsets might focus directly on impacts to humans and the environment. For example, the community might agree to bear some risk from PCB-contaminated sediments if the responsible party agree to offset that risk by reducing other risks to human health. In such a case, a community might opt to have a catch-and-release program for its contaminated fish, if the responsible party restores a contaminated river bank rather than dredging deep-water sediments. A plan to restore wetlands or avoid the anticipated loss of a critical habitat area might be used to offset ecological impacts from PCB contamination.

If all parties cooperate and think creatively, offsets can result in solutions that reduce overall impacts to the community and simultaneously reduce costs to the responsible party. Again, the key is to implement offsets within the context of a negotiation process that identifies solutions that are preferred by all affected parties.

The design of this type of system for offsets requires several steps. First, all potential PCB impacts must be identified. Next, a set of feasible actions must be identified to offset each type of impact. All affected parties should be involved in identifying impacts and the set of potential offsetting actions, and all affected parties should be satisfied that the proposed set of actions at least offsets the PCB-related impacts. Finally, alternative sets of actions that both offset all PCB-related impacts *and* result in a cost savings relative to PCB risk management can then be identified. The least costly set of actions that fully offset PCB-related impacts is then implemented, or an alternative set may be chosen if agreed to by all affected parties.

Source Control

Many of the contaminated-sediment problems faced by the nation are a legacy of the poor control of industrial and municipal effluents in the period prior to the passage of the Clean Water Act. With respect to PCBs, much of the problem stems from effluents and activities before to the ban on PCB usage and also before effective control of PCB sources. Because of the historical nature of the PCB-contamination problem, sediments now often serve as a major contributor of PCBs to overlying water rather than as a sink for external sources. In some areas, however, uncontrolled land-based sources of PCBs continue to release contaminants to the water and, due to their strongly sorptive nature, to the sediments. The first goal of any management activity is to conduct a critical assessment of the sources of the sediment contamination and control sources that are significant relative to the sediment cleanup goals. Significant source control might have been achieved already, however, and further source control might be more difficult to manage than the residual sediment contamination that might result. If a significant external source is not identified or is allowed to persist, then efforts to reduce risk through other management options might not be successful. Controlling on-going sources is consistent with the EPA contaminated sediment management strategy (EPA

1998), which has as its first goal "preventing the volume of contaminated sediment from increasing."

Identification and control of remaining sources of PCBs to a water body and to the sediments is often difficult but is a necessary part of developing an accurate conceptual model of a site. Full development of an accurate, verifiable, material-balance-based mathematical model of the site remains one way to identify other as yet unidentified sources. A detailed database and understanding of the fate and transport processes of the contaminant are a necessary prerequisite to the development of such a model. However, these requirements are unlikely to be met except in the case of the most studied sites where the expense of sophisticated model development can be justified by the potential risks and costs associated with risk management of the site. As shown in Box 7-2, even sites that have received a great deal of study might have significant sources that go unrecognized. Although finding all external sources might be difficult, the long-term cost of inadequate source control is that site remediation might have to be repeated at some time in the future.

A continuing source of sediment PCB contamination often results when the water body is adjacent to a contaminated-soil site. Runoff of surface contamination, continued seepage of groundwater contaminants, or movement of nonaqueous-phase liquids can pose a continuing source of PCBs to the sediments. Lack of source control might make sediment management efforts unsuccessful. In other cases, a continuing source, if not significant, might limit the cleanup levels that are achievable. In addition, if it is not possible to control the migration of contamination from the soil site, it might be appropriate to manage the sediments to lessen the impact of the contamination on the water body. In such a situation, continuous monitoring and periodic remediation of the contaminated sediments might be alternative management options. In any event, no "walk-away" solution is possible without adequate source control.

One of the most difficult sources to quantify and control is urban stormwater runoff and combined sewer overflows. Since the overflows are episodic in nature, it is often difficult to monitor and characterize these releases effectively. Due to improved control and management of other potential sources of PCBs and over time, the amount of PCBs associated with storm water runoff or combined sewer overflows is likely to be much less than that in decades past, but it still might be important relative to target sediment cleanup goals. PCBs are strongly sorbing, and even low water-borne loads might lead to significant sediment concentrations. King County, Washington, is eliminating combined-sewer overflows at Denny Way, where sewers overflow from the south end of the city of Seattle, partly to avoid recontamination of remediated sediments (King County 2000).

BOX 7-2 Location of Additional Sources of PCBs on the Hudson River

From the late 1940s until 1977, two General Electric capacitor manufacturing plants near Fort Edward and Hudson Falls, New York, discharged PCBs into the Hudson River. Much of the sediments accumulated behind a dam at Fort Edward that was removed in 1973. Subsequent high-flow events on the river led to dispersal of the PCBs throughout the Hudson River. The site was proposed for listing on the National Priorities List in 1983. In 1984, on the basis of a lack of a technologically feasible, cost-effective management alternative, in-place containment of remnant shoreline deposits was selected to mitigate the most significant threats to human health and the environment. No action was taken on dredging the river system. In the late 1980s, significant decreases in fish-tissue PCB levels were observed.

Despite extensive study of the river prior to the issuance of a record of decision, however, another source of PCBs to the river was found in 1991. In that year, a failure of a gate in an abandoned paper mill at Hudson Falls led to a significant increase in PCB loading to the Hudson River. PCB loadings into the lower portions of the river, which had decreased to about 2 pounds (lb) per day by 1990, increased to 5-10 lb/day in 1991 and 1992 (Farley et al. 1999). A large pool of PCBs was found entering the river through the abandoned paper mill and via fractures in the surrounding bedrock. Containment and removal actions taken since 1991 have led to the removal of 136 tons of PCBs from this source. This represents a quantity of PCBs equal to about half of all of the PCBs estimated to be within the river system at the time of the Record of Decision in 1984. Groundwater extraction systems have been installed in an attempt to contain the PCBs in the fractured bedrock but the current PCB load due to this source remains 0.2-0.4 lb/day or 10-20% of that from other sources (Farley et al. 1999).

The significance of the continuing source in slowing natural attenuation of the surficial sediments and decreasing fish-tissue PCB levels remains a subject of debate and analysis. It is clear, however, that this source was and continues to be sufficiently significant that it must be included in the assessment and evaluation of the potential effectiveness of management options for the river. The continuing source might have gone largely unnoticed without the gate failure, but seepage from the bedrock could have posed a long-term source that would have at least slowed any natural attenuation of the system and could have substantially reduced the effectiveness of any management option undertaken prior to its identification.

Natural Attenuation

EPA (1999b) defines natural-attenuation processes for soil or groundwater as including "a variety of physical, chemical, or biological processes that, under favorable conditions, act without human intervention to reduce the mass, toxicity, mobility, volume, or concentration of contaminants in soil or groundwater. These in situ processes include biodegradation; dispersion; dilution; sorption; volatilization; radioactive decay; and chemical or biological stabilization, transformation, or destruction of contaminants." Since natural-attenuation processes will affect all contaminated sites to some extent, an evaluation of the change in risk with the passage of time posed by natural-attenuation processes should be a component of all risk-management proposals. Some natural-attenuation processes, such as dispersion, dilution, and volatilization, might transfer the risk at one location to another location, which might or might not reduce overall risk. Other processes, such as biodegradation and sorption, and stabilization through burial, might produce overall risk reduction. However, in water bodies with PCB-contaminated sediments, stable burial by deposition of clean sediment (i.e., sedimentation) is often the major natural-attenuation process that can lead to further recovery of the water body. Natural-attenuation processes will be a part of any management strategy, because some residual PCBs are expected to remain at a site despite efforts to remove all contamination. These residual PCBs might be in marginally contaminated areas outside the area receiving more active management efforts, or they might be the residual contamination remaining within the remediated zone.

The question then is not whether natural attenuation should be considered a part of a risk management strategy, but how much should it be relied upon for reduction of risk to humans and the environment, and how can this risk reduction be measured? (See Chapter 4 for a discussion of community concerns regarding natural attenuation.) Because many of the processes that might attenuate contaminants are different in sediments than in other environmental media, the direct application of existing protocols for natural attenuation of contaminants in soils and groundwater is inappropriate. Reduction of the PCB content in fish through time might be used as one indicator of the effectiveness of natural attenuation, because fish consumption frequently is the major pathway of PCB exposure to humans. Sloan (1999) provided data on changes in lipid-based PCB content in Hudson River fish. The concentration in all species combined dropped over a 15-year period from 1983 to 1998 in both the upper Hudson and lower Hudson, in spite of a major increase of PCBs in fish that appears to have occurred in 1991 as a result of a major upper Hudson PCB discharge, which resulted from a gate failure in an abandoned paper mill. Using the regression line provided (Sloan 1999), one can calculate a first-order rate of PCB decrease in fish of about 3% per year for the upper Hudson and about 4% per year for the lower Hudson. The rates of reduction might have been greater if the 1991 spill had not occurred.

As indicated in Box 7-1, however, the spill allowed identification of a continuing source of PCBs to the river, which became more obvious as sediment contaminant levels were reduced by natural-attenuation processes. As another measure of risk reduction, source control with the reduction in PCBs through time might be useful. For example, the PCB discharge from the Saginaw River to Saginaw Bay, Michigan, decreased from about 4,500 kilograms (kg) in 1972, when point-source PCB discharges were first regulated, to 110 kg in 1991, a rate of decrease of about 20% per year (Verbrugge et al. 1995). That relatively high rate undoubtedly reflects the combined effect of source control and natural attenuation, a decrease of 14% since 1979 is perhaps more indicative of the portion attributable to natural attenuation, although the data available are inadequate for good statistical analysis. A similar decrease in PCB discharges from the upstream reaches of the lower Fox River in Wisconsin is discussed in Box 7-3.

Natural Attenuation by Biodegradation

Of the natural-attenuation processes, biodegradation is generally considered the most desirable because it can result in elimination of risk. However, biodegradation does not always result in complete destruction of a compound; it results more often in the transformation of one compound into another. The daughter products can at times be even more hazardous than the parent compound. Thus, a knowledge of biodegradation pathways is needed to properly evaluate whether biological processes are likely to be beneficial in risk reduction. The biodegradation pathways of the collection of chlorinated compounds that comprise the PCBs differs depending upon whether the degradation occurs in the presence of oxygen (aerobic processes) or in its absence (anaerobic processes). A detailed discussion of the various pathways is given in Appendix E. Generally, anaerobic processes are most effective in removing chlorine atoms from PCB molecules containing more than three chlorine atoms, particularly when they are located at what are termed the meta and para positions rather than the ortho positions. Aerobic processes are more effective at complete destruction of the PCB molecule when it has three or less chlorine atoms. (See Appendix E for a more detailed discussion of the fate of PCBs in the environment.) A combination of anaerobic and aerobic processes can result in complete destruction of PCB molecules. However, the rate of natural biodegradation of PCBs is often slow. The slow rate is due partly to the poor

BOX 7-3 Simulation of Natural Attenuation of Upstream Reaches of Lower Fox River, Wisconsin

The lower Fox River extends 39 miles from Lake Winnebago to Green Bay, Wisconsin. Beginning in the mid-1950s, paper mills along the river began using PCBs, primarily in carbonless copy paper. Recycling of this paper discharged substantial quantities of PCBs into the river and out into Green Bay. In 1990, over 4,000 kg of PCBs remained within the upstream reaches of the Fox River between Lake Winnebago and the DePere Dam, 7 miles upstream of Green Bay. Studies of the upstream reaches demonstrate the importance of sediment stability, partially through burial by depositional processes, on containment of the PCBs and reduction of PCB fluxes within that reach of the river.

Because natural-attenuation processes are generally slow, there are rarely sufficient time-series data to demonstrate the rate and extent of natural attenuation. Instead, sophisticated mathematical models that simulate future behavior are required to predict the effects of natural recovery. If natural attenuation is included as part of the management strategy, continued long-term monitoring should provide the time-series data needed to validate and calibrate the model and improve its predictions.

During the early 1990s, the Wisconsin Department of Natural Resources sponsored the development of such a model for the upstream reaches of the lower Fox River (WDNR 1995). The model was able to simulate the available observations of a variety of PCB congeners while modeling a 1-year period between May 1989 and April 1990. The upstream Fox River model predicted that 143 kg of PCBs were transported over the DePere Dam during this period and 19 kg were released to the air by volatilization. Of this 162 kg total, all but 2 kg were estimated to have entered the water column by release from the bed sediments.

Using the calibrated model to predict future behavior, the mass of PCBs estimated to be transported over the DePere Dam was expected to decrease by a factor of two every 5 years primarily due to burial of contaminated sediments by clean sediments. Although these predictions suggested that PCB transport during normal flows over the DePere Dam would decrease dramatically, similar simulations in the downstream reaches would be required to predict PCB transport into Green Bay. Separate simulations also showed that a major storm event during 1989 and 1990 could have caused significant increases in transport of PCBs. A simulation of such an event suggested that 140 kg of PCBs could be resuspended, leading to 81 kg being transported over the DePere Dam, 6 kg being volatilized, and the remaining 53 kg being distributed over the upstream reaches. The significance relative to sediment-management decisions depends on the probability of such an event occurring during the early years of natural attenuation as well as the potential consequences. In either case, prognostic modeling remains the only tool available for assessment in the absence of sufficient time-series observations.

bioavailability of PCBs because of (1) their strong tendency to sorb to sediments, (2) their generally low concentrations, especially in water where they are more readily available to biological attack, and (3) their complex structure. Probably no single organism is capable of complete destruction of PCBs. A single organism might be capable of degrading or transforming only a few of the 209 different possible PCB isomers or congeners. Thus, many different organisms are generally required to effect more complete PCB biodegradation, and all those required might not be naturally present at any given time or place. In addition, there is yet no good evidence that microorganisms benefit from PCB transformations. The transformations occurring might be a fortuitous process, carried out by enzymes present in microorganisms for other purposes, a process termed cometabolism. If no direct benefit is received, the process might not be reliable and might depend upon other factors affecting the viability of the PCB-transforming organisms.

Given the above, what in general can be said about the potential for biodegradation to reduce PCB risk? Anaerobic biodegradation tends to be most effective for the more-chlorinated PCB congeners. The dioxin-like PCB congeners tend to be modified most readily by anaerobic transformations, and thus the arylhydrocarbon receptor (AhR)-mediated toxicity of the critical PCB congeners can be reduced significantly by anaerobic processes. The proportion of nondioxin-like PCBs present is then increased. Thus, anaerobic biodegradation might cause a change from a prominence of dioxin-like toxicity to a prominence of nondioxin-like toxicity. Anaerobic transformations also reduce the chlorine content of PCB molecules, making them more mobile, less sorptive, and more susceptible to aerobic biodegradation. Frequently, "weathered" PCBs are noted to be more toxic per unit total PCB mass, than "nonweathered" PCB mass (Giesy and Kannan 1998). That difference is probably the result of the loss through diffusion and volatilization of the lesser chlorinated PCBs, leaving behind the more highly chlorinated and toxic congeners. Thus, some natural-attenuation processes might render the remaining PCBs mixture more toxic per unit mass, and others, such as biodegradation, might render them less toxic. The particular processes involved in change in PCB mass must be clearly understood by site risk assessors to draw valid conclusions about risk reduction.

Natural Attenuation by Sedimentation

One of the most important natural-attenuation processes affecting risk reduction of PCB-contaminated sediments is sedimentation itself. Historically, contaminated sediments often developed in net depositional environments. If the hydraulic conditions of the water body have not changed, natural attenuation is often assisted by the continuing deposition of clean sediments. For PCBs, which are persistent and subject to only limited degradative processes in the environment, burial by deposition of clean sediment may be the dominant natural-attenuation process. Because this process does not change the contaminant levels at depth in the sediments, it is often viewed with concern and distrust, particularly by affected communities. Changes in hydraulic forces, for example, by dam removal or major flooding or storm events, might also cause erosion of the sediments and mobilization and transport downstream of contaminants. Stable deposition and burial, however, can result in substantial recovery of resources at risk, for example, fish in the overlying water. Natural-attenuation processes might or might not lead to recovery of the water body or resource at risk. Whether natural-attenuation processes can be viewed as natural recovery depends on whether risk-management goals are met and maintained over the long term. That is, natural recovery depends on the ability of attenuation processes to maintain risk reduction to humans and ecological health over an extended time and not necessarily on the reduction of concentration levels in the contaminated sediments.

Key factors affecting sedimentation and benthic stability are the energy of the overlying flow and whether a particular sediment deposit is net erosional or depositional. Under high-flow conditions, a bed sediment deposit tends to be coarse grained and noncohesive with little sorptive capacity and low depositional rates. Substantial amounts of sediment and associated contaminants can be suspended in the water. Because most persistent sediment contaminants such as PCBs are associated with the solid phase, any mobilization of the solid phase dramatically increases contaminant mobility. However, high suspended sediment levels in a water body does not necessarily mean that the bed and its associated contaminants have been resuspended. High suspended sediment loads may result from surface runoff or simply transport from upstream.

In low-energy environments, deposits are typically fine-grained, providing high sorptive capacity and significant slowing of advection and oxygen transport. Low-flow conditions might stem from low runoff during dry periods, from a widening of a stream to a pool in a pool-riffle system, or from hydraulic controls, such as a dam, which contributes to sediment accumulation immediately upstream. Under stable sediment conditions, even relatively thin layers of sediment pose important barriers to contaminant transport and release to the overlying water. Deep within the sediments, the release or fate of contaminants from the bed sediment is largely governed by physicochemical and microbial processes. Important physicochemical processes include advection, diffusion and sorption and desorption. Because PCBs are highly sorbing, however, advection and diffusion processes in the pore water are severely retarded because of accumulation on the sediment solids.

Generally, the most important process within stable sediment deposits is bioturbation, the mixing of sediments associated with the normal life-cycle activities of sediment-dwelling organisms. This process is inherently faster than pore-water based advective and diffusional processes. Many organisms, especially head-down deposit feeders, prefer fine-grained, organic-rich sediments, enhancing uptake and bioturbation in those areas where PCB accumulation is most likely and where advective and diffusive processes are normally suppressed. The presence of these organisms tends to produce a relatively well-mixed zone of sediment near the surface. This biologically active zone largely governs the extent of exposure to benthic organisms and overlying water. It is important to note that the contaminant concentration in the biologically active zone might not be represented well by a depth-averaged composite concentration which is typically measured (see Chapter 10, Box 10-1). Freshwater benthos, for example, might populate in large numbers only the upper 5-10 centimeters (cm) of sediments. In marine sediments, animals living at the sediment-water interface tend to be larger and influence a larger sediment depth. The region of sediment heavily affected by benthic organisms, however, tends to remain 5-15 cm, and occasional deeper excursions by organisms generally do not significantly affect the overall mass of contaminants or the exposure of animals living in overlying waters. More than 90% of the 240 observations of bioturbation mixing depths in both fresh and salt water reported by Thoms et al. (1995) were 15 cm or less, and more than 80% were 10 cm or less. Almost all of those estimates were based on measurements of the vertical distribution of various radionuclides, measurements that have proved to be useful tools in identifying erosion and deposition rates and the degree of mixing within the upper layers of sediment. The observed effective particle diffusion coefficients fell within the range of 0.3 to 30 cm^2 per year more than two-thirds of the time. Recognizing the strong sorption of PCBs to particles, that observation suggests that bioturbation can move contaminants at least 10-1,000 times faster than molecular diffusion in the pore water. Thus, depositional processes that bury contaminants below the biologically active zone can provide effective containment of contaminants unless episodic flooding or storm events are capable of returning the deep contaminated layers to the surficial sediments and the biologically active zone.

In summary, it is clear that some natural-attenuation processes, notably burial in a stable system, can reduce risk, and others, such as bed erosion, dilution, dispersion, and volatilization, may simply transport the risk elsewhere. Because different natural-attenuation processes have such different impacts on risk management, it is not sufficient to quantify natural attenuation without adequate characterization of the particular processes involved at a given site. The important processes effecting natural attenuation at a given site must themselves be characterized to understand the long-term effectiveness of a given management option, the rate at which risk is likely to be reduced, and the potential for transporting risk to some other location. Natural attenuation by deposition is most effective in areas that are hydrodynamically stable and that have sedimentation resulting in the burial of contaminated sediments. In contrast, this process might not be effective in areas that are subject to continuous or episodic erosion and exposure of deep sediments or in areas where sources are not yet controlled. To characterize the natural-attenuation processes properly, mathematical models are often required because of the complexity of the many processes involved.

Recommendations for evaluating natural-attenuation processes in groundwater (NRC 2000) can be applied similarly to PCB contamination in sediments:

- Evidence of natural-attenuation processes should be used to document which mechanisms are responsible for observed decreases in contaminant concentration.
- A conceptual model of sites being considered for natural attenuation should be prepared to show where the contaminants are moving.
- Field data on natural attenuation should be analyzed at a level commensurate with the complexity of the site and the contaminant type.
- A long-term monitoring plan should be specified for every site at which natural attenuation is approved as a formal remedy for contamination.

In Situ Treatment Options

In situ treatment options are often designed to enhance natural-attenuation processes. Risk reduction as a result of natural-attenuation processes in sediments is often due to depositional processes that carry the contaminant below the biologically active zone. Similarly, in situ containment by capping with clean sediments has been proposed as a sediment management option. The extremely slow transport processes of diffusion or advection in the region below the biologically active zone provide effective containment of sorbing contaminants, such as PCBs (e.g., Palermo et al. 1998).

In situ treatment options that do not rely upon physical separation from the biologically active zone have been considered but have not yet been demonstrated at the field scale for PCBs. Biological-degradation processes could be implemented in situ, but most PCBs degrade only slowly or to a limited extent in sediments, as discussed above. In principle, biological degradation could be enhanced by the addition of required substrates and nutrients or reagents or catalysts (Koenigsberg and Norris 1999), but no effective in situ delivery system has yet been developed for contaminated sediments. An effective delivery system would likely involve mixing of the sediment, encouraging resuspension and loss of both sediments and contaminants. The lack of an effective delivery and homogenization system has also hindered the application of in situ stabilization systems. A demonstration in Manitowoc Harbor, Wisconsin, also encountered difficulties in the management of pore water released by the solidification process (Fitzpatrick, W., Wisconsin Department of Natural Resources, Madison, WI, personal communication, 1994 as cited in EPA 1994a). As a result of these limitations, in situ treatment and stabilization technologies are unlikely to be used except when the contaminated sediment can be isolated from the water body, for example, through sheet piling or temporary dams. In situ treatment options that do not involve delivery of chemicals to the sediments have also been proposed. In situ vitrification uses electricity to raise sediments to sufficiently high temperatures to produce a glass-like product (EPA 1995). The energy costs of heating high-moisture-content sediments to glass-formation temperatures are formidable, and the technology is unlikely to be used except for small volumes of highly contaminated sediments.

In the absence of other in situ contaminated-sediment treatment systems that have been demonstrated to be effective, attention is focused here on capping. For ease of placement, sand or other coarse media are normally used as capping material. Geomembrane material may be used beneath a cap in soft sediments to aid in the support of the cap and stones, or other large material may be used as armoring on top of the cap to reduce cap resuspension and erosion. The purpose of a cap is to separate contaminated sediment from organisms living at the sediment-water interface, isolate the chemical contaminants from the overlying water, and provide protection from breaching as a result of cap erosion. A cap may be placed upon undisturbed contaminated sediments or upon dredged material that has been discharged back to the water column in a separate location from where it was generated. Due to the potential for contaminant losses during removal and subsequent placement, capping dredged material is generally only considered for the management of marginally contaminated sediments in regions where navigation needs necessitate sediment removal.

Capping contaminated sediment is usually implemented by either

- Thin-layer capping, also referred to as enhanced natural attenuation.
- Thick-layer capping and armoring.

Thin-layer capping or enhanced natural attenuation is the process most closely related to natural sedimentation processes. By placing a thin layer of clean sediment over the contaminated sediment, the process is potentially less disruptive of the benthic community. Because the sediment-water interface tends to approach an equilibrium state, the small modification provided by a thin-layer cap is potentially more stable than a thick-layer cap without additional armoring. A layer of only 5-15 cm will generally isolate the bulk of the contaminants from the benthic community and the overlying water. Isolated penetrations of a thin-layer cap can still occur, however, but are unlikely to lead to aquatic organism exposure to significant contaminant mass. As indicated earlier, the depth effectively mixed by benthic organisms rarely exceeds 15 cm, even in the marine environment. Primary concerns associated with thin-layer capping is the long-term stability of the capping layer without armoring and the ability to accurately place a thin layer of sediment. Accurate placement of a thin-layer cap requires shallow water depths for better control of material placement and sediments of sufficient static strength to support a cap without intermixing with the bottom sediments.

Thick-layer capping is the conventional approach to containment of contaminated sediments. Cap thicknesses are normally 20 cm to as much as 1 m. The larger depths help ensure that an isolating cap layer remains even if there is significant heterogeneity in placement thickness or small amounts of post-placement erosion. However, the larger depth of capping material might result in load-bearing problems for an underlying soft sediment or require placement in multiple layers to allow the underlying sediment to consolidate and develop sufficient strength to support the cap layer. Once placed, however, the load of the cap can contribute to the consolidation and strengthening of the underlying sediment, aiding in its stability even in the event of major storms that might remove the cap. A membrane layer can also be used to provide strength for the underlying sediment during placement, but the requirement for a membrane layer has been demonstrated in only a few cases. A number of examples exist of the placement of sand caps of 20-50 cm directly over sediments with a surface undrained strength of less than 1 kilopascal (Palermo et al. 1998). A capping pilot-scale demonstration (1 hectare) over soft sediments was conducted by the National Water Research Institute of Canada in 1995. Although full-scale capping projects have been implemented, the pilot-scale project benefitted from the collection of extensive measurements before, during, and after capping to assess effectiveness and implementability. The results and the implications for cap effectiveness are discussed in Box 7-4.

Capping can also be used to promote wetlands and restore habitat. The lower layers of a cap may be used for chemical containment and physical separation of contaminants from organisms that might burrow or contact the sediments. The materials used for this portion of the cap should be selected to maximize their effectiveness in those roles—for example, sorbing amendments might be used to aid chemical containment, or grain size might be selected to control the type and extent of bioturbating organisms. The upper layers of a sediment can be selected to enhance habitat and tailored to the

BOX 7-4 In Situ Capping Demonstration at Hamilton Harbour, Lake Ontario

Hamilton Harbour, located on the extreme west end of Lake Ontario, is contaminated with a variety of organic and metal compounds, including PCBs. The sediment is soft and exhibits low shear strength, but water depths of 12-17 m protect an area that was capped from even severe storm events. The 50-cm sand cap used was chosen to contain contaminants, and monitoring programs were developed to evaluate mixing during placement and release after placement. Capping sands were placed via tremie pipe from a feed hopper. Accurate placement was assured by cable anchoring of the sand barge with a winch to control horizontal movement. Three layers of sand were placed. The capping operation and its results are described by Zeman and Patterson (1997).

Observations after placement indicated that the initial lift of sandy material intermixed with the underlying sediment, strengthening the sediment for subsequent sand placement. Subsequent sand lifts did not significantly intermix with underlying sediment. Turbidity plumes observed during placement were determined to be associated with a small fraction of fine-grained material with the sand and not by resuspension of the underlying sediment. The cap and underlying sediment consolidated 6-8 cm within a period of days after placement, consistent with laboratory assessment prior to the demonstration. As a result of consolidation as well as inaccuracy in cap placement and horizontal spreading of the cap material, the final cap depth was about 34 cm rather than the design depth of 50 cm. Coring subsequent to cap placement showed no sediment intermixing between the second and third sand lifts and the underlying materials. Core analyses also showed that contaminant levels of insoluble metals decreased from 100 to 1,000 of mg/kg in the sediment to effectively zero in the capping material. PCBs decreased from a maximum of about 0.2 mg/kg in the underlying sediment to about 0.01 mg/kg in the capping material. Soluble metals showed penetration into the capping material as a result of cap placement and consolidation, but pore-water levels at the cap-water interface were still effectively zero.

species for which improved habitat is desired. Box 7-5 illustrates how a capping project can be used to enhance habitat. (See Appendix D for a more detailed description of the St. Paul Waterway problem area.)

Guidance exists for the design, placement and monitoring of a cap as a sediment management option (Palermo et al. 1998). This guidance includes quantitative information on design of armoring layers, design for contaminant containment and stability analysis during cap placement. A capping layer even 20 cm thick has been demonstrated to be an effective means of reducing

BOX 7-5 Capping and Habitat Restoration at St. Paul Waterway, Tacoma, Washington

The St. Paul Waterway problem area was a 17-acre contaminated-sediment site located at the mouth of the Puyallup River near Tacoma, Washington. The area was contaminated by pulp and paper-mill operations over 6 decades. Discussions between the site owners; the Puyallup American Indian Tribe; various environmental groups; interested citizens; and federal, state, and local government officials led to the emergence of a comprehensive environmental cleanup and restoration approach. The approach included

- Source control, including storm-water collection and treatment, plant process modifications, and a new outfall for the pulp and paper mill's secondary wastewater treatment plant.
- Isolation of the contaminated sediments by capping with clean sediments from the Puyallup River.
- Habitat restoration and enhancement of near-shore and intertidal areas.
- Preventive measures to reduce the potential for recontamination of the sediments.
- Long-term monitoring and adaptive management planning.

The cleanup action and habitat restoration was initiated in 1987 and completed in 1988 (Weiner 1991). It involved a very deep cap in shallow water (4-20 feet in thickness depending on the area being capped and the desired tide flat habitat elevations) (Sumeri 1989, Sumeri et al.1994). The deep cap in shallow water allowed integration of the cleanup action with the creation of new intertidal and shallow-water habitat capable of supporting a diverse benthic community and higher organisms that could use that benthic community for food. Commencement Bay, in which the area is located, had lost about 90% of such habitat over the last 100 years. Monitoring 10 years after placement of the cap showed minimal redistribution of cap materials and the development of a productive habitat where none had previously existed (Parametrix 1999). There now exist diverse biological communities of benthic and epibenthic organisms. Shorebirds and salmon use the site for feeding and rearing, and tidal pools are abundant with invertebrates.

The cost-effective project demonstrated the ability of capping to contain contaminants and provide viable habitat. The project remains one of the few sediment management efforts for which sufficient time-series monitoring (over 10 years) has been accomplished to demonstrate its success. The project was also the first completed Superfund cleanup in U.S. marine waters and the first natural resource damages settlement in the United States without litigation and with all federal, state and tribal trustees.

chemical flux and effectively isolates the contaminated sediment from much of the mixing associated with benthic organisms (Palermo et al. 1998). Thus, the ultimate effectiveness of capping as a sediment management alternative is generally reduced to a question of the stability of the capping layer. Some applications have depended on natural features to contain material to be capped (Commencement Bay, Washington) or spreading of cap material (Sumeri 1989) beyond the bounds of the material to be capped (New York Bight, New York). Subsequent monitoring in both cases has shown little movement of cap material or movement of contaminants into cap material (Parker and Valente 1988). A more general approach, however, is to engineer an armored cap to ensure containment of capping material under a particular disturbance (Palermo et al. 1998). That containment might be especially important in that a cap might disrupt the equilibrium surface of a riverine system and might encourage further erosion.

Assessment of capping in any particular situation depends on the ability to describe the hydraulics of the water body in which the cap resides and the resulting potential for cap resuspension and erosion. Because most caps consist of noncohesive sandy materials, the ability to predict the onset of sediment resuspension and erosion is comparatively good. Quantitative evaluation of the potential for cap loss under a variety of flow conditions, including episodic storm events, can be conducted. The stability of a cap can be further enhanced by the addition of an armoring layer composed of stones or other material currently used for bank or levee protection. With proper design and armoring it is possible, in principle, to build a cap that would be stable in any storm or flow situation. As with any engineered structure, however, the potential for failure in a catastrophic event remains, and plans for monitoring and long-term maintenance of a cap must be part of any application of this approach.

One potential cause for a reduction in the effectiveness of a capping layer is substantial seepage of groundwater through the cap into the overlying water. In such a situation, PCBs present in the sediment pore water can be transported into the overlying body of water. As with any pore-water transport process, sorption in the cap material can slow this process. For hydrophobic organic compounds, such as PCBs, the addition of organic material in the capping layer can enhance sorption and retard the movement of the contaminant through the cap. Ultimately, however, continued groundwater movement can cause the migration of sufficient PCBs to saturate the overlying capping layer, resulting in breakthrough of contaminants into the overlying water. Substantial groundwater seepage might occur in near-shore areas of lakes and along the coast and in river systems not underlain by low permeability clays or bedrock. Solutions to this problem include hydraulic control of groundwater (e.g., a slurry wall or trench system to divert water around the contaminated sediments) or control of the permeability of the capping layer to minimize seepage. Clay materials that expand upon placement in water or sediment solidification might also reduce seepage to manageable levels. Commercially available capping materials, such as Aquablok, a clay mineral-based capping material, also exhibit low permeability upon saturation with water and high resistance to erosion. The use of such material as a cap tends to divert groundwater seepage away from the contaminated sediments.

Although it is possible in principle to place and maintain a cap over contaminated sediments, a cap can be inappropriate at a given site for other reasons. The presence of a cap might hinder navigation or in the case of shallow environments, create unwanted wetland or upland conditions. The presence of a cap might also be incompatible with current or future uses of the water body or region. However, the cap can be used beneficially to create desired wetland areas or appropriate habitat at the sediment-water interface, as indicated previously, or to reclaim land by building the cap above the air-water interface. Sediment management options that include beneficial outcomes exhibit many advantages and are often more acceptable to the community than those that are viewed strictly as remediation efforts. Capping with clean sediments, with proper design and implementation, is a widely used and highly effective means of ensuring isolation of contaminated sediments in areas for which the resulting reduction in water depth is acceptable or desired. As with any engineered structure, however, provisions for monitoring and maintenance of a cap is necessary to ensure long-term containment.

As discussed above, in situ treatment options include capping and enhanced biological degradation. The use of capping is limited to sites where adequate placement and maintenance of the cap is feasible. For example, in situ containment by thick-layer capping and armoring can be an effective means of reducing risks where the cap can be maintained because of (1) a hydrodynamically stable environment, (2) adequate design of protective structures, and (3) adequate monitoring and maintenance of the containment system.

Removal Technologies

Options that involve removal of contaminated sediments from a water body are much more complicated than in situ approaches. Removal options generally require

- Controls to minimize contaminant loss during sediment removal and transportation.
- Pretreatment of produced dredged material for dewatering (i.e., removing water from the dredged material) and use of equalization basins to

assist in control of the rate of dredged material handled in subsequent steps in the treatment train.

- Treatment or transport and disposal of the dredged material.
- Management of the residual contamination left in the sediment as well as effluent streams from pretreatment or treatment operations (e.g., contaminated water).

Thus, removal options involve not only dredging but also several other component technologies to manage the dredged material. This discussion of removal options will consider the applicable component technologies separately:

- Sediment removal via mechanical or hydraulic dredging.
- Dredged material pretreatment technologies, including dewatering and particle-size separation.
- Dredged material extraction, stabilization, or destruction technologies.
- Dredged material disposal technologies, including upland and subaqueous disposal.
- Options for management of residual contamination.

Sediment Removal Via Dredging

Key factors affecting the selection and performance of dredging or excavation technologies include

- Production rate.
- Solids content of produced dredge material.
- Resuspension of sediments and associated contaminants.
- Dredging accuracy and residual contamination.
- Operational limitations.
- Availability.

Selection and design of removal technologies depend on those factors. The performance of some available technologies with respect to those factors is discussed below.

Hydraulic and Mechanical Dredging

Table 7-1 summarizes the key characteristics of the most common dredging types for the subaqueous removal of contaminated sediment. These dredges fall into one of two basic categories: (1) hydraulic dredges that primarily

TABLE 7-1 Summary of Operating Characteristics of Common Dredges

Dredge Type	Operational Depths, m	Production Capacity, m^3 hr	Solids Fraction, % by wt	Accuracy, m (horizontal/ vertical)	Resuspension Minimization	Turbidity Generating Units, kg/m^3 [a]	Debris Handling	Maintenance Dredging Costs, $\$/m^3$ [b]	Comments
Hydraulic									
Cutterhead	1.2-15[c]	25-2,500	37,183	1/0.3	Fair to good	1.4-45.2	Fair to good	36,956	Widely available in pipe sizes 6-30 inches
Horizontal auger	0.5-5	46-120	37,193	0.15/0.15	Fair		Fair to poor	36,956	
Matchbox	36,910	18-60	37,025	1/0.3	Fair		Fair to poor	36,956	Enclosed horizontal auger
Dustpan	2.5-19[c]	19-3,800	37,183	1/0.15	Fair		Poor	36,956	Partially enclosed pure suction dredge
Plain suction	2-19[c]	19-3,800	37,178	1/0.3	Fair	7.1-25.2	Poor	36,956	
Mechanical									
Clamshell	0-48[d]	23-460	~in situ	0.3/0.6	Poor	17.6-55.8	Good	37,018	Enclosed bucket (e.g. cable arm dredge)
Excavator	36,964	20-150	~in situ	0.3/0.3	Poor	11.9-89	Good	37,019	
Dry dredge	36,956	50-80	0.2-0.5	0.3-0.3	Fair to good		Good		Excavator with hydraulic pump

[a]Kilogram of sediment resuspended per cubic meter of sediment dredged (from Nakai, 1978, as cited in Herbich and Brahme 1991)
[b]Remediation dredging costs typically 2-3 times navigation dredging costs.
[c]Greater depths achievable through use of submerged pumps.
[d]Demonstrated depth of operation; greater depth is theoretically possible.
Sources: EPA (1994a); Foster Wheeler Corp. (1999).

use suction and hydraulic action to remove sediments, and (2) mechanical dredges that remove sediments by direct mechanical action. Hydraulic dredges typically exhibit high production rates and minimize sediment resuspension. Mechanical dredges are applicable for high solids content, low water production, improved performance in the presence of debris and obstructions, and greater accuracy. Hybrid dredges have also been used that are predominantly mechanical in action but also withdraw water to control migration of a resuspension plume. The selection of a particular dredging technology, or the risks associated with dredging relative to other management options, is dependent upon site-specific factors, and no general guidance can be provided. Some of the site-specific factors include sediment grain size and cohesiveness, the presence of debris, and the conditions controlling the relationship between the contaminant release and the exposure and risks faced by sensitive organisms. A more complete list of factors is included in Table 7-2.

There are a variety of specific dredge technologies other than those included in Table 7-1, but they have seen limited use in environmental dredging, and no evidence exists that they have significant, consistent advantages over the dredges listed (e.g., McLellan and Hopman 2000). In addition, there are a variety of minor variations on the types of dredges listed in Table 7-1 that have merits for particular applications.

One of the most important factors in the selection of dredges for removal of PCB-contaminated sediments is the resuspension potential. PCBs are largely associated with the solid phase in sediment beds, and therefore resuspension of particles results in resuspension of contaminants. Sediment characteristics, such as grain size, largely control resuspension. Fine-grained sediments settle the most slowly and result in the most resuspension in and around a head of the dredge. Dredging effectiveness is also limited by residual sediment contamination not targeted or captured by the dredging operation and the influences of debris, sediment heterogeneity, and dredge type. In the presence of large debris, hydraulic dredges can be ineffective or have increased resuspension rates. Hard, consolidated sediment layers, or hardpan, might make dredging overlying contaminated sediments extremely difficult and of limited effectiveness. Sediments also tend to settle back into the cuts of mechanical dredges, resulting in increased resuspension rates.

Hydraulic dredges operated slowly and with care to avoid unnecessary resuspension generally give rise to less resuspension than mechanical dredges or dredges operated to maximize production rate. Nakai (1978, as cited in Herbich and Brahme 1991) estimated releases of 5-45 kg of suspended solids per cubic meter of sediment dredged using hydraulic dredges in silt and clay sediments. That amount represents between 0.5% and 4.5% of the sediments dredged, assuming a typical sediment bulk (dry) density of about 1,000 kg/m^3. Nikai also estimated releases of 25-90 kg of suspended solids per cubic meter

TABLE 7-2 Factors That Affect Contaminant Loss During Dredging

Category	Factors
Sediment type and quality	Grain size Sediment cohesion Organic matter content Sediment density Volatiles concentration
Dredging equipment and methods	Type of dredge Dredge production rate Condition of equipment Equipment reliability Operating precision of equipment Sediment loss during operations Training and skill of operators
Hydrodynamic conditions	Water depth Morphology of shoreline Flows and suspended solids Waves, tides, and currents Hydraulic effects of dredging operations
Water quality	Temperature Salinity Density

Source: St. Lawrence Centre 1993 (as cited in EPA 1994a).

from mechanical dredges in silt and clay (i.e., approximately 2.5-9% of the sediments dredged). Resuspension from sandy sediments was as much as an order of magnitude less for either dredge type. Nikai refers to the production normalized resuspension rate as a turbidity-generating unit, and some of those are listed in Table 7-1. The use of the term turbidity-generating unit is misleading in that turbidity does not directly indicate resuspended sediment concentrations. Values shown in Table 7-1 are based on sparse data and should be used with caution. Kauss and Nettleton (1999), for example, measured water concentrations and estimated that an enclosed cable-arm mechanical dredge lost only 0.1-1.3% of the contaminants hexachlorobenzene and hexachlorobutadiene in a particular application. Unfortunately, the estimation of a mass release rate in such cases must be inferred from a dispersion model and cannot be estimated directly by evaluation of mass flows. Additional information on expected resuspension characteristics of dredge can be found in McLellan et al. (1989) and Hayes et al. (2000a,b).

In some cases, resuspension can be much greater than those estimates suggest. Palermo et al. (1990) estimated that 20-30% of the sediment from a clay and silt bed was spilled from the clamshell during hoisting through the water in a particular situation. In another study, Kauss and Nettleton (1999) noted that the negative impact of debris on a cable arm hindered its ability to close and enhanced resuspension. Enclosed clamshells can reduce losses by a factor of about 2, and similar variations in resuspension potential can be realized by changes in operation of a particular dredge (McLellan et al. 1989). Debris, the presence of hardpan, poor control of operations, fine-grained, fluffy sediments, and use of nonoptimal dredging equipment can all cause sediment resuspension and residual contamination to be much greater.

There have been important improvements in hydraulic and mechanical dredging technologies in the past 10 years, but improvement has been largely limited to improvements in production rate and location and depth accuracy (McLellan and Hopman 2000). In general, improvement (i.e., reductions) in sediment resuspension and contaminant release come at the expense of volumetric efficiency and production rates. The low resuspension rates of the enclosed cable-arm dredge noted by Kauss and Nettleton (1999) were aided by continuous monitoring and in-water cycle times of 2-6 minutes during normal operation, much slower than would be expected during navigational dredging. During hydraulic hot-spot dredging in New Bedford Harbor in 1994 and 1995, efforts to control resuspension led to the capture of 160 million gallons of water, which had to be decanted and treated, while targeting the dredging of only 10,000 yd^3 of sediments (an average solids concentration based on targeted sediments of little more than 1%) (Foster-Wheeler 1999)

Although resuspension-related contaminant loss can result in residual surficial sediment concentrations, the overall effectiveness of dredging is also reduced by contaminated sediments that are targeted but not captured by the dredge. Because historically contaminated sediments are generally found in net depositional environments, the highest concentrations are often found at depth in the sediment column. Difficulties in removing all the sediments can result in surficial concentrations that might increase after dredging because of the exposure of the more contaminated sediments at depth. Sediments that are difficult to dredge include those underlain by bedrock or hardpan. In such a situation, it is not possible to use overdredging or "overbite" to improve removal efficiency. Debris and boulders can also reduce removal efficiency. The difficulties of obtaining specific cleanup levels through dredging are illustrated in Box 7-6.

The effectiveness of dredging is also reduced by the presence of contaminated sediments that are not targeted by the dredge. Hot-spot dredging, for example, only targets contaminants within the region containing elevated concentrations. The success of such an effort, even if 100% effective at capturing targeted sediments, depends on the extent to which the hot spot contrib-

BOX 7-6 Effectiveness of Dredging PCB Contaminated Sediments at Massena, New York

Between 1959 and 1973, hydraulic fluids used at the General Motors facility in Massena, New York, contained PCBs, a portion of which were ultimately released to the sediments in the adjacent St. Lawrence River. In 1995, approximately 13,000 yd^3 of sediment adjacent to the facility were dredged for removal of PCBs to a target level of 1 mg/kg. The management activities and results are described in BBL (1996). Removal of boulders and debris was conducted by mechanical excavation, and subsequent sediment removal was conducted by horizontal auger dredge. The selection of a horizontal auger dredge was based on expected removal efficiency and minimization of sediment resuspension. Containment of sediments resuspended by dredging operations was originally attempted with silt curtains, but high flows in the river forced the placement of sheet piling to separate the dredging area from the river flow. The contaminated sediments were underlain with dense glacial till, which made it impossible to use overdredging to increase sediment removal efficiency.

In areas in which initial concentrations exceeded 500 mg/kg,15-18 dredge passes were required to reduce sediment concentrations below 500 mg/kg. In one particular area that initially exceeded 500 mg/kg, eight additional attempts, including multiple dredge passes, were conducted to reduce sediment concentrations. After as many as 32 dredge passes, the contractor had concluded, with EPA concurrence, that attainment of target cleanup levels in this quadrant was not possible with dredging alone. It was decided that capping the residual contamination was the most effective means of reducing surficial sediment concentrations and risk. A 6-inch sand cap with an additional 6 inches armoring by 2-inch gravel was placed over all sediment (75,000 ft^2) that contained in excess of 10 mg/kg.

A review of this management activity indicates that dredging alone was not sufficient to achieve desired cleanup goals due to resuspension within the sheet-pile walls and the inability to remove all sediments as a result of underlying dense glacial till. A cap with an armoring layer, however, effectively separated at-risk species from the contaminants and was largely responsible for risk reduction for the most highly contaminated sediments at the site. The combination of dredging and mass removal for control of long-term risk and capping to control short-term risk is a highly effective, although expensive, means of managing contaminated sediments.

utes to the risk to the resource of concern. If the elevated contamination levels are at depth in the sediment column and the risk is largely controlled by the surficial sediment concentrations in the surrounding area, the hot-spot dredging will not be successful at risk reduction. Hot-spot dredging of the Grasse

River by Alcoa, Inc., for example, has apparently not significantly reduced fish body burdens, despite removal of 84% of the targeted PCBs within the hot spot and an 86% reduction in average PCBs concentrations in the top 12 inches of sediment (Smith 1999).

Thus, residual sediment contamination can result from (1) leaving sediments nontargeted by dredging; (2) exposing previously buried sediments; or (3) contaminant losses during dredging. Therefore, surficial sediment concentrations may increase or decrease less than expected despite obtaining high mass removal rates. Because at least short-term exposure and risk is related to surficial sediment concentrations within the biologically active zone, mass removal itself might not achieve risk-management goals. Although the contaminated-sediment management strategy (EPA 1998) has a goal of reducing the volume of existing contaminated sediment, a goal more consistent with the proposed framework is reducing the volume solely to the extent that it reduces the broadly defined risks of the contaminated sediments.

The assessment of the effectiveness or ineffectiveness of dredging is a strong function of site-specific conditions. It is not possible to state generally that dredging is appropriate or inappropriate. At this time, the only guidance that can be provided is to identify conditions that hinder the successful application of a removal technology. As indicated above, these conditions include the presence of buried high contaminant concentrations near hardpan or bedrock, and the presence of significant quantities of debris or stone.

The current database on the success or failure of dredging is not sufficient to draw strong general conclusions as to its applicability in particular situations. This situation is changing rapidly, however, as a result of improved monitoring of dredging operations and increasing scrutiny of the success or failure of such operations. As an illustration, the nongovernmental organization, Scenic Hudson, has prepared a report on the effectiveness of environmental dredging (Scenic Hudson 1997). The General Electric Co. has supported the development of a database on sediment management projects, including those that used dredging (General Electric 1999). Other firms and an industrial group, the Sediment Management Workgroup, have supported the investigation of dredging activities in the Fox River and in Manistique Harbor, Michigan (Brown and Doody 2000). These reports are largely based on limited recent data sets, and several of the sites are still undergoing additional management. Furthermore, as will be discussed in Chapter 10, there appears to be some discrepancy in terms of the actual goals against which the various reports compared the results. It appears, however, that the greater scrutiny and oversight of dredging projects has caused more extensive monitoring to be conducted at sites undergoing management. It is expected that the resulting data sets will soon substantially improve our ability to understand the positive and negative consequences of removal actions.

Controls on Sediment Resuspension Losses

The impact of sediment and contaminants resuspended at the point of dredging can be reduced by the addition of sheet piling or silt curtains around the area to be dredged. Sheet piling provides the greatest control of both particulate and contaminant resuspension but might not be feasible or required at many sites. Silt curtains are designed to increase the residence time of suspended solids around the dredgehead, encouraging settling and reducing the amounts of resuspended sediments reaching the main body of water. Silt curtains are normally constructed of vinyl or polyurethane and are not capable of eliminating flow between the zones inside and outside the silt curtain. Generally, they also cannot be placed and maintained in the presence of any significant current. Where applicable, however, loss of suspended solids and particulate-bound contaminants might be effectively reduced by properly installed and maintained silt curtains.

Although potentially effective on suspended particles, silt curtains are not normally expected to reduce contaminant loss in dissolved form. As noted by DiGiano et al. (1993), the concentration of the dissolved-contaminant can be estimated by assuming local equilibrium for hydrophobic organic compounds between the suspended sediment particles and the water within the silt curtain. A statement of this equilibrium can be written

$$C_W = \frac{C_S W_S}{1 + K_{SW} C_S},$$

where C_w is the dissolved-phase concentration in the water, C_S is the suspended-solids concentration in the water within the silt curtain, W_s is the chemical loading on the sediment being dredged, and K_{sw} is the sediment-water partition coefficient. At low suspended-solids concentrations ($C_S << 1/K_{sw}$), the dissolved-phase concentration in the water is just the mass of contaminant resuspended by the dredge (i.e., all of the resuspended PCB is in dissolved form and not effectively contained by silt curtains). At high suspended-solids concentrations ($C_S >> 1/K_{sw}$), the total water-borne concentration is higher, but the bulk of the contaminants are associated with suspended solids. Under these conditions, much of the PCBs are sorbed to suspended-sediment particles, but the dissolved-phase concentration is at its maximum. At high suspended-sediment concentrations within the silt curtain, the dissolved PCB concentrations would approach those found in the pore water of the sediment.

The effectiveness of a silt curtain and other means of controlling contaminant releases during dredging and shore operations are illustrated in Box 7-7, which describes a hot-spot dredging operation in the Grasse River, New York.

BOX 7-7 Effectiveness of Control Measures During Removal and Treatment of Dredged Material, Grasse River, New York

The Alcoa, Inc. facility in Massena, New York, has historically discharged storm water and treated wastewater to the Grasse River. As a result of the past use of PCBs at the facility, sediments within the Grasse River were ultimately contaminated with PCBs. During 1995, a Non-Time-Critical Removal Action (NTCRA) was implemented on sediment in the most upstream contaminated areas. The monitoring conducted during the NTCRA allows an assessment of the effectiveness of various control actions that were implemented (Alcoa 1999; Thibodeaux et al. 1999).

Mechanical excavation equipment was used to remove 390 yd^3 of rocks and debris to expose the contaminated sediment. A total of 2,640 yd^3 of sediment were removed using a horizontal auger dredge. Approximately 84% of the targeted sediment was removed by the action, but because of the presence of rocks and debris and a hard bottom, or hardpan, the ability to dredge to overcut to improve removal efficiency was limited. Despite removal of as much as 98% of the PCBs from the sediment column, the average PCB concentrations in surficial sediments (upper 8 inches) were reduced by only 53% (Thibodeaux et al. 1999). The entire dredging operation was conducted within a three-layer silt curtain. The dredged sediments were dewatered onshore and the water was discharged back to the Grasse River after treatment with a sand filter and granular activated carbon.

Monitoring downriver showed that the silt curtain was effective at containing sediment resuspended during the dredging operations. Alcoa (1990) estimated that 8 yd^3 of sediment were released through the silt curtains, which equals approximately 0.3% of the sediment dredged. They also estimated PCB losses through the silt curtain at 5-30 lb (2-14 kg), which is similar in percentage to that of sediment losses. Such correlation between PCB losses and sediment losses depends on the particle concentration in the escaping water and the partitioning of the PCBs between the particles and water. Silt curtains are not designed to contain water or dissolved contaminants, and no such correlation would be expected if the majority of the PCBs were not particle-bound.

The dewatering operations produced 11,667,000 gallons of water that required treatment. Thibodeaux et al. (1999) estimated that 0.16 kg of PCBs were lost by volatilization, 11.7 kg of PCBs, presumably that fraction associated with particulate matter, were removed by sand filtration, and 0.33 kg of PCBs were taken up by the granular activated carbon. Less than 0.0045 kg of PCBs were returned to the Grasse River following treatment by the activated carbon treatment system. However, significant savings in water treatment costs would have been obtained if untreated water were returned to the area within the silt curtains. If it is assumed that the particulate-bound PCBs in the returned water were contained as effectively as the sediment resuspended during dredging, the net increase in PCB release through the silt curtains would have been only about 0.37 kg, far less than the 2-14 kg estimated to have been released due to resuspension during dredging itself.

The illustration also shows the utility of mass-flow analysis in determining the most significant factors contributing to contaminant loss, in this case showing that water treatment can result in minor improvements in contaminant containment while dramatically increasing the cost.

Dry Excavation

In addition to conventional dredging approaches, it is sometimes possible to temporarily dam a water body, remove the overlying water, and conduct the contaminated-sediment removal via dry excavation. This approach has important advantages with respect to control of resuspension and minimization of residual contamination. Through this approach, the degree of control afforded land excavation can be applied to contaminated sediments. The hydraulic conditions of the waterway can make isolation and dewatering infeasible. Direct exposure of the contaminated sediments will also result in significant increases in volatilization of PCBs (Valsaraj et al. 1995). These evaporative losses must be assessed to ensure that the risk of exposure during the removal action is not unacceptable. However, in some situations, higher short-term risks may be considered acceptable to reduce long-term risks. This decision should be made in conjunction with all affected parties, including local community organizations. Evaporative losses are very high in freshly exposed sediments and may be controllable in a particular situation by working with small areas at a time. As with resuspension losses during subaqueous removal, evaporative losses during dry excavation are essentially negligible after completion of the risk management process, and the long-term risk is determined by untargeted or uncaptured residual contamination. Examples of dry excavation have been largely limited to near-shore sediments that can be readily isolated and might already be exposed under low-water or tidal conditions.

Summary of Removal Technologies

Ex situ remediation technologies, such as dredging and dry excavation, might be most effective for exposed and accessible hot spots that pose significant risks. Removal options, such as dredging and dry excavation, require pretreatment (dewatering and volume equalization) and appropriate treatment and disposal options for the excavated sediments (landfilling, treatment, incineration, or placement in a confined disposal facility) and for any separated liquids.

Although no dredge can remove all PCB contamination, dredging can remove a substantial mass of PCBs from contaminated areas. However, even with substantial mass removal, sufficient PCBs might be left behind to cause water-quality and, hence, ecosystem risks. The importance of these residual PCBs depends to a great degree on natural attenuation processes, including deposition. In addition, the dredging process can result in the exposure of high PCB concentrations buried in the sediments directly to the water column and the dispersal of PCBs to other areas through resuspension. The effective removal of contaminated sediment with less dispersal can best be achieved through a relatively controlled dry excavation. For a contaminant such as PCBs, however, that must be done carefully and with small areas at a time to avoid unacceptable volatilization losses. Methods exist for prediction of volatile losses that could be used to assess the risks of this exposure pathway.

Wet excavations of contaminated sediment using either hydraulic or mechanical dredges, however, remains the more common approach. Mechanical dredges are preferred near submerged structures and when a large amount of debris exists in the sediment. Hydraulic dredges tend to produce less resuspended sediment if they are operated carefully in sediments with little or no debris. Silt curtains can aid in limiting dispersion of resuspended sediment and in evaluating the amount of resuspended sediment and contaminants generated by the dredge but cannot reduce the amount of dissolved contaminants dispersed in the water body. During wet dredging operations in silty clay sediments, loss and resuspension of 0.5-5% of the sediments can be expected during a single pass in the absence of debris or heterogeneities in sediment characteristics. The significance of such losses depends on the distribution of contaminants. Multiple dredging passes may be required to achieve desired sediment and contaminant recoveries even under ideal conditions.

Pretreatment Technologies

Dredged material removed from a contaminated-sediment site normally requires pretreatment prior to ultimate treatment or disposal. The purpose of pretreatment is normally twofold:

- Remove excess water to reduce volume and aid subsequent treatment or disposal.
- Provide a volume equalization basin to allow matching of dredging rates with subsequent treatment or disposal rates.

In navigational (not environmental) dredging, the material produced by mechanical means is close to in situ sediment density, and hydraulic means introduce much more water. In environmental dredging, the slower production rate and operational procedures to reduce contaminant and sediment resuspension generally increases the produced water content. It is not unusual to have mean water contents that exceed 90% for hydraulic dredging when attempting to minimize solids loss and resuspension (Foster Wheeler 1999; General Electric 1999). Reduction of these high water contents is normally necessary for subsequent cost-effective treatment or disposal of the dredged material.

Dredging is normally subject to wide variations in production rate. Even when pumping hydraulically dredged material as a slurry through a pipeline, wide variations in production rate result because of sediment heterogeneity and the presence of debris. Subsequent treatment or disposal steps often cannot maintain effectiveness if the feed rate is widely variable, and so a temporary storage system is normally required to serve as a basin for watering and volume equalization.

Variations in sediment conditions and production rates cause difficulties when adding coagulants within a dredged material pipeline to aid dewatering (Jones et al. 1978). Similar problems would be expected with the addition of nutrients or reagents to aid decontamination of dredged material or with any effort to feed the dredged material to a process unit for dewatering or other pretreatment operation. Among other pretreatment operations that have been considered is hydrocyclone separation of fine from coarse sediments (EPA 1994a). The organic fraction that contains the bulk of the PCBs can be separated from the clean sands in this fashion. No advantage is realized, however, unless the sand fraction is sufficiently clean that it can be returned for open water disposal or used as clean fill on land. The ability of a hydrocyclone operation to achieve these goals is dependent upon the sediment characteristics, particularly the particle-size distribution, and the distribution and potential for separation of the organic-matter fraction.

Because of these difficulties, pretreatment is normally limited to use of primary settling basins for dewatering. Potential contaminant concerns in such systems are evaporation of PCBs from the exposed sediment and overlying water (Valsaraj et al. 1995) and carryover of dissolved and suspended contaminants with the effluent water. The residual contamination in the effluent water is normally treated before discharge back into the water body. Due to low solubility and high sorptivity of PCBs, however, as well as the relatively high Henry's Law constant or volatilization constants of many PCBs, the mass of PCBs normally associated with the produced water is very low compared with the mass of PCBs on the sediment. Despite that, regulations normally require treatment of the water produced from the dewatering cycle. A mass-flow analysis might suggest, however, that return of the produced water to the point of dredging, or within the silt curtain, can add a mass of PCBs that is negligible compared with the mass lost by resuspension by the dredge (see Box 7-7). In such cases, the expense of water treatment is difficult to justify. To address such concerns, the committee encourages regulatory agencies to retain maximal flexibility to adapt to site-specific challenges and opportunities—for example, not requiring treatment of residual effluent streams that contain a small mass of PCBs.

Treatment and Disposal Technologies

In most sediment management activities that have been completed or are under way, the dewatered dredged material is either left in a confined disposal facility or transported to a landfill. In a confined disposal facility, the ultimate disposal is typically at the same facility in which primary dewatering has taken place. In an upland landfill, the partially dewatered dredged material may be further dewatered, for example, via filtering, and then transported for ultimate disposal.

If disposed of in an upland landfill, the dredged material is not normally subjected to further treatment. A confined disposal facility, however, can be useful as a treatment facility. Particular technologies that have been considered in a confined disposal facility include biodegradation, phytoremediation and solidification and stabilization. Problems include the heterogeneity of the dredged material and the difficulty of applying biodegradation and phytoremediation to the entire column of dredged material, which may be tens of feet thick. A completely confined disposal facility may be capped to control leachate production and vaporization and to provide a physical barrier to direct contact by terrestrial animals and birds. Additional development and field testing are required before these approaches will receive widespread acceptance.

In principle, the public is more supportive of treatment technologies that permanently destroy the contaminants, but the costs of these treatment technologies or disposal have generally not been competitive with landfill or disposal-facility placement. A recent review of eight technologies (PIANC 2000) suggested that contracts of 10 or more years involving the treatment of a million or more cubic yards of dredged material per year were required for sufficient economies of scale to make the technologies commercially viable. At sites where a large volume of sediment and effluent must be managed, technologies that generate a product that can be sold to offset the costs of the technology, might receive greater acceptance by some of the affected parties. These technologies would require development of regulatory standards that establish the safety of these product. These volumes are available only in large harbors subject to navigational dredging of sediments unable to be disposed of in open water (e.g., New York/New Jersey Harbor) or in a few large contaminated-sediment sites. It might also be possible to build centralized facilities capable of processing the contaminated dredged materials from multiple sites although public acceptance would have to be gained and regulatory barriers would have to be overcome. Even treatment technologies that permanently destroy PCB-containing wastes, however, might generate residuals that are released to the environment or disposed of in licensed landfills.

There are a number of technologies that might be appropriate for contaminated sediment and competitive with landfill disposal under certain conditions, or they might significantly reduce the volume or toxicity of the material that could be placed in a landfill. It is not possible to address all of these technologies in this report, and the reader is referred to the more comprehensive technology summaries, which include costing information, such as PIANC (2000), EPRI (1999), General Electric (1999), and EPA (1994a,b,c).

Essentially all dredged-material treatment technologies can be characterized into one of three categories:

- Extractive technologies that seek to separate the contaminants from the sediment, producing a residual material that is smaller in volume and has a greater variety of disposal options.
- Stabilization or containment technologies that seek to minimize the mobility of contaminants and thereby reduce exposure and risk.
- Destructive technologies that seek to eliminate the contaminant while producing an innocuous residual material.

Extractive technologies include thermal desorption, solvent extraction (including supercritical gaseous solvents), and soil washing for either particulate-size separation or contaminant removal. Stabilization or containment technologies include disposal in secure landfills as well as chemical processing to bind contaminants in a stable matrix, such as concrete. Destructive technologies are generally limited to thermal processes, such as incineration, vitrification, or high-temperature desorption followed by reduction. Nonthermal destruction technologies, such as biodegradation, are generally not suitable for PCBs because of slow or limited biological-degradation rates, as discussed previously in the section on natural attenuation.

Soil-washing technologies serve to reduce contaminant levels by partial removal of fine-grained particles and organic material that contain the majority of the contaminants. The net result is small reductions, by factors of 2 to10, in the more-soluble contaminants in the sediments. Reductions in contaminant levels of less-soluble components, such as PCBs, are likely to be by a factor of 2 or less. Therefore, the sediment treated by washing might still require disposal in a secure landfill. The goal of some soil-washing technologies is production of a manufactured soil. In such cases, the ability to use the soil as fill depends on the availability of use-dependent quality standards and the ability of the washed sediment to meet those standards. The effectiveness of selected soil-washing processes is shown in Box 7-8. As illustrated by these

BOX 7-8 Effectiveness of Sediment Washing

Sediment washing may be used to remove PCBs when fine-grained silt and clay fractions that generally contain the bulk of the PCB contamination represent only a small fraction of the sediments. Sediment washing can remove the fine-grained contaminated materials from the larger and cleaner sands and gravels, thereby reducing significantly the volume of contaminated material for subsequent treatment and disposal. The effectiveness of these processes is illustrated by two pilot-scale demonstration projects, a sediment washing and classification system at Saginaw River, Michigan (EPA 1994c) and a sediment washing and treatment system in New York/New Jersey Harbor (PIANC 2000).

Saginaw River: Washing of Saginaw River, Michigan, sediments using a system designed by Bergman USA was demonstrated by the EPA Assessment and Remediation of Contaminated Sediments (ARCS) and Superfund Innovative Technology Evaluation (SITE) programs in 1991 and 1992. A total of 800 yd^3 of Saginaw River sediments were dredged via open clamshell bucket for use as feed material during the demonstration. The feed material was relatively homogeneous and contained PCBs at 1.2 ± 0.23 mg/kg. The process involved a separation of oversize material followed by three hydrocyclone stages to size segregate the sediments. Observed concentrations of PCBs in washed sand were reduced by more than 80% to 0.21 ± 0.07 mg/kg, and concentrations in two fine-grained solid streams averaged 3.9 ± 2 mg/kg and 2.2 ± 0.4 mg/kg, respectively. Produced water contained PCBs at 1.34 ± 0.54 micrograms per liter PCBs and was recycled back into the process. Successful application of this treatment process requires that the produced sand be sufficiently clean to allow alternative uses or disposal and that the volume of the more contaminated fine-grained solids be significantly smaller than the starting volume.

New York/New Jersey Harbor: The sediment washing process developed by BioGenesis Enterprises was evaluated in a pilot-scale test by EPA Region 2, U.S. Army Corps of Engineers—New York District, and Brookhaven National Laboratory. The process used an initial washing step, a separation system to remove floating organic materials, a second washing step using a collision chamber, and a two-stage cavitation and oxidation system to reduce sediment contaminant levels. The resulting stream was separated using hydrocyclones and a centrifuge, and the resulting partially dewatered sediment was available for beneficial uses. During the treatment of 700 yd^3 of material dredged from the Seabord/Koppers Coke Site in Kearny, New Jersey, the process removed about 45% of the PCBs from an average sediment concentration of 0.398 to 0.22 mg/kg. Successful application of the process requires that the produced material be sufficiently clean to allow alternative uses or disposal. In addition, water and a floating organic stream require treatment prior to discharge.

examples, product streams from a soil washing system can still contain high contamination, and the disposal or subsequent treatment of these streams must be included in the identification, analysis, and screening of sediment management options.

Other extractive technologies, such as thermal desorption and solvent extraction, are generally more effective than simple sediment washing. As with any extraction process, however, product streams and residuals contain the contaminants, and these streams generally require further treatment before disposal. Generally, the product streams are more concentrated in contaminants than the original sediment, since the desire is to reduce the volume of material for subsequent treatment or disposal. If the product stream is a fluid, either air or water, subsequent treatment might be more easily accomplished than treatment of the original sediment. If the product stream is a solid, subsequent treatment options might be identical to those available for the original contaminated sediment, but the reduction in volume might result in reduced overall costs. Because of the cost and complexity, solvent extraction has seen limited application to either sediment or soil remediation and is unlikely to be used except for specialized, small-volume applications. Thermal desorption has received somewhat wider usage and has been used in sediment management programs. Box 7-9 illustrates two examples of the effectiveness of thermal desorption and the residuals that might be produced. One example is low-temperature desorption, which has relatively low-energy requirements when applied to wet sediment but is limited in effectiveness. The second example is high-temperature desorption, which exhibits high effectiveness with correspondingly greater energy costs and process complexity. Both processes illustrate a key feature of extractive technologies: the products and residuals require further treatment or disposal.

Stabilization technologies involve introduction of additives to the dredged material to prevent mobility of contaminants, providing a more secure material for disposal, or a reusable product, such as flowable or solid fill for construction. The contaminant levels are normally unchanged except for dilution due to mixing with the various additives required to prepare the product. Stabilization is expected to significantly reduce the potential for leaching of the contaminants. An important barrier, however, is the lack of regulatory standards for use of the product. Fill standards based on total contaminant levels are not suitable for stabilized materials, while fill standards based on leachate tests, such as EPA's toxicity characteristic leaching procedure (TCLP), may not be sufficiently protective or acceptable to the community. Examples of solidification and stabilization processes that are being applied to contaminated sediments are the flowable fill technology marketed by Pohlman Materials Recovery and the solidification process of OENJ Cherokee. In the flowable fill process, partially dewatered sediment is blended, after debris removal, with

BOX 7-9 Effectiveness of Thermal Desorption

Thermal desorption enhances the evaporation of contaminants from soils and sediments. The resulting air stream can be treated for removal and destruction of contaminants prior to release to the atmosphere. Low-temperature thermal desorption has lower energy requirements, and high-temperature thermal desorption has the advantage of a high removal and destruction efficiency, which is typical of high-temperature processes. High-temperature processes, however, also involve high-energy costs associated with treatment of wet sediments and might result in the formation of dioxins and furans. The effectiveness of the process is illustrated with a pilot-scale demonstration on sediment dredged from Ashtabula, Ohio (EPA 1994b) and a full-scale management effort in Waukegon, Illinois (EPA 1995).

Ashtabula River, Ohio: The Remediation Technologies (RETEC) Inc., process was evaluated in a pilot-scale demonstration on 36-55 gallon drums of sediment from Ashtabula, Ohio, in 1992 (EPA 1994b). This is a low-temperature (190-250°C) desorption process that collects particulate matter carried by exhaust gases with a cyclone separator and collects the vapors with a multistage condenser system. Average PCB concentrations in the sediments were 2.2 mg/kg before treatment and 0.4-0.8 mg/kg after treatment. The average removal efficiency was 82.8%, although efficiencies as high as 95.1% were observed. With low-temperature desorption, the PCBs are not destroyed. The bulk of the desorbed PCBs (as much as 55%) were collected in the condensate system. An activated carbon-polishing step collected as much as 6% of the desorbed PCBs. The remainder of the desorbed PCBs were not accounted for.

Waukegon, Illinois: High-temperature (up to 1207 °F) thermal desorption was used for the full-scale remediation of the Outboard Marine Corporation site in Waukegan, Illinois, during 1992 (EPA 1995). The process, developed by Soiltech ATP, involves a rotary kiln desorption under anaerobic conditions and with a mean solids residence time of 30-40 minutes followed by exhaust gas and water treatment. A total of 12,755 tons of soil and sediment containing an average of 10,484 mg/kg PCBs were treated, and 255 tons of soils and sediments containing extractable organic halides at 1,900 mg/kg were processed during a site process evaluation at the same site during June 1992. The treated soil contained an average of 2.2 mg/kg for greater than a 99.98% treatment efficiency. Because desorption was conducted under anaerobic conditions, PCBs in the effluent gases were removed with particulate collection devices and with an oil absorption system. Approximately 50,000 gallons of PCB-contaminated oil was produced that required subsequent disposal. PCBs released to the atmosphere met destruction and removal efficiency (DRE) targets (99.9999%), but meeting dioxin and furan emission targets required process modification.

proprietary silicate binders and fine aggregate. Contaminants are not destroyed but are stabilized by incorporation into the physical matrix of the product. Reduction in mobility by 2-3 orders of magnitude is expected. Conventional stabilization of the New York/New Jersey Harbor sediments with Portland cement was conducted by OENJ Cherokee and used as a cap on a municipal landfill in Elizabeth, New Jersey. A similar project plan was to use 4.5 million yd^3 of dredged harbor sediment as a cap on a 38-acre municipal landfill and a 97-acre industrial site in Bayonne, New Jersey. Because of the lack of quality standards for either the stabilized material or the leachate that it might produce, these processes are likely to be applied only to relatively clean sediments and only for use as fill in industrial or landfill applications.

As indicated previously, destruction processes are primarily thermally based. Conventional incineration has a high cost, results in the formation of dioxins and furans, and is subject to considerable community resistance. Alternative destructive processes might include the production of a reusable product, such as blended cement, lightweight aggregate, or glass from the dredged material. The blended-cement process includes the use of dredged material along with other components to produce cement. It would be necessary, however, to develop regulatory standards that establish the safety of these products. Thermal destruction in cement kilns to make cement, however, can raise air-emission permit and community-acceptance issues similar to those that arise for a conventional incinerator. The production of lightweight aggregate also uses rotary kiln technology for the destruction of contaminants and production of the aggregate. The production of glassy products from dredged material using a plasma torch has been proposed. Very high temperatures (in excess of 5,000° C) are achieved in the plasma torch, causing vaporization and degradation of organic materials in the dredged material. The final product is of high quality and essentially contaminant free. In situ vitrification is a similar process in which very high temperatures are achieved in situ with the accompanying potential for contaminant vaporization and degradation. High-temperature degradation processes can also be conducted under reducing (oxygen-free) conditions. Under these conditions, the generation of dioxins and furans might be minimized. The Eco Logic process (Eco Logic 1998) is a high-temperature (more than 850 °C) desorption process followed by a gas-phase reduction reaction using hydrogen. The process can achieve high destruction efficiencies with minimal production of dioxins, furans, and other products of incomplete reaction (Eco Logic 1998). The high temperatures, associated energy costs, and complexity of the process, however, have limited its application to contaminated-sediment treatment.

The various nonthermal and thermal technologies, while providing permanence, are costly and not likely to compete on a cost basis with direct disposal of the dredged material in a landfill. The useful products that some of the above processes produce have the opportunity to offset part of the cost of treatment if their introduction to the marketplace in large volume will not overly disrupt the market. The costs of these processes are also likely to be large except when large volumes are dredged because of economies of scale. It has been estimated, but not demonstrated, that many of these technologies can be implemented for $20-$60 per cubic yard of dredged material if amounts greater than 100,000 yd^3 per year for 10-20 years can be guaranteed (PIANC 2000). The success of the various technologies and the products they produce currently depends on community and regulatory acceptance of their operation and the proposed usefulness of the products. The factor common to all treatment technologies for dredged material is that 100% effectiveness cannot be realized and that residual and effluent streams containing significant contamination might require further treatment or disposal or both. The contaminant losses, treatment and disposal of the residuals, and the risks involved need to be considered when identifying, evaluating, and selecting sediment management options. Inadequate consideration of such problems can give rise to inaccurate and misleading comparisons between removal and nonremoval sediment management options.

CONCLUSIONS AND RECOMMENDATIONS

A summary of the various risk-management options and the areas of potential risks that must be assessed is given in Table 7-3. Ultimately, the ability to make decisions, the next step in the management framework, requires that the process of identifying and evaluating options collects the information necessary to complete such a table. It should be emphasized that the table is a conceptual framework in which to consider the various options. The information needed to make decisions requires far more information and data about the individual options than can be presented in a simple table.

On the basis of the discussion in this chapter, a number of considerations should be kept in mind when identifying and evaluating options.

- The current ability to reduce health and environmental risks from PCB-contaminated sediments through technical options alone is limited. Successful contaminated-sediment management and risk reduction requires a combination of technical and institutional options as well as natural attenuation.
- There should be no preferred or default PCB-contaminated-sediment management option. The optimal option for a particular site is dependent upon site-specific factors and conditions and should be selected as a result of active participation by all affected by the decision.

TABLE 7-3 Sediment Management Options and Associated Risks

Option	Component	Goal	Feasibility	Cost	Risk of Imple-menting	Short-Term Risk	Long-Term Risk
Socioeconomic	Institutional controls	Sever exposure pathways					
Source control		Eliminate source					
In situ management	Natural attenuation	Containment and degradation					
	Thin-layer capping	Containment					
	Thick-layer capping	Containment					
Ex situ management	Mechanical dredging	Removal					
	Hydraulic dredging	Removal					
	Dry excavation	Removal					
	Pretreatment	Dewatering, size separation					
	Treatment and disposal	Separation or destruction					

- The first goal of any management activity for PCB-contaminated sediments should be to identify and, where possible, control the point and nonpoint sources that have caused and will continue to cause the contamination problem. The sources include, but are not limited to, run-off from contaminated soils, combined sewer overflows, and atmospheric inputs.
- Effectively responding to the contaminated sediment at a site generally requires using options that involve multiple technological and institutional components, and the evaluation, screening, and selection of these options must consider all the components, their interrelationships, and their impacts. Seven broad rules govern the analysis of management options:

1. All sites require a conceptual model of the system, and the interaction of the management options with the sediments and contaminants is required.
2. The use of mass flows can assist in developing and testing the

conceptual model and can identify components of the management option that require additional review and analysis

3. Evaluation of options at large, complicated sites requires the development of a sophisticated understanding of the specific system dynamics and the ability to predict future contaminant behavior and risks that are likely to result from the application of each of the various management options.

4. The reduction of risks to human health and the environment at one location can often result in the creation of additional risks at other locations. The impact of these transferred risks and their acceptability to all concerned should be identified and fully explored.

5. Natural attenuation is a component of all contaminated-sediment management options. No remediation technology is effective in removing all sediment contaminants from a site, and all remediation technologies result in the production of contaminated residuals and effluents that cannot be eliminated by known or likely technology. Such contaminated residuals and effluents left in place must ultimately be subjected to natural-attenuation processes.

6. Removal options might produce short-term risks and, due to residual contamination from resuspension losses or leftover contaminated sediment, long-term risks that must be managed.

7. Nonremoval options might produce long-term risks due to the potential of exposure to the remaining contaminants, and provisions for long-term monitoring and maintenance is required.

- The committee recommends that opportunities to restore or create critical habitats not be overlooked in the development and implementation of PCB-contaminated sediment risk-management strategies. Cleanup projects, such as the St. Paul Waterway Area Remedial Action and Habitat Restoration Project, demonstrate how sediment risk management can be successfully coupled with natural-resource restoration.
- Research should be directed toward improving understanding of the acute and chronic exposure and risk associated with the various management options. Because the appropriateness and effectiveness of the various options are dependent upon site-specific characteristics, such research should be directed toward defining models that can be used to project mass flows for a particular management option as a function of site conditions.

— A particular area of uncertainty is the long-term stability of sediments and sediment caps and the mode of failure if destabilized. Specifically, an ability is needed to quantitatively predict the extent of failure and the resulting exposure and risk in a low-probability storm or flow event.

— Due to the inability to target or capture all contaminated sediments, a second particular area of uncertainty is the assessment of residual contamination mass and concentrations. Under what conditions will dredging be unlikely to reduce risk?

— A third particular area of uncertainty is the assessment of the financial costs of a management alternative. Part of this uncertainty is due to the inability to adequately describe site conditions that influence the effectiveness and cost of management options. Part of this uncertainty is also due to the lack of a standard approach to accounting for present and future costs and inadequate representation of the entire life-cycle of management options.

References

Alcoa. 1999. Analysis of Alternatives Report. Grasse River Study Area, Massena, New York. December.

Averett, D.E., B.D. Perry, E.J. Torrey, and J.A. Miller. 1990. Review of Removal, Containment and Treatment Technologies for Remediation of Contaminated Sediment in the Great Lakes, Final Report. Paper EL-90-25, U.S. Army Engineer Waterways Experiment Station, Vicksburg, MS.

Belton, T., B. Ruppel, K. Lockwood, S. Shiboski, G. Bukowski, R. Roundy, N. Weinstein, D. Wilson, and H. Wholan. 1985. A Study of Toxic Hazards to Urban Recreational Fisherman and Crabbers. Trenton, NJ: New York Department of Environmental Protection, Office of Science and Research . September 15.

BBL (Blasland, Bouck & Lee). 1996. St. Lawrence River Sediment Removal Project Remedial Action Completion Report. Prepared for General Motors Powertrain, Massena, New York. Blasland, Bouck & Lee, Syracuse, New York. June.

Brown, M.P, and J.P. Doody. 2000. A Dredging Effectiveness Review - Case Studies and Lessons Learned. 16th Annual International Conference on Contaminated Soils, Sediments and Water, University of Massachusetts, Amherst, MA. October 16-19.

Cropper, M.L., and P.R. Portney. 1990. Discounting and the evaluation of lifesaving programs. J. Risk Uncertainty 3(4):369-379.

DiGiano, F.A., C.T. Miller, and J. Yoon. 1993. Predicting release of PCBs at the point of dredging. J. Environ. Eng. 119(1):72-89.

Eco Logic. 1998. PCB-Contaminated Soil and Sediment Treatment Using Eco Logic's Gas-Phase Chemical Reduction Process. Rockwood, Ontario: ELI Eco Logic International Inc. July 18.

EPA (U.S. Environmental Protection Agency). 1994a. Assessment and Remediation of Contaminated Sediments (ARCS) Program: Remediation Guidance Document. EPA 905-R-94-003. Great Lakes National Program Office, Chicago, IL. October.

EPA (U.S. Environmental Protection Agency). 1994b. Assessment and Remediation of Contaminated Sediments (ARCS) Program: Pilot Scale Demonstration of Thermal Desorption for the Treatment of Ashtabula River Sediments. EPA 905-R-94-021. Great Lakes National Program Office, Chicago, IL.

EPA (U.S. Environmental Protection Agency). 1994c. Assessment and Remediation of Contaminated Sediments (ARCS) Program: Pilot Scale Demonstration of Sediment Washing for the Treatment of Saginaw River Sediments. EPA 905-R94-019. Great Lakes National Program Office, Chicago, IL.

EPA (U.S. Environmental Protection Agency). 1995. Remediation Case Studies: Thermal Desorption, Soil Washing, and In Situ Vitrification. EPA 542/R-95/005. Federal Remediation Technologies Roundtable, U.S. Environmental Protection Agency, Washington, DC.

EPA (U.S. Environmental Protection Agency). 1996. Assessment and Remediation of Contaminated Sediments (ARCS) Program: Estimating Contaminant Losses from Components of Remediation Alternatives for Contaminated Sediments. EPA-905-R-96-001. Great Lakes National Program Office, Chicago, IL.

EPA (U.S. Environmental Protection Agency). 1998. EPA's Contaminated Sediment Management Strategy. EPA-823-R-98-001. Office of Water, U.S. Environmental Protection Agency, Washington, DC. April.

EPA (U.S. Environmental Protection Agency). 1999a. Treatment Technologies for Site Cleanup: Annual Status Report, 9th Ed. EPA 542-R-99-001. [Online]. Available: http://clu-in.org/products/asr/index2.html. [May 26,1999].

EPA (U.S. Environmental Protection Agency). 1999b. Use of Monitored Natural Attenuation at Superfund, RCRA Corrective Action, and Underground Storage Tank Sites. OSWER Directive 9200.4-17P. Office of Solid Waste and Emergency Response, U.S. Environmental Protection Agency, Washington, DC. April 1999. [Online]. Available: http://www.epa.gov/OUST/directiv/d9200417.htm [June 16, 1999].

EPRI (Electric Power Research Institute). 1999. Review of Sediment Removal and Remediation Technologies at MGP and Other Contaminated Sites. Report No. TR-113106. Electric Power Research Institute, Palo Alto, CA. September.

Farley, K.J., R.V. Thomann, T.F. Cooney, D.R. Damiani, and J.R. Wands. 1999. An Integrated Model of Organic Chemical Fate and Bioaccumulation in the Hudson River Estuary, Manhattan College, Riverdale, NY. Final Report to the Hudson River Foundation. [Online]. Available: http://www.hudsonriver.org/

Foster Wheeler (Foster Wheeler Environmental Corporation). 1999. New Bedford Harbor Cleanup Dredge Technology Review, Final Report. Prepared for U.S. Army Corps of Engineers, New England District. March.

General Electric Company. 1999. Major Contaminated Sediment Site Database. [Online]. Available: http://www.hudsonwatch.com/mess/.

Giesy, J.P. and K. Kannan. 1998. Dioxin-like and non-dioxin-like toxic effects of polychlorinated biphenyls (PCBs): implications for risk assessment. Crit. Rev. Toxicol. 28(6):511-569.

Hahnenberg, J. 1995. Presentation at the Workshop on Interim Controls held July 31, 1995. Committee on Contaminated Sediments, National Research Council. EPA Headquarters, Chicago.

Harkness, M.R., J.B. McDermott, D.A. Abramowicz, J.J. Salvo, W.P. Flanagan, M.L. Stephens, F.J. Mondello, R.J. May, and J.H. Lobos. 1993. In situ stimulation of aerobic PCB biodegradation in Hudson River sediments. Science 259:503-507.

Hayes, D.F., T. Borrowman, and T. Welp. 2000a. Near field turbidity observations during Boston Harbor Bucket comparison study. Pp. 357-370 in Proceedings of the Western Dredging Association 20th Technical Conference and 32nd Annual Texas A&M Dredging Seminar, R.E. Randall, ed. CDC Report 372. Center for Dredging Studies, Texas A&M University.

Hayes, D.F., T.R. Crockett, T.J. Ward, and D. Averett. 2000b. Sediment resuspension during cutterhead dredging operations. J. Waterw. Port Coastal Ocean Eng.126(3):153-161.

Herbich, J.B., and S.B. Brahme. 1991. Literature Review and Technical Evaluation of Sediment Resuspension During Dredging. Contract Report HL-91-1. Vicksburg, MS: U.S. Army Engineer Waterways Experiment Station. January.

Jones, R.H., R.R. Williams, and T.K. Moore. 1978. Development and Application of Design and Operation Procedures for Coagulation of Dredged Material Slurry and Containment Area Effluent. Technical Report D-78-54. Vicksburg, MS: U.S. Army Engineer Waterways Experiment Station.

Kadak, A.C. 2000. Intergenerational risk decision making: a practical example. Risk Anal. 20(6):883-894.

Kauss, P.B., and P.C. Nettleton. 1999. Impact of 1996 Cole Drain Area Contaminated Sediment Cleanup on St. Clair River Water Quality. Technical report. Ministry of the Environment, Toronto, Ontario. NTIS MIC-99-07256INZ.

King County. 2000. King County's Combined Sewer Overflow (CSO) Control Program. Department of Natural Resources, Wastewater Treatment Division, Seattle, WA. [Online]. Available: http://dnr.metrokc.gov/WTD/cso/. [August 17, 2000].

Koenigsberg, S.S., and R.D. Norris, eds. 1999. Accelerated Bioremediation Using Slow Release Compounds: Selected Battelle Conference Papers: 1993-1999. San Clemente, CA: Regenesis Bioremediation Products.

McLellan, T.N., and R.J. Hopman. 2000. Innovations in Dredging Technology: Equipment, Operations, and Management. ERDC TR-DOER-5. Vicksburg, MS: U.S. Army Corps of Engineers, Engineer Research and Development Center.

McLellan, T.N., R.N. Havis, D.F. Hayes, and G.L. Raymond. 1989. Field Studies of Sediment Resuspension Characteristics of Selected Dredges. Technical Report HL-89-9. Vicksburg, MS: U.S. Army Engineer Waterways Experiment Station.

Nakai, O. 1978. Turbidity generated by dredging projects. Pp. 31-47 in Management of Bottom Sediments Containing Toxic Substances: Proceedings of the Third U.S./Japan Experts Meeting. EPA-600/3-78-084.

NRC (National Research Council). 1997. Contaminated Sediments in Ports and Waterways: Cleanup Strategies and Technologies. Washington, DC: National Academy Press.

NRC (National Research Council). 2000. Natural Attenuation for Groundwater Remediation. Washington, DC: National Academy Press.

NYSDOH (New York State Department of Health). 2000. Health Consultation 1996 Survey of Hudson River Anglers: Hudson Falls to Tappan Zee Bridge at Tarrytown, NY. Final Report. CERCLIS No. NYD980763841. Troy, NY: New York State Department of Health, The Center for Environmental Health.

Palermo, M.R., J. Homziak, and A.M. Teeter. 1990. Evaluation of Clamshell Dredging and Barge Overflow, Military Ocean Terminal, Sunny Point, NC. Technical Report D-90-6. Vicksburg, MS: U.S. Army Engineer Waterways Experiment Station.

Palermo, M.R., S. Maynord, J. Miller, and D.D. Reible. 1998. Assessment and Remedation of Contaminated Sediments (ARCS) Program: Guidance for In situ Subaqueous Capping of Contaminated Sediments. EPA 905-B96-004. Great Lakes National Program Office, Chicago, IL. [Online]. Available: http://www.epa.gov/glnpo/sediment/iscmain/index.html [December 09, 1998].

Parametrix. 1999. St. Paul Waterway Area Remedial Action and Habitat Restoration Project. Final 1998 Monitoring Report. Prepared for Simpson Tacoma Kraft Company, Tacoma, WA, and Champion International, Stanford, CT, by Parametrix, Inc., Kirkland, WA. February.

Parker, J.H. and R. Valente. 1988. Long-Term Sand Cap Stability: New York Dredged Material Disposal Site. Contract Report CERC-88-2. Vicksburg, MS: US Army Engineer Waterways Experiment Station.

PIANC (Permanent International Association of Navigation Congress). 2000. Innovative Dredged Sediment Decontamination and Treatment Technologies, U.S. Section PIANC Specialty Workshop, May 2, 2000, Waterfront Plaza Hotel, Oakland, CA. [Online]. Available: http://www.wes.army.mil/el/dots/training/pianc.html [March 2000].

Portney, P.R., and J.P. Weyant, eds. 1999. Discounting and Intergenerational Equity. Washington, DC: Resources for the Future Press.

Scenic Hudson. 1997. Advances in Dredging Contaminated Sediments, New Technologies and Experience Relevant to The Hudson River PCBs Site. Prepared by J. Cleland for Scenic Hudson, Poughkeepsie, NY. April.

Sloan, R.J. 1999. Hudson River Fish and the PCB Perspective. Paper presented to the National Research Council Committee on Remediation of PCB-Contaminated Sediments November 8.

Smith, J.R. 1999. Non-time-critical removal action (NTCRA) pilot dredging in Grasse River. Pittsburgh, PA: Alcoa Inc. November 8.

St. Lawrence Centre. 1993. Selecting and Operating Dredging Equipment: a Guide to Sound Environmental Practices. Prepared in collaboration with Public Works Canada and the Ministere de l'Environement du Quebec, and written by Les Consultants Jaques Berube Inc. Cat No En 40-438/1993E.

Sumeri, A. 1989. Confined Aquatic Disposal and Capping of Contaminated Bottom Sediments in Puget Sound. Proceedings of the WODCON XII, Dredging: Technology, Environmental, Mining, World Dredging Congress, Orlando, FL, May 2-5, 1989.

Sumeri, A., T.J. Fredette, P.G. Kullberg, J.D. Geermano, D.A. Carey and P. Pechko. 1994. Sediment Chemistry Profiles of Capped Dredged Sediment Deposits Taken 3 to 11 Years After Capping. Dredging Research Technical Note. DRP-5-09. Vicksburg, MS: U.S. Army Engineer Waterways Experiment Station. May.

Thibodeaux, L.J., D.D. Reible, and K.T. Valsaraj. 1999. Effectiveness of Environmental Dredging. Final Report to Alcoa, Massena, NY. Hazardous Substance Research Center/ South and Southwest, Louisiana State University, Baton Rouge, LA.

Thoms, S.R., G. Matisoff, P.L. McCall, and X. Wang. 1995. Models for Alteration of Sediments by Benthic Organisms. Project 92-NPS-2. Alexandria VA: Water Environment Research Foundation.

Valsaraj, K.T., L.J. Thibodeaux, and D.D. Reible. 1995. Modeling air emissions from contaminated sediment dredged materials. Pp. 227-238 in Dredging, Remediation, and Containment of Contaminated Sediments, K.R. Demars, G.N. Richardson, R.N. Young, and R.C. Chaney, eds. Philadelphia: American Society for Testing and Materials.

Verbrugge, D.A., J.P. Giesy. M.A. Mora, L.L. Williams, R. Rossmann, R.A. Moll, and M. Tuchman. 1995. Concentrations of dissolved and particulate polychlorinated biphenyls in water from the Saginaw River, Michigan. J. Great Lakes Res. 21(2):219-233.

WDNR (Wisconsin Department of Natural Resources). 1995. A Deterministic PCB Transport Model for the Lower Fox River Between Lake Winnebago and De Pere, Wisconsin. Prepared by J. Steuer, S. Jaeger, and D. Patterson. Publication WR 389-95. Wisconsin Department of Natural Resources.

Weiner, K.S. 1991. Commencement Bay Nearshore/ Tideflats Superfund Completion Report for St. Paul Waterway Sediment Remedial Action. Submitted to the U.S. Environmental Protection Agency for Simpson Tacoma Kraft Company and Champion International Corporation. January.

Zeman, A.J., and T.S. Patterson. 1997. Results of in situ capping demonstration project in Hamilton Harbour, Lake Ontario. Pp. 2289-2295 in Engineering Geology and the Environment: Proceedings: International Symposium on Engineering Geology and the Environment, Athens, Greece, June 23-27, 1997, P.G. Marinos, ed. Rotterdam, Brookfield: A.A. Balkema.

8

Making Decisions

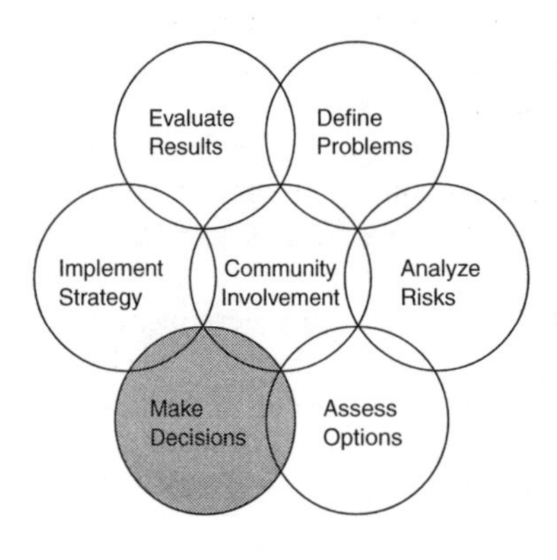

This chapter presents the fourth stage in the risk-based framework for managing the risks associated with PCB-contaminated sediments. At this point in the process, general risk- management goals would have been established, the site would have been characterized in terms of hydrodynamics, geology, and other physical and chemical factors, and the nature and extent of the PCB contamination at the site should have been delineated. The human health, ecological, social, cultural, and economic risks for the site have also been identified. Possible management options, both technical and nontechnical, have been explored, and the risks for each option have been assessed as shown in Table 7-3. Now a decision must be made on which management option or combination of options is most likely to achieve the risk-management goals established in stage one of the framework (Chapter 5), based on the assessment of risks posed by the contamination (Chapter 6) and those associated with each option (Chapter 7).

The difficulty in selecting the best risk-management strategy for a site stems from many factors, including the differing levels of contamination at the site, the lack of uniformity in the sediments themselves (e.g., depth and debris cover), the varying ecology (human and environmental) at different locations around the site, and the sheer magnitude of many PCB-contaminated sites, some of which can cover many square miles. Because of the unique character of each site, determining the most appropriate course of action can

Consideration of Options

- The paramount consideration for PCB-contaminated sediment sites should be the management of overall risks to humans and the environment rather than the selection of a remediation technology (e.g., dredging, capping, or natural attenuation). Indeed, the lack of consideration of these risks is one of the major impediments to the adoption of a successful management strategy.

- There should be no presumption of a preferred or default risk-management option that is applicable to all PCB-contaminated-sediment sites. A combination of technical and nontechnical options is likely to be necessary at any given site.

be difficult, not only for the designated decision-makers but also for the other affected parties who will have to implement and live with the decision. Affected parties might assume that the availability of a wealth of site-specific data, well-conducted human health and ecological risk assessments, and a thorough review of the risks associated with potential risk-management options will automatically lead to a determination of the most-effective and acceptable risk-management strategy. However, that assumption is often not the case, because decision-makers are dealing with a multivariate system, which involves a diverse community with differing expectations, interests, risk tolerance, and willingness to pay. This complexity has been a major contributor to the slow progress on risk management at many PCB-contaminated-sediment sites to date. New procedures are needed to accommodate the multivariate complexity at contaminated sites.

In this chapter, the committee summarizes the regulatory environment in which any decision-making process on a PCB-contaminated site must take place, reviews the various components of the decision-making process, discusses how these processes relate to the risk-management framework described in this report, and shows how the process may be facilitated.

REGULATORY REQUIREMENTS

The need to comply with various federal and state regulations when managing a PCB-contaminated-sediment site cannot be ignored. The regulations involved in the risk-management strategy for PCB-contaminated sediments can complicate its development and impede its implementation. The NRC report (1997) on contaminated marine sediments analyzed the regulatory and legal challenges posed by contaminated sediments and provided a summary of the current regulatory requirements for the management of contaminated marine sediments. The analysis and conclusions provided in this NRC report are equally relevant to the risk management of PCB-contaminated sediments and are discussed briefly below.

Regulatory Overlap

As explained in the 1997 NRC report, several federal laws (CERCLA, CWA, RCRA, MPRSA[1]) have been enacted that deal with health and environmental issues surrounding the presence of contaminated sediments in navigable and nonnavigable waters. These laws were not designed specifically to manage contaminated sediments but rather deal with issues such as water quality and the disposal of hazardous waste. These laws are implemented through a variety of often contradictory or overlapping federal and state regulations. In addition to federal agencies, such as EPA and the U.S. Army Corps of Engineers (USACE), states also exercise important authority related to water-quality certification, coastal-zone management, and use impairments for water bodies. Many states also have their own Superfund and RCRA-type laws which may not be entirely consistent with federal laws. In some cases, local laws might also apply. State environmental, health, and transportation agencies might play a role in the management of contaminated sediments.

The principal federal agencies involved in management of contaminated sediments are the following:

- EPA, which is responsible for implementing the Superfund program under CERCLA and has major site designation, regulation development, and veto responsibilities under the CWA and MPRSA.
- U.S. National Oceanic and Atmospheric Administration, which assesses the potential threat of Superfund sites to coastal marine resources and has important responsibilities for research under the MPRSA and review and comment under CWA and MPRSA.

---

[1]In CERCLA is the Comprehensive Environmental Response, Compensation and Liability Act (Superfund); CWA is the Clean Water Act; RCRA is the Resource Conservation and Recovery Act; MPRSA is the Marine Protection, Research, and Sanctuaries Act.

- USACE, which assists in the design and implementation of remedial actions under Superfund and has responsibilities for dredged material under the CWA, MPRSA, and Rivers and Harbors Act.
- U.S. Fish and Wildlife Service, which acts as a federal trustee at contaminated sites under the Natural Resource Damage Act.

If a decision is made to dredge contaminated sediments, several laws affect their disposal on land or in the water. It is currently necessary to secure different types of permits for the placement of sediment in navigation channels or ocean waters as part of the construction of land or containment facilities (under the Rivers and Harbors Act), the dumping of sediments in the ocean (under MPRSA), the discharge of sediments in inland waters or wetlands (CWA), and the containment of contaminated sediments on land (RCRA).

Regulatory Constraints

Existing regulations complicate and can impede, but do not necessarily prevent, implementation of the risk-based framework for the risk management of PCB-contaminated sediments. Regulatory differences need to be acknowledged and, if possible, resolved early in the management process so that all participants are working toward a common goal. For example, although EPA may be the lead regulatory agency for risk management at a site, the U.S. Fish and Wildlife Service may be the trustee for a natural resource damage assessment and has very different regulatory concerns and needs. However, it may be necessary to complete the risk assessments and determine what management options before all applicable laws are identified. For example, RCRA may not be an applicable law if natural attenuation is the selected management option or the MPRSA may not be applicable for the dredging of river sediments with land disposal. The risk-management framework is not intended to supplant existing regulatory processes; rather it provides a consistent and comprehensive approach to facilitate their implementation. Using this framework will not only satisfy the requirements of the CERCLA remedial-investigation and feasibility-study (RI/FS) process but will also enhance it by encouraging the use of all available data on a site. The community surrounding the site will be included early and consistently in the process, and a broader assessment of the various risks the contamination poses to humans and the environment will be provided.

The current regulatory regime focuses on the type of sediment management activity—removal, containment, or treatment—and not on the management of all risks to humans and the environment. As noted in the 1997 NRC report, the mechanisms of the regulatory process in a given situation depend on where the sediments are located; where they will be placed; the nature and extent of the contamination; and whether the purpose of removing or manipulating the sediment is navigation dredging, environmental cleanup, site development, or waste management.

Each set of laws uses a different approach to assess management options for contaminated sediment, and none considers fully all the risks (NRC 1997). The regulatory process under MPRSA places primary emphasis on the intrinsic toxicity of the constituents of dredged material without full consideration of site-specific conditions (e.g., proximity to shellfish beds, other sensitive receptors, food-chain carriers, or the containment of contaminants by an engineered clean cap) that might influence the impact on various organisms.

Risk reduction is the goal of the Superfund remedial-action program. Site-specific remedies are chosen on the basis of exposure assessments during the feasibility study, and management options are identified on the basis of their capability to reduce risks of exposure to an acceptable level. The final selection of a risk-management strategy now involves choosing the most cost-effective alternative, not necessarily the alternative that will have the broadest community acceptance.

Finally, the current regulatory regime places a presumption on action-oriented approaches to risk management at PCB-contaminated sites. Decisions under MPRSA and CWA are made when sediments are proposed for dredging and placement in inland or ocean waters. Remedy selection under Superfund places a preference on management approaches that remove contaminants from the environment (EPA 1991). In addition to Superfund's nine criteria for evaluating management options (Box 8-1), there is a general statutory preference for treatments that "permanently and significantly reduce the . . . toxicity or mobility" of contaminants (section 121(b) of CERCLA). Section 121 also establishes off-site transport and disposal of untreated contaminated sediments as the least desirable alternative (Renholds 1998). Natural attenuation or the in-place or ex situ capping of contaminated sediments does not have the requisite quality of "permanence"—in the same sense that the destruction or detoxification of sediment contaminants have permanence.

COMPARATIVE DECISION-MAKING

Science has a critical role to play in the decision-making process. Technical considerations should not be excluded from the decision-making process nor should they totally control the process. To make rational decisions, decision-makers need to have accurate estimates of the potential exposures; effects; and subsequent human, ecological, social, economic, and cultural risks. Thus, it is the responsibility of scientists (including economists and other social scientists) and engineers to present their risk-assessment results and their estimates of engineering feasibility and effectiveness in an unbiased, clear, and systematic manner.

One of the inherent problems when comparing risks of different types (e.g., economic risks with health risks) is that there is no scientific methodology to do so (Davies 1995). Implicit in any such comparisons of different types of risk are the values of the decision-makers and other affected parties. The regulatory environment, in which such values are considered and risk-management decisions are made, must also be considered, as noted above. Decision analysis, and ultimately, decision-making, at PCB-contaminated sites might require several types of analyses to assess the costs, benefits, effectiveness, and risks associated with each potential management strategy. In this chapter, the committee presents a discussion of decision-making with a description of the RI/FS process as practiced by EPA. This process is followed by EPA and other regulatory agencies at Superfund sites. The committee then provides a brief description of some approaches that may be used by EPA in the RI/FS process and that may be used in a more general sense by all affected parties to develop a risk management strategy. These approaches include cost-benefit analysis (NRC 1997); risk tradeoffs (Graham and Weiner 1995; NRC 1997), and risk ranking (Graham and Hammitt 1995). Finally, the decision-analysis process, which incorporates information from the cost-benefit, risk tradeoffs, and risk-ranking exercises is discussed (NRC 1997). The committee recognizes, as with other stages in the risk-management framework, that selection of an approach to decision-making is dependent on the needs and values of the affected parties and the specific situation at a PCB-contaminated-sediment site.

The RI/FS Process at EPA

At present, the primary decision-making process used by EPA for managing National Priority List (Superfund) sites, is the RI/FS process (as described in Chapter 3 of this report) (EPA 1991). Although written specifically for Superfund sites, this process is also used by states and other responsible parties and can be applied to non-Superfund sites as well. In the RI/FS process, EPA, another federal or state agency, or a potentially responsible party (PRP) conducts a screening, a baseline risk assessment, or both (as described in Chapter 6) and characterizes the site to develop and analyze a preliminary list of management options. These options (usually four to five of them) are described in following terms: (1) volume of material to be remediated; (2) implementation of requirements and timetables; (3) method of remediation and general response actions; (4) remediation technologies (treatment or containment) and process options; (5) monitoring procedures; (6) capital, operational, and maintenance costs; (7) need for 5-year review; and (8) the applicable or relevant and appropriate requirements (ARARs) that are triggered by this option, such as land-disposal restrictions. A detailed analysis of the alternatives is then conducted, and the alternatives are compared with nine EPA evaluation criteria (see Box 8-1). The alternative that best meets the criteria is then presented to the public for a comment as the proposed risk-management plan. After a public comment period, the risk-management plan, modified as necessary to accommodate the comments received, becomes a Record of Decision if the site is on the Superfund National Priorities List.

However, as heard by this committee during the public sessions, this process has not received general approval. Many affected parties stated that decisions appear to be made arbitrarily by the regulatory agencies with little or no input from those most affected, that public comments received during the public comment period do not appear to have much, if any, influence on the decisions, and that all available information and all risks (human health, ecological, social, economic, and cultural) do not appear to be considered by the regulatory agencies during the RI/FS process. Nevertheless, the committee believes that the RI/FS process is a generally useful mechanism for

BOX 8-1 EPA Evaluation Criteria for Remedial Action Alternatives

1. Overall protection of human health and the environment
2. Compliance with ARARs
3. Long-term effectiveness and permanence, including magnitude of residual risk and adequacy and reliability of controls
4. Reduction of toxicity, mobility, or volume through treatment.
5. Short-term effectiveness
6. Implementability
7. Costs
8. State (support agency) acceptance
9. Community acceptance

Source: EPA (1991)

gathering the necessary site-specific technical information at a contaminated site and for providing a blueprint for the data collection and analysis efforts. The RI/FS process is used consistently by EPA and many other regulatory agencies, and the regulated industries are familiar with the process as well. However, EPA guidance for the RI/FS and ARARs processes has been criticized as being frequently too vague or inconsistently applied. The committee concludes that it also might also be too restrictive to accommodate the various risks that must be assessed and evaluated at PCB-contaminated sites and to accommodate the need for involvement of all affected parties in the risk-management process. To address many of the criticisms of the RI/FS process, the committee recommends that increased efforts be made to provide the affected parties with the same information that is to be used by the decision-makers and to include, to the extent possible, all affected parties in the entire decision-making process at a contaminated site. In addition, such information should be made available in such a manner that allows adequate time for evaluation and comment on the information by all parties.

Cost-Benefit and Cost-Effectiveness

Other decision-making tools are available to the decision-makers and affected parties: cost-benefit analysis and cost-effectiveness analysis. Cost-benefit analysis is based on first identifying all positive (benefits) and negative (costs) impacts of an action, with the impacts quantified in a common (dollar) unit. The set of actions that leads to the largest net benefit is identified. Cost-benefit analysis is effective for evaluating management strategies as it combines risk and cost information to determine the most efficient allocation of resources; however, this type of analysis may be extremely difficult to use when the benefits and costs (risks as well as monetary) are not easy to estimate or calculate (NRC 1997).

Cost-effectiveness analysis identifies the least costly set of actions that achieves a given goal or, rarely, the set of actions that leads to the highest benefit for a given cost. Cost-effectiveness analysis is a more general approach than cost-benefit analysis. Any risk-management strategy that maximizes net benefits is considered to be a cost-effective solution. If one strategy can achieve the same level of benefit as another strategy but at a lower cost, the former strategy has a higher net benefit. Such analysis does not require that the costs and benefits be expressed in common units.

The advantage of cost-benefit and cost-effectiveness analyses for contaminated-sediment sites is that they allow the comparison of very different risk-management options (e.g., institutional controls versus dredging) on an equal basis across sites. They also have the advantage of transparency of analysis and many assumptions, thus making them accessible to examination, critique, and reassessment through sensitivity analysis when some parameters are uncertain. These types of analyses are best viewed as providing input into a more complete decision-making process rather than providing the single decision criteria. The RI/FS process discussed above includes a consideration of cost-effectiveness analysis.

Risk Ranking

Any decision on how to manage a PCB-contaminated site should follow from an effort to achieve frank and open discussions among the affected parties after the risks identified for each risk-management option have been reviewed. As discussed in Chapter 7, it is helpful if the various risks for each of the options have been presented in a consistent fashion, as shown in Table 7-3. This table provides a mechanism by which each of the risks for the management options may be compared and ranked by the affected parties and the decision-makers.

Several methods are available for ranking risk (Graham and Hammitt 1995). These methods include the following:

- Ranking risks associated with management options and baseline risks.
- Ranking risks according to the expected value of risk reduction.
- Using uncertainty as a source of information.
- Ranking risks according to target risks (e.g, health risks from PCB-contaminated sediments) and competing risks (e.g., health risk from not eating fish).
- Ranking risks according to resource costs and savings (cost-effectiveness analysis).

However, to do this ranking in the most objective manner, the criteria for assigning values to the various risk categories (e.g., what constitutes a high short-term ecological risk or what cost is important) should have been established in stage one of the framework when the risk-management goals were identified and priorities were set, or, at the very least, in stage two of the framework during the risk assessment. It might be helpful to have each of the affected parties rank each management option and then compare and discuss the rankings with the goal of reaching a consensus on the best options or combination of options to minimize the overall risks at the site. The NRC Committee on Contaminated Marine Sediments (NRC 1997) presented, as an example of one approach to decision-making, comparative analysis of the remediation technologies and options that might be used for marine sediments (Table 8-1). When considering this table, however, it must be remembered that these rankings are relative only and that the values are subject to change (i.e., a pilot technology in 1997 might be in commercial use in 2001) and site-specific conditions. As that committee noted, no single technology scored the best in all categories. Furthermore, unlike Table 7-3, Table 8-1 does not consider any of the risks associated with the technologies.

Risk Tradeoffs

Even when risk-management efforts involve relatively nondisruptive activities, possible health, ecological, societal, cultural, and economic risks should be considered. One limitation of the current RI/FS process as practiced by EPA is that risks other than human health and ecological risks receive little consideration in the decision-making process. Furthermore, limiting involvement of all affected parties in the process also means that a thorough comparison of the total risks at a site might not influence the risk-management decision. The methods by which various types and levels of risks might be compared have been studied (Graham and Wiener 1995). With regard to PCB-contaminated-sediment sites, risks to human health and wildlife are considered to be target risks. Risks, such as economic or cultural risks, that stem from the target risks, are considered to be competing risks. Affected parties must be able to weigh and tradeoff these risks to make a decision on what is the best risk-management strategy. For example, although it might seem prudent to recommend avoidance of consumption of PCB-contaminated fish, there are potential adverse health impacts associated with the recommendation. Epidemiological evidence suggests that consumption of fish improves human health (TERA 1999). In addition, a representative of the Mohawk tribe stated to the committee that the avoidance of fish consumption had disrupted the tribe's cultural tradition of fathers training sons to fish and had led to adverse health effects by reducing the physical activity of male tribal members (Ken Jock, personal commun., 1999). Thus, it seems reasonable to ask if avoiding consumption of fish with low concentrations of PCBs is beneficial or detrimental to overall human health.

When considering possible risk tradeoffs, several comparisons may be made between the target risks and the competing risks based on whether the countervailing risks are to the same or different populations and of the same or different type (Graham and Wiener 1995). Risk transfer occurs if the target risks are of the same type but moved from one medium to another—for

TABLE 8-1 Comparative Analysis of Technology Categories

Approach	Feasibility	Effective	Practicality	Cost
INTERIM CONTROL				
Administrative	0	4	2	4
Technological	1	3	1	3
LONG-TERM CONTROL				
In Situ				
Natural recovery	0	4	1	4
Capping	2	3	3	3
Treatment	1	1	2	2
Sediment Removal and Transport	2	4	3	2
Ex Situ Treatment				
Physical	1	4	4	1
Chemical	1	2	4	1
Thermal	4	4	3	0
Biological	0	1	4	1
Ex Situ Containment	2	4	2	2
SCORING				
0	<90%	Concept	Not acceptable, very uncertain	$1,000/yd^3$
1	90%	Bench		$100/yd^3$
2	99%	Pilot		$10/yd^3$
3	99.9%	Field		$1/yd^3$
4	99.99%	Commercial	Acceptable, certain	<$1/yd^3$

Note: For each control and technology, the four characteristics were rated separately on a scale of 0 to 4, with 4 representing the best available (not necessarily the best theoretically possible) features. The effectiveness rating is an estimate of contaminant reduction or isolation, and removal efficiency scores represent a range of less than 90% to nearly 100%. The feasibility rating represents the extent of technology development, with 0 for a concept that has not been verified experimentally and 4 for a technology that has been commercialized. The practicality ranking reflects public acceptance; 0 means no tolerance for an activity, and 4 represents widespread acceptance. The cost ranking is inversely related to the cost of using the control or technology (excluding costs of monitoring and so forth).
Source: NRC (1997).

example, dredging might remove PCBs from the sediments but increase PCBs in the water column. Risk offsets are when the risks are the same in the same target populations as might occur when either dredging or capping kill a benthic community—population mortality occurs with either management option. The fish-consumption example above illustrates risk substitution where one type of outcome (risk of health effects from consumption of PCB-contaminated fish) is replaced by another outcome (risk of heart disease from lack of fish consumption). A framework for comparing the health risks associated with eating fish contaminated with such materials as PCBs versus not eating fish has recently been published; the public-health implications of such risk substitution are also discussed (TERA 1999). Finally, target risks can be transformed, resulting in a different outcome to a different population. Transformation might occur if the risks to reproduction of birds at one contaminated site are compared with the risks of immunological effects in humans who lived near a sediment disposal site. Another example might be the economic risks to farmers if their dairy cows are believed to produce PCB-tainted milk. The concept of voluntary versus involuntary risks and risks to special groups, such as children and endangered species, must also be factored into consideration of risk tradeoffs. Affected parties will be required to consider each type of risk for a PCB-contaminated-sediment site and possibly several permutations for each risk tradeoff type. By considering the various risks in broad terms of public and ecosystem health, it should be possible to make decisions that have net benefits on the community.

An important factor in achieving a decision that works for all the affected parties is the consideration of the willingness of different affected parties to accept the various risks. Not all parties are equally affected by different risks, and some parties are disproportionately affected and thus might be unwilling to accept a risk that they primarily must bear. For example, many local communities are unwilling to accept a "beneficial" use or application of sediment removed from a contaminated river even if the sediment has been treated to reduce contaminant levels. Such uses might include inclusion of sediment in construction material or the building of a confined disposal facility that may provide new habitat. The communities where the sediment would be burned or where sediment-containing construction materials are used are often unwilling to accept the risks of the air emissions or the risks of contaminants leaching from the construction material even if the risk is considered low or "acceptable." They are likely to oppose such plans especially if they are not included in the discussion of options and tradeoffs before a decision is made. If such tradeoffs are to be considered, representatives of those directly affected by the risks, such as the people who live downwind of the cement kiln, should be included in the discussion of the risks and tradeoffs.

Decision Analysis

Decision analysis is a computational tool that might be of use when evaluating, comparing, and ultimately selecting a risk-management strategy. It can be used to integrate the various aspects of the risk-management process described above. Aspects include risk and site assessment, economic assessment, and technical feasibility studies to estimate the outcome of possible management strategies (NRC 1997). Decision analysis also incorporates the results of the comparative risk-assessment analyses described above: cost-benefit and/or cost-effectiveness analyses, risk ranking, and risk tradeoffs. It has the advantage of formally accounting for uncertainties that are inherent in the risk assessments at the very least and of being reproducible. Furthermore, it allows users to modify the computations to ascertain how slight changes in the various risk parameters might affect the outcome. Such changes in scenarios might have profound impacts on the decision-makers and affected parties as they determine the most effective and acceptable risk-management strategy.

The NRC report on contaminated marine sediments (NRC 1997, Appendix E) provides a hypothetical test case applying decision analysis to the risk management of a hot spot at a contaminated site. Although technical in concept and complex in execution, decision analysis is a valuable approach for evaluating competing management options at a PCB-contaminated-sediment site. Decision analysis is also more suitable for decision-making in the multivariate context generally associated with the management of PCB-contaminated-sediment sites.

FRAMEWORK CONSIDERATIONS

The committee has concluded that all risk-management strategies will be multifaceted. An advantage to using the Presidential/Congressional Commission on Risk Assessment and Risk Management's (PCCRARM 1997) framework recommended by this committee is that affected parties are encouraged to consider a broad array of risk-management options, often resulting in solutions incorporating a variety of technological and nontechnological methods at a site rather than seeking the elusive "silver bullet." Dredging or other technologies should be considered when developing a risk-management strategy, but focusing them without proper consideration of all options can lead to further issues at the site with a resulting delay in implementation. Therefore, although a particular technology might pose high risks in one area (e.g., dredging might have large, short-term ecological impacts), it might be appropriate for use in a high-risk situation, such as a large hot spot. It also may be used for a limited amount of time if the technology is very effective in reducing the immediate risks or meets other criteria, such as low cost.

It might be possible for all affected parties to discuss each of the risks at a site and identify the best options for managing both short- and long-term risks. At this point in the decision-making process, tradeoffs might be required to achieve consensus. Some affected parties might feel that no cost is too much to achieve a complete cleanup of a site but they might accept the reality that funds are not unlimited and that achieving a zero level of contamination is impossible. Others might be willing to forgo eating fish from the area if habitat for the fish can be restored and a catch-and-release program is instituted. At other sites, it might be imperative to reduce the contamination in the sediment to address immediate human health or ecological risks, and some risk-management strategy using a combination of technologies and other options must be implemented quickly. In that case, a phased approach consisting of an immediate response followed by a long-term response might be best.

There is a presumption that removal of PCBs from a site will reduce a source of exposure of PCBs to the aquatic environment. But that is true only if PCB contamination at a site provides the major source of exposure. If there are other environmental sources contaminating the site, site remediation might have little or no impact on total risks. For example, while 40% of the Great Lakes shoreline is considered to be impaired by PCB contamination of sediments, almost 50% of the shoreline is contaminated by atmospheric deposition of PCBs and 30% is impaired by land disposal of PCBs (Muno 1999). Nonpoint sources, such as atmospheric deposition or land runoff, can have a substantial impact on the effectiveness of particular risk-management strategies to reduce total exposure.

Any stage of the framework may be revisited as new significant information becomes available. Therefore, the greater the understanding and documentation of the site and the risk-management goals and the risks associated with the PCBs at the site and those associated with their management, the more readily new information may be incorporated into the framework without the need to "start over." However, this new information must be understood to have an impact on the process or site—for example, simply identifying a new animal study on PCBs that does not affect the hazard assessment is not sufficient to delay the development and implementation of the risk-management strategy. Modifications to the risk-management strategy may be considered in light of any new compelling information either to increase or to decrease the strategy requirements (e.g., to require more monitoring or a decrease in the area to be dredged).

The committee recognizes that this risk-based framework is ideally used at a newly identified contaminated site; in reality, however, most PCB-contaminated sites are already in the process. At sites where the risk-management process has not been completed and is the subject of contention, the committee urges the lead federal or state agency, to consider whether the participants involved at a contaminated-sediments site might use this framework to resolve a stalemate over site management. Conflict resolution might be a large factor in the decision-making process. The committee cautions that the use of the framework or other risk-management approach should not be used to delay a decision at a site if sufficient information is available to make an informed decision. Particularly in situations in which there are immediate risks to human health or the ecosystem, waiting until more information is gathered might result in more harm than making a preliminary decision in the absence of a complete set of information. The committee emphasizes that a "wait-and-see" or "do-nothing" approach might result in additional or different risks at a site.

Most PCB-contaminated sites have had some risk-management activity, even if only an acknowledgment that PCBs are present in the sediments. The risk-management framework (Chapter 3) can be used at any of these sites. For some sites, interim decisions have been or will be made. Again, the committee emphasizes that even though decisions might have been made at a site and a remediation technology is being implemented, this framework will still provide a mechanism by which all affected parties can assess the effectiveness of the risk-management strategy and make any necessary modifications for improvement. Such corrections can range from redefining the risk-management goals to gathering more sediment or fish samples, to identifying new PCB sources, to assessing whether an experimental technology has field or commercial potential.

The committee cautions that interim decisions should not be construed as a final solution for the site and stresses that the process must be iterative. Further evaluations and decisions by the affected parties will be required until the risk-management goals are met. The committee also recommends that before a management option is selected, a public comment period be allowed to provide an opportunity for those who may not have been involved in the framework process to express their concerns before a decision is made and a risk-management strategy implemented.

FACILITATING THE PROCESS

Decision-making with respect to PCB-contaminated sediments can be adversarial, in part because of the huge costs associated with many risk-management strategies at contaminated sites (see Table 5-1a,b). This adversarial relationship is detrimental to the development of societally acceptable decisions, and conflict resolution must be an inherent part of any decision-making process. In the CERCLA process, this dissension can arise during preliminary discussions at a site to identify potentially responsible parties who might be at substantial financial risk. For example, EPA has recently estimated the costs to General Electric of dredging hot spots in the upper Hudson River to be approximately $490 million. Parties might take opposing views to allow themselves negotiating room. This process can result in lengthy and divisive negotiations by the affected parties. This protracted debate can cause the community to distrust both industry and regulatory agencies and regard the information that they provide as self-serving. The early establishment of partnerships among all parties, as discussed in Chapter 4, might circumvent or overcome the development of adversarial positions by some parties.

Decision-making might require the use of an outside party, such as a facilitator, to help all the affected parties express themselves and understand the points of view and preferences of the others. Facilitators can be of use at any site, but they can be particularly valuable at sites where there is substantial distrust among affected parties and obstacles to the risk-management process appear to be intractable. The committee hopes, however, that all affected parties, including the regulatory agencies, will arrive at the decision-making stage with a willingness to listen to others, to consider their concerns, and to address their concerns as quickly as possible. The committee recognizes that, in general, the ultimate decision for managing a site resides with a federal or state regulatory agency, such as EPA. However, the committee emphasizes that these agencies do not operate in a vacuum and that they, the community, and the PRPs are subject to public scrutiny and societal and legislative pressures. In situations in which the affected parties feel that their concerns and needs have been acknowledged and addressed, the risk-management process can be more effective (see Box 8-2).

CONCLUSIONS AND RECOMMENDATIONS

To supplement the RI/FS process, the committee recommends that increased efforts be made to provide the affected parties with the same information that is to be used by the decision-makers in a form allowing for impartial and comprehensive review and evaluation. All affected parties must be involved in the data review and have an understanding of the regulatory constraints under which the decision-makers must operate. The affected parties must also have sufficient time to thoroughly review and comment upon the information.

BOX 8-2 Waukegan Harbor Area of Concern

Waukegan Harbor, located on the shore of Lake Michigan, was identified as polluted in 1972 (EPA 1997). PCBs were the primary pollutant of concern and had resulted in degradation of the benthos, restrictions on dredging activities, beach closing, degradation of plankton populations, and loss of fish and wildlife habitat. Fish consumption advisories had also been posted. A Citizens Advisory Group (CAG) was organized in 1990 with members from industry, fishing interests, environmental interests, and residents. The CAG assisted in obtaining cooperation from local interests on additional investigations, such as groundwater monitoring. The CAG was able to obtain access from business and federal grant money to install the monitoring wells. The CAG is also working with a local bank to resolve environmental concerns regarding a defunct salvage yard. In 1997, the fish consumption advisory was removed for Waukegan Harbor. Stage one of the remedial action plan was successfully completed in 1993, and subsequent risk-management efforts are still under way to address the other impairments.

When considering the best ways to manage risk at a PCB-contaminated site, a broad variety of options should be considered individually and collectively (see Table 7-3). Focusing too early on possible remediation technologies to the exclusion other options and having particular preference for a single-faceted solution can curtail consideration of potentially viable options.

The risks posed by PCB-contaminated sediments and management efforts extend beyond human health and ecological effects to economic, cultural, and societal impacts that must be factored into any risk-management strategy. The committee appreciates that it might be more difficult for the regulatory agencies to factor in risks other than human health or ecological impacts, let alone quantify them. At most sites, management goals focus on reducing risks to humans who may consume contaminated fish (see Table 5-1). The committee found that regulatory agencies do not give sufficient attention to other risks, including ecological effects, impacts on the local economy, or cultural traditions. Consequently, acceptance of the proposed risk-management strategy is often lacking among affected parties for whom those risks are major concerns.

There is no preferred or default risk-management strategy for all PCB-contaminated sites. The optimal strategy for a particular site is dependent upon site-specific factors and risks and such conditions as sediment depth and composition, ecosystems, extent of contamination, and the presence of co-contaminants. The array of risks to be assessed are also site-specific and management-option-specific. When selecting a risk-management strategy for a site, decision-makers must be sensitive to affected parties' views on the short-term and long-term risks, not only from the PCBs themselves but also from the implementation of any management option. Risk tradeoffs might be required to achieve the risk-management goals.

REFERENCES

Davies, J.C.. 1995. Comparing Environmental Risks: Tools for Setting Government Priorities. Washington, DC: Resources for the Future.

EPA (U.S. Environmental Protection Agency). 1991. Guidance on Oversight of Potentially Responsive Party Remedial Investigations and Feasibility Studies. Office of Solid Waste and Emergency Response (OSWER) Directive No. 9835.1(c). [Online]. Available: http://es.epa.gov/oeca/osre/910701-1.html.

EPA (U.S. Environmental Protection Agency). 1997. Waukegan Harbor Area of Concern. Great Lakes Area of Concern. U.S. Environmental Protection Agency. [Online]. Available: http://ww.epa.gov/grtlakes/aoc/waukegan.html. [November 21, 1997].

Graham, J.D., and J.K. Hammitt. 1995. Refining the CRA Framework. Pp. 93-109 in: Comparing Environmental Risks: Tools for Setting Government Priorities, J.C. Davies, ed. Washington, DC: Resources for the Future.

Graham, J.D., and J.B. Wiener, eds. 1995. Risk Versus Risk: Tradeoffs in Protecting Health and the Environment. Cambridge, MA: Harvard University Press.

Muno, W. 1999. A Great Lakes Perspective of Sediment Contamination. U.S. EPA Region 5. Paper presented at Public Meeting Sections, University of Wisconsin at Green Bay, September 28, 1999.

NRC (National Research Council). 1997. Contaminated Sediments in Ports and Waterways: Cleanup Strategies and Technologies. Washington, DC: National Academy Press.

PCCRARM (Presidential/Congressional Commission on Risk Assessment and Risk Management). 1997. Framework for Environmental Health Risk Management. Final Report. Washington, DC: The Commission.

Renholds, J. 1998. In Situ Treatment of Contaminated Sediments. Technology Innovation Office, Office of Solid Waste and Emergency Response, U.S. Environmental Protection Agency, Washington, D.C. December. [Online]. Available: www.epa.gov/swertio1/products/renhold.htm.

TERA (Toxicology Excellence for Risk Assessment). 1999. Comparative Dietary Risk: Balancing the Risks and Benefits of Fish Consumption. Final Report. Toxicology Excellence for Risk Assessment, Cincinnati, OH. [Online]. Available: http://www.tera.org/pubs. August.

9

Implementing the Strategy

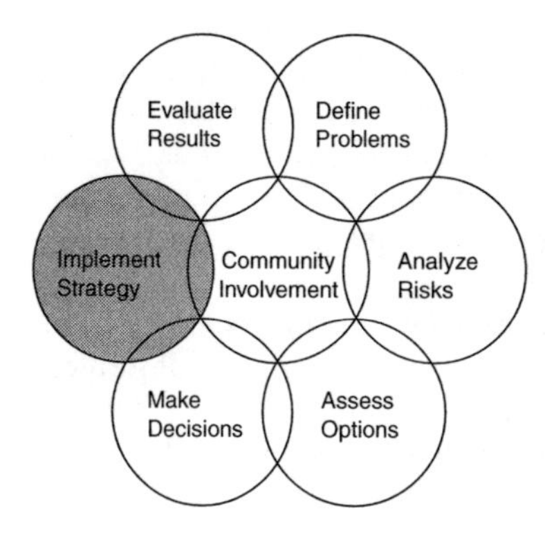

The preceding four components of the risk-management framework established the foundation necessary for the implementation of a risk-management strategy at PCB-contaminated sites. The nature of the PCB-contamination problem and its subsequent risks have been defined, and potential management options have been evaluated. As a result of these analyses, a risk-management strategy has been developed on the basis of input from all affected parties. The fifth stage of the framework is the implementation of the strategy.

The committee recognizes that implementation is often the stage in a risk-management project where a substantial disconnect occurs between previous planning and risk-management objectives and the subsequent field operations. Decisions that look good on paper or in the meeting room can be more problematic when they are put into practice, and communication and trust among the interested parties can break down. For this reason, implementation must involve all affected parties. Implementation requires detailed planning, funding, and evaluation throughout its entire process. Furthermore, the implementation plan must be flexible to respond to unforseen situations or problems. For instance, estimated risks might change during the implementation process because of unanticipated environmental factors (e.g., floods or storms) or the development of new remediation technologies, new regulatory requirements, or the availability of new scientific data. In other situations, short-term risks that were not factored into the initial risk assessments (e.g., risks to workers at the contaminated sites or risks of transporting the dredged materials through a community) have to be considered. These considerations might require that the implementation plans be modified or revised completely. As shown in Box 9-1, New Bedford Harbor offers a specific example of mid-course correction occurring during management, stemming from concerns of local residents about risks from exposures to contaminated sediments and from exposures that might result from the management of the sediments.

BOX 9-1 Midcourse Evaluation at New Bedford Harbor

From the 1940s until the late 1970s, the New Bedford Harbor area (the Achusnet River estuary) received wastes from industries using PCBs in the manufacture of capacitors and other electrical components. Following several years of assessment (1976-1981), EPA added this site to the National Priorities List. Sediments in the upper part of the harbor (Figure 9-1) contained PCBs at concentrations of as much as 100,000 ppm or greater on a dry-weight basis. Management action concerns were focused on the hot spots; the upper and lower Harbor, and the Buzzards Bay area. In 1983, EPA began to evaluate alternatives for addressing the PCB contamination. Initial remediation included PCB-contaminated sediments with concentrations in excess of 4,000 ppm with a sediment volume of about 10,000 m^3. An environmental feasibility study (EFS) and pilot studies for remediation were completed in 1989, and EPA recommended the use of a cutterhead dredge connected to a hydraulic pipeline, which would deliver sediment to a confined disposal facility (CDF), where, dewatering, treatment of the water, and incineration of the PCBs in the sediment would occur (EPA 1997).

However, concerned community groups raised strong objections to the incineration of the PCBs, and the incineration plan was modified. The revised plan called for the temporary storage of the dewatered sediment in a CDF along the edge of the upper harbor near the hot spot (Figure 9-2) until a final decision for treatment of the dredged sediments could be made.

Dredging of the hot spot was completed in 1995. Subsequently, after consideration of several alternatives, including pilot studies of solidification and chemical destruction of the sediments, the sediments were further dewatered and moved to an offsite Toxic Substances Control Act-permitted landfill.

This example shows how mid-course correction might be necessary in certain stages of the framework and illustrates the importance of involving affected parties early in the process to achieve a desirable outcome.

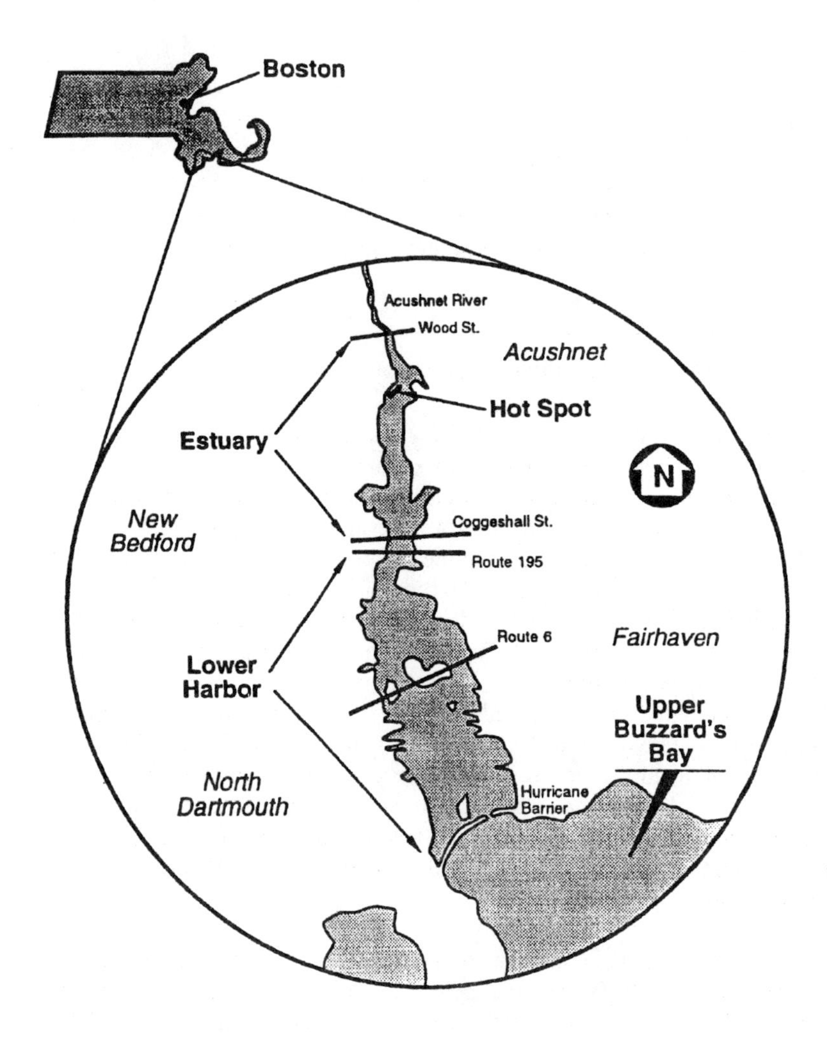

FIGURE 9-1 Site map of New Bedford Harbor. Source: EPA (1992).

Throughout the implementation process, there must be continual feedback between the application of the management option and the subsequent monitoring conducted to establish if risk-management goals and engineering objectives (if applicable) are being met (see Chapter 10 for a discussion of monitoring). This feedback is extremely important, as monitoring should be used to determine whether any course modification is necessary. Feedback is also particularly important for PCB-contaminated sites because of the biopersistence and changing toxicity of PCBs over time. Because PCBs are not static contaminants, it is imperative that affected parties know both the short-

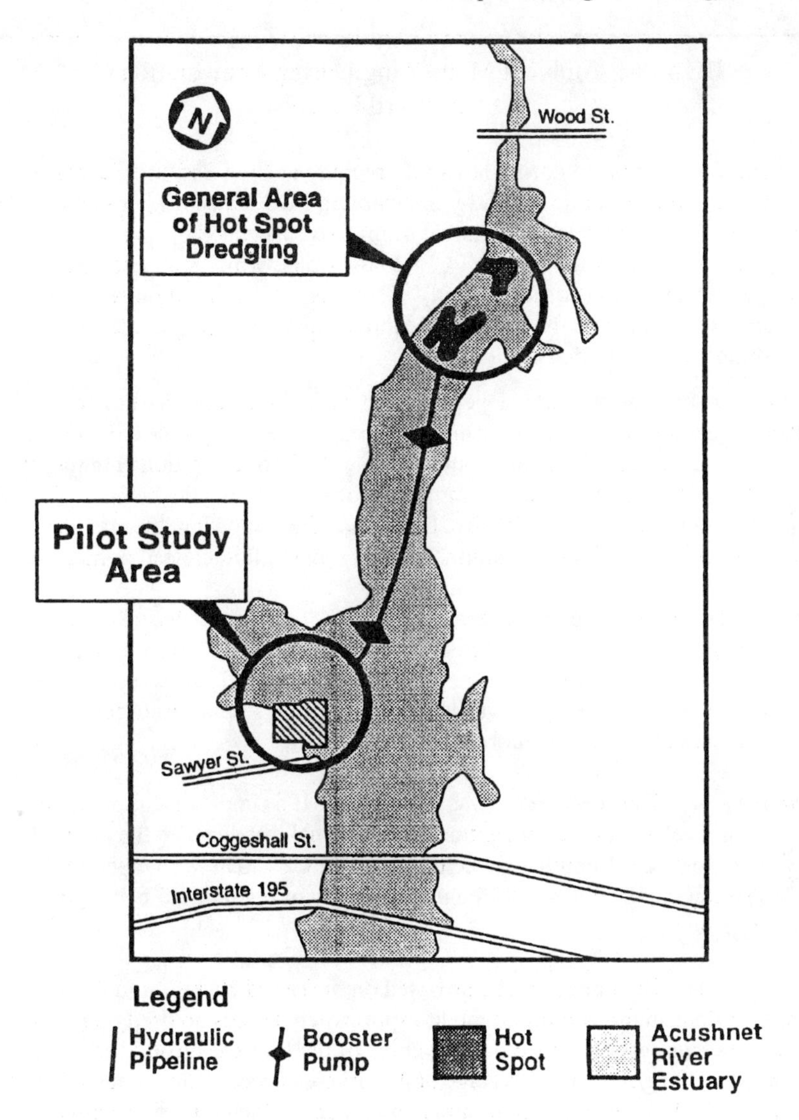

FIGURE 9-2 Selected alternative for hot-spot sediments. Source: EPA (1992).

term and long-term effectiveness of a management option. Feedback between implementation and monitoring can also be used to evaluate the impacts of the risk-management strategy on ecological resources. Box 9-2 illustrates the important role played by ambient-air monitoring at the New Bedford site, where high ambient-air PCB concentrations forced a delay in the management process.

BOX 9-2 Ambient Monitoring During Remediation of New Bedford Harbor

Ambient-air monitoring during hot-spot dredging at New Bedford Harbor was designed to protect public health by maintaining PCB concentrations below certain levels. There was concern that sediments suspended during the dredging operation would release PCBs to the water, which was relatively shallow, especially at low tide (1-3 m). That release in turn might result in exchange of PCBs from the water to the atmosphere, resulting in increased atmospheric concentrations.

Sixteen air-monitoring stations were located around the upper harbor area, and initial concentrations before remedial dredging of the hot spots were collected. Initial background assessments found average PCB concentrations ranging from 10 to 50 ng/m^3, depending on the sampling location around the harbor, with maximum concentrations of 100 to 200 ng/m^3. The following "trigger" levels were agreed upon and published in a bulletin sent to all interested parties:

Shut-Down Level—The shut-down level of 1,000 ng/m^3 was based on the recommendation of the National Institute for Occupational Safety and Health (NIOSH) for toxic substances in the workplace. If a single air sample exceeded the shut-down level, dredging was to be immediately stopped until the source of the PCBs was identified and corrective action taken.

Action Level—The action level was 500 ng/m^3. If a single air sample exceeded the action level, the dredging contractor had to make changes in the dredging process to reduce the emitted concentrations of PCBs. These changes could include reducing the period of dredging each day or operating the dredge at lower speeds.

Notice Level—The notice level was based on measured background PCB concentrations in the vicinity of the hot spot, when there is no dredging. The notice level was estimated from the average of the four background sampling events plus 30 ng/m^3. If the average daily sample exceeded the notice level, the contractor had to notify the government and identify the corrective action.

Table 9-1 contains the record of the averages and maximums for the entire hot spot. This figure illustrates that during the dredging operations, 26% of the samples exceeded the notice level, 1% of the samples exceeded the action level, and 0.25% (or 10 samples) exceeded the shut-down level (EPA 1997).

Air monitoring did result in the shut down of the dredging operations and was important for identifying the potential health risks to the local community. This example illustrates the importance of feedback between the implementation phase of the strategy and the monitoring of its outcomes.

TABLE 9-1 Summary of Airborne PCB Sampling Results: New Bedford Harbor Hot-Spot Dredging Remedy

Sample Location	Total No. of Samples Collected	Average Concentration (ng/m^3)	Action Level Exceedances >50 ng/m^3	>500 ng/m^3	>1,000 ng/m^3
		ON-SITE			
1	204	33.19	37	0	0
2	204	44.18	55	0	0
3	200	180.11	107	22	7
3D	161	147.70	87	9	2
4	131	14.04	6	0	0
5	162	20.06	18	0	0
6	206	51.98	68	0	0
17	30	52.29	13	0	0
		NEAR SITE			
7	86	9.45	0	0	0
8	94	8.45	0	0	0
9	94	14.92	11	0	0
		DREDGE			
10	311	29.17	52	0	0
11	313	174.12	251	12	1
12	313	28.69	50	0	0
13	313	80.99	143	4	0
13 D	282	77.54	121	2	0
14	314	11.16	13	0	0
15	313	23.08	28	0	0
16	310	10.09	3	0	0
		TOTAL			
	4,041		1,063	49	10

Note: Summary of data from March 11, 1994 to September 5, 1995.
Source: EPA (1997).

Coupling implementation with monitoring will also provide valuable information on the effectiveness of specific remediation technologies and management options. This information can be used to compile a database, such as the General Electric Corporation's *Major Contaminated Sediment Sites Database* (General Electric 1999), that would serve as a catalogue of demonstration projects. Such a comprehensive database would be useful as a risk communication tool and could also serve to identify areas for further research, including new or improved remediation technologies and monitoring techniques.

Many of the issues relating to implementation of a risk-management strategy have been addressed in the NRC report *Contaminated Sediments in Ports and Waterways: Clean Up Strategies and Technologies* (1997). The report emphasizes several salient issues pertaining to contaminated sediments. The issues addressed include the need for the development of a project plan and the subsequent components, including affected and interested party involvement, interim controls, and source control.

PROJECT PLAN

Successful implementation of the selected risk-management strategy begins with a project plan. A project plan should be developed that details the schedule of activities that will occur, what additional activities need to be conducted to ensure the management options are as effective as possible, who will be involved in conducting and orchestrating these activities, how they will be carried out, and what the costs are. Elements of the project plan can include a more extensive site characterization, specific to the risk-management options chosen; necessary source control; and specifications of performance standards to be met during the risk-management process. A key factor in the development of a project plan is the involvement of the affected parties. The project plan may be implemented in phases, particularly if the site is very large, complex, several management options are to be used, or a new technology is to be tried. In the latter case, the project plan may call for a pilot test of the technology before it is used on a field scale. The plan should then specify what will constitute a successful pilot test and what activities are anticipated if the test is a success or failure.

A critical element of the implementation project plan is the request for proposal (RFP). A well-written, comprehensive RFP in which the project objectives, technology specifications, and desired schedule are clearly specified is essential to satisfactory project completion. The development of an RFP requires a more detailed understanding of the site characteristics than was obtained during the initial site and risk assessments. Greater emphasis should be placed on the spatial distribution of PCBs; the physical, chemical, and environmental factors affecting contaminant dispersal and degradation and human and wildlife exposure; and any applicable regulatory requirements, such as permits.

During implementation, it must be recognized that many considerations can potentially and profoundly affect project protocols, technology applications, scheduling, and costs. These considerations include seasonal factors that might dictate the management option to be used (e.g., the presence of ice cover), concerns regarding the impact of the management option on the ecosystem (e.g., loss of spawning grounds), and limitations on the recreational uses of the contaminated site (e.g., the presence of silt curtains preventing boat passage). These considerations will vary depending on the type and location of the water body that is being managed (e.g., ice cover is a problem at northern sites and wide rivers might not be obstructed by the placement of silt curtains). Ideally, respondents to the RFP will have a thorough understanding of the physical and environmental characteristics of the site.

Because the project plan should be viewed as the time to clarify all expectations of the management outcomes and to put in place procedures that acknowledge the "high visibility" of most management projects and the need for success, there must be flexibility to allow for modifications in the project plan. Too often emphasis is placed more on administrative protocols than on engineering or scientific end points. The strict timelines typical of many engineering projects are often not appropriate with sediment-management projects. At these sites, there is a need to minimize or eliminate adverse environmental impacts or human exposure, and these considerations dictate procedures. There might be numerous uncertainties regarding either the site or the technology characteristics. For instance, at the General Motors facility in Massena, New York, a specific area had PCB concentrations that exceeded 500 mg/kg. After as many as 32 dredge passes, it was determined that target cleanup levels (ranging between 0.1 and 1.0 mg/kg for sediment) in this area would not be achieved with dredging alone and that capping would be necessary (Chapter 7, Box 7-5; BBL 1996). This situation highlights the need for both a flexible project plan and a contract that can be modified if the original strategy is shown to be ineffective or otherwise in need of revision.

Innovation should be encouraged throughout the management process, not only for the actual implementation of the management option but also for the site and risk assessment and monitoring stages as well. It is also important that the evaluation be conducted frequently so that problems can be identified and addressed early in the process. Contracts should include economic incentives for the development and use of new technologies and protocols to stimulate innovative risk-management efforts, including, when appropriate, the beneficial use of contaminated materials (e.g., as construction aggregate) or habitat restoration. The committee suggests that CDFs can be used as experimental sites to test innovative methods for treating sediments. Incentives for innovative but protective approaches can be based on a variety of performance metrics, including production rates or remedial endpoints, such as contaminant concentrations or bio-indicators (e.g., species diversity). These incentives have the potential to contribute substantially to an increase in the rate of successful site risk management.

A partnership between a broad-based group of affected parties can also assist in implementation of the risk-management strategy. Success at such projects as the St. Paul Waterway (Box 7-5 and Appendix D) and Waukegan Harbor (Box 8-2) highlight this. Partnering is particularly important in management of sites where costs and risks are high and there is a good potential for contention and liability.

Affected parties who are involved in the implementation process will have a better appreciation of the progress and problems that can occur at the site. These partnerships are particularly important for larger systems, such as the Fox River or the Hudson River. In such large, multifaceted, and dynamic regions, contaminant management upstream could result in adverse effects in downstream regions. Management efforts to satisfy needs of one region might not coincide with the needs or desires of the residents in an adjoining region, as is evident among communities along the upper and lower Hudson River. Under such conditions, identification of the requisite affected parties might be best served by the formation of a watershed-wide committee to bring together the disparate and geographically separated groups and their respective points of view. Although it might be difficult or impossible to satisfy all factions of this larger group, successful implementation of the risk-management strategy requires their active participation in the process and consideration of their opinions and concerns.

SOURCE CONTROL

As discussed in Chapter 7, source control should be the first milestone in any risk-management strategy. The failure to control a source and prevent recontamination often represents one of the major impediments to overall site risk management. The possibility of continuing contaminant inputs, at even extremely low concentrations, makes it substantially more difficult to specify engineering endpoints for contractual and operational purposes and complicates subsequent evaluations of both the efficiency and efficacy of contaminated sediment removal or containment and the desired long-term benefit of the management strategy. Beyond those operational effects, the potential for continuing contamination can also affect acquisition of the required state and federal permits. Without some assurance that the primary PCB source is eliminated or negligibly small, regulatory agencies might find it difficult to issue project permits, arguing that the need for continuing risk-management and the associated disruption might be more of a risk than the undisturbed presence of the PCBs. Therefore, it is imperative that sources be controlled before proceeding with management measures.

Despite the importance of source definition and control and its influence on the implementation of a risk-management strategy, it might be difficult to achieve. At some sites, sources are difficult to identify and may be even more difficult to eliminate and control. Their origin and associated transport might be distant from the ultimate sink. That has been illustrated at several sites, including the Housatonic River in Massachusetts and the Fox River in Green Bay, Wisconsin, where there was a failure to identify the source early in the management process, resulting in wasted expenditures of money and time. At other sites, sources might be even more elusive or diffuse. Allocating responsibilities and costs among all sources might expedite implementation of the risk-management strategy.

CONCLUSIONS AND RECOMMENDATIONS

As discussed in Chapter 8, overlapping and frequently contradictory regulations and goals can lead to delays in implementation of the risk-management plan and the subsequent frustration of affected parties. Such delays can result in increased costs, increased contamination levels, larger areas of contamination, and other problems. The committee recommends that the appropriate federal, state, and local regulatory agencies coordinate the various legal and regulatory requirements under which they operate to expedite the implementation of the selected risk-management strategy.

REFERENCES

BBL (Blasland, Bouck & Lee). 1996. St. Lawrence River Sediment Removal Project Remedial Action Completion Report. Prepared for General Motors Powertrain, Massena, New York. Blasland, Bouck & Lee, Syracuse, New York. June.

EPA (U.S. Environmental Protection Agency). 1992. Superfund Program, New Bedford Harbor Site, New Bedford, Massachusetts, Proposed Plan. Region 1, U.S. Environmental Protection Agency. January.

EPA (U.S. Environmental Protection Agency). 1997. Report of the Effects of the Hot Spot Dredging Operations New Bedford Harbor Superfund Site, New Bedford, Massachusetts. October.

General Electric Company. 1999. Major Contaminated Sediment Site Database. Available: http://www.hudsonwatch.com/mess/.

NRC (National Research Council). 1997. Contaminated Sediment in Ports and Waterways: Cleanup Strategies and Technologies. Washington, DC: National Academy Press.

10

Evaluating Results

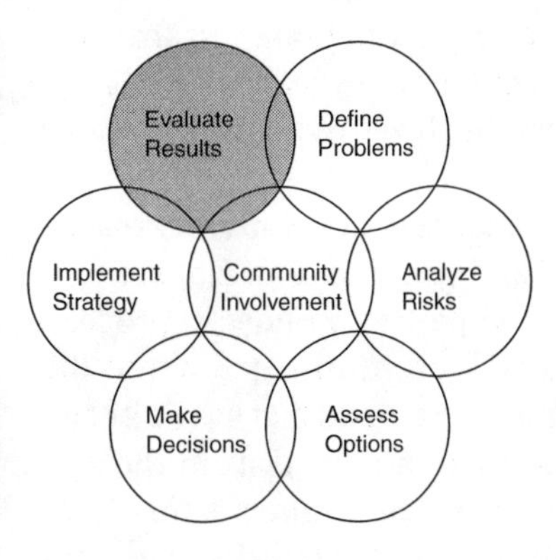

The controversies surrounding the management of PCB-contaminated sediments, especially sediments with high concentrations of PCBs and large volumes of sediments, have been intense. In part, these controversies have arisen because of a lack of evaluation of the effectiveness of management options that have been applied in large pilot studies or at full-scale contaminated sites. Fortunately, in the past 2 years, there has been progress in the evaluation of some aspects of remediation efforts at several sites.

This chapter presents the rationale for an effective evaluation program, drawing on the overall recommendations for evaluating the suggested framework (Chapter 3). The chapter also describes the committee's reviews of some of the recent evaluations conducted for selected remediation sites and then presents its conclusions and recommendations. The actual details of all aspects of site monitoring are not presented in this chapter, because there have been several excellent reports and reviews about the nature of monitoring in general (e.g., NRC 1997), methodologies for PCB analyses (Appendix F and Chapter 6), and biological effects measurements (Appendix G and Chapter 6).

FRAMEWORK GUIDANCE

In its report *Framework for Environmental Health Risk Management*, the Presidential/Congressional Commission on Risk Assessment and Risk Management stresses the need for an evaluation of any risk-management strategy. An evaluation provides the decision-makers and other affected parties with valuable information, including the following:

- Whether the actions are successful, whether they accomplish what was intended, and whether the predicted benefits and costs are accurate.
- Whether any modifications are needed to the risk-management strategy to improve success.
- Whether any new information has emerged that indicates a decision or stage of the framework should be revisited and possibly revised.
- Whether the framework process is effective and how the involvement of the affected parties contributes to the outcome.
- What lessons can be learned to guide future risk-reduction strategies at PCB-contaminated sites to improve the decision-making process.

The framework (Chapter 3) identifies several information tools that will assist in the evaluation process: environmental monitoring, health monitoring, research, disease surveillance, analyses of costs and benefits, and discussions with affected parties.

Of particular relevance to the committee's charge to develop a risk-based framework are learning lessons from the past and ongoing remedial actions. It is difficult to build consensus among affected parties on how to proceed with future remedial actions when there is a lack of information about the effectiveness, costs, and risks of risk-management strategies that have already be implemented. Thus, it is crucial that information gained from the few evaluations of remedial actions for PCB-contaminated sediments that have been conducted be thoroughly and impartially assessed in a process that involves all affected parties. As the commission (1997) notes,

> In the past, evaluation, when conducted, has been performed by the regulatory authority itself. As with other stages of the risk management process, evaluation will benefit if stakeholders are involved, helping to:
>
> - Establish criteria for evaluation, including the definition of 'success.'
> - Assure the credibility of the evaluation and the evaluators.
> - Determine whether an action was successful.
> - Identify what lessons can be learned.
> - Identify information gaps.
> - Determine whether cost and benefit estimates made when evaluating the risk management options were reasonable.

REVIEWS OF EVALUATIONS OF REMEDIAL ACTIONS

Among the reviews of remedial actions at PCB-contaminated-sediment sites that have been conducted over the last several years by various organizations, two studies were recently completed—General Electric (2000) and Scenic Hudson (2000). The General Electric report lists 54 sites for which data were assembled on the type of site, methods of remediation and disposal, volume of sediment removed (if removed), total costs, and per unit cost. A summary of information for several sites is presented in Table 10-1.

Among the key findings of the General Electric report are the following:

- There has not been a systematic experience-based review of the capabilities and limitations of dredging technology in reducing risks posed by contaminated sediments. Thus, an opportunity exists to apply lessons learned from the current base of experience that can help to guide future decision-making.
- Based on evaluation of projects in the United States, information is now available on the capabilities and limitations of dredging technology. The data on post-dredging residual contaminant levels in surface sediments, production rates, and costs need to be more rigorously used in the evaluation of dredging technology in sediment remedy decisions.

The Final Observations and Conclusions of the General Electric report include, among nine bulleted points, the following:

> The experience at completed projects needs to be considered in making future decisions. Adequate monitoring data and formal plans for pre- and post-remediation evaluation of risk reduction are essential elements in sediment remediation projects. These types of essential data can reduce uncertainty and allow one to draw sound conclusions regarding the relative effectiveness of remedial activities.

EPA Region 5, in a presentation to the committee during a public session, indicated that there were six sites (five in the United States and one in Sweden) where monitoring data for PCBs in fish indicated a 2- to 9-fold reduction of PCB concentrations as a result of remedial dredging of PCB-contaminated sediments. Appendix C of the General Electric Report, contributed by the Fox River Group, takes exception to EPA Region 5's contention that sediment removal (dredging or dry excavations) has been successful in reducing risks as evidenced by decreases in PCB concentrations in fish sampled in the areas of the remedial action. General Electric contends that "EPA presents its findings as conclusive without properly qualifying those conclusions based

TABLE 10-1 U.S. Sediment Remediation Projects Implemented (>10,000 cubic yards)—Primary Goal Versus Outcome

Project	Primary Goal	Basis for Primary Goal	Sediment Remedial Target	Relationship of Target to Goal	Remediation Method	Achievement of Remedial Target	Achievement of Primary Goal
Manistique River, MI (97,050 cubic yards)	Reduce PCB in fish levels; reduce carcinogenic and noncarcinogenic risks to $<10^{-4}$ and <1, respectively, except for high-end subsistence and some high-end recreational exposure from fish consumption	Human-health risk assessment	10 ppm PCBs	Default level after using biota to sediment accumulation factor (BSAF) to estimate a target sediment level, then increasing the estimate to 10 ppm PCBs, which EPA justified based on cleanup levels at other EPA projects, the likelihood of achieving <10 ppm, and future natural burial	Hydraulic dredging	In progress; consistent achievement of 10 ppm or less proving difficult	Too soon to tell. Remediation still in progress in Year 5. No postmonitoring program defined as of yet

New Bedford Harbor, MA (14,000 cubic yards)	Remove PCB mass at an optimum "residual concentration to volume removed" ratio and reduce PCB flux to the water column (interim measure)	Mass removal calculations; flux modeling studies conducted by PRPs; water column data	4,000 ppm PCBs in 5 acres of hot spots	Direct	Hydraulic dredging	Achieved based on a limited number of verification samples (15 composite samples ranging from 67 to 2,068 ppm PCBs)	Achieved mass removal; water column data post-dredging (if collected) not obtained. PCBs in surface sediment samples in the Upper Harbor increased 32% on average, following hot-spot dredging
General Motors (Massena), NY (13,800 cubic yards)	Reduce PCB levels in fish	Human health risk assessment	Achieve 1 ppm PCBs and remove as much sediment as technically feasible	Vague; 0.1 ppm PCBs desired, but 1 ppm selected based on technical feasibility	Hydraulic dredging	Not achieved. Average residual PCB levels at completion in six dredged quadrants across 11 acres ranged from 3 to 27 ppm with a maximum of 90 ppm	Two annual post-dredging fish monitoring programs completed; no discernible trends other than a slight increase in fish PCB concentration in Year 2 versus Year 1

Sources: Manistique River (Scenic Hudson 2000; BBL 2000); New Bedford Harbor, Massachusetts (NRC 1997; Bergen et al. 1998); GM Massena, New York (BBL 1996; Scenic Hudson 2000; General Electric 2000).

on known uncertainties and limitations of the underlying data" (General Electric 2000).

The Scenic Hudson report reviews many of the same remediation activities and monitoring data as the General Electric report. The Scenic Hudson (2000) report, which is intended to inform the EPA's decision regarding remediation of PCB-contaminated sediments in the Hudson River, states,

> Dredging is still the preferred remedy for sediment contamination. . . . [M]onitoring data show reductions in sediment and fish contamination following sediment cleanups. . . . and options for large-scale sediment dewatering, water treatment, and disposal are well developed.

The committee carefully reviewed these reports and the supporting documents to ascertain why these groups could assess the same sets of evaluation data and arrive at different general conclusions. The committee determined that there are three principal reasons for this disagreement. First, in some instances, there is disagreement about the remediation goals and the measures by which achievement of the goals can be assessed. Second, in some cases, the available post-remediation monitoring data are sparse and incomplete compared with pre-remediation data and control data. Third, in some cases, it is the intention of reviewers, agencies, and industries to support their preferences, and that might lead to more conflict. The committee acknowledges that objective evaluations are difficult to organize and even more difficult to produce; however, that does not preclude the affected parties from striving for objectivity.

The lack of a comprehensive set of data on site conditions prior to remediation limits the ability to provide compelling evidence of post-remediation risk reductions. A plot of monitoring data for PCBs in fish collected at the Queensbury, New York, site on the Hudson River (Figure 10-1) illustrates that point. Niagara Mohawk completed the first phase of remediation on this site between 1993 and 1996. That phase involved removal of onshore contamination and dry excavation of a portion of the near-shore sediments in 1996. The 1992 data in Figure 10-1 are for fish that might be larger in size than the other fish samples, and the fish were sampled at a different time of the year (Scenic Hudson 2000). The Scenic Hudson report states, "According to the New York State Department of Environmental Conservation (DEC), the pre-cleanup fish contamination levels at the site are best represented by the 1993 data" (Scenic Hudson 2000). The inclusion or exclusion of the 1992 data set makes a major difference in interpretation of the influence of the remediation on the current concentrations of PCBs in fish. Similar difficulties arose for assessments of

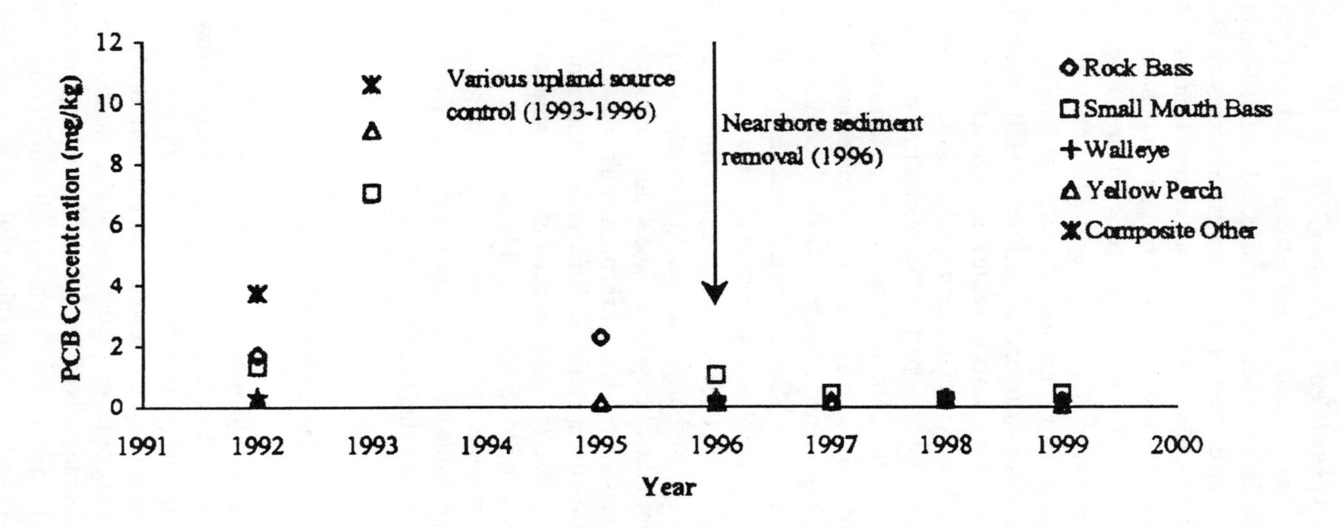

FIGURE 10-1 Average PCB concentrations in fish collected at the Queensbury site. Source: Niagara Mohawk (2000). Reprinted with permission; copyright 2000, Niagara Mohawk Corp.

monitoring data for PCB concentrations in fish for the General Motors, Massena, New York, site (General Electric 2000).

The committee has previously emphasized that achieving an understanding of the issues, goals, and risks among affected parties for a project is critical. Also, critical is agreement on how success in achieving these goals will be monitored and evaluated. The need for agreement becomes clear when two different perspectives on possible risk-management goals are considered for a hypothetical area with PCB-contaminated sediments. One goal might be to decrease concentrations of PCBs in surface sediments and thereby decrease PCB concentrations in fish at the site. As discussed in Chapter 6, PCB concentrations in the surface sediments can influence PCB concentrations in fish. However, the degree of certainty of achieving a reduction in PCB concentrations and the time allowed to make such a reduction in the surface sediments and/or fish should be stated as part of the overall risk-management goals.

In a second situation, PCB concentrations in surface sediments were found to be higher after remediation than the goal concentrations established before the remedial dredging (General Electric 2000). (See Box 10-1 for a possible explanation of this increase.) Some affected parties might view that phenomenon as a failure to achieve a management goal. However, if the management goal was to reduce the overall sediment burden (mass) of PCBs, including those in those in deeper sediments that might continue to diffuse low concentrations of PCBs into the overlying surface sediment (i.e., the mass removal option), then the long-term and certain benefit might outweigh the temporarily increased risk associated with the higher concentrations of PCBs in surface sediments. It is important to have an accurate assessment of short-term and long-term risks associated with each possible management option to select the risk-management strategy that will maximize the risk reductions for human health and the ecosystems or that will best achieve the risk-management goals established in the earlier stage of the framework.

ASSESSMENT MONITORING

In the following section, monitoring data collected after the remediation of PCB-contaminated sediment sites are excerpted from a few of the reports submitted to the committee. These data illustrate the type of monitoring that can and should be conducted and also some of the pitfalls that can arise from incomplete monitoring or incomplete interpretation of the data. The specific types of samples and methods of analysis for monitoring depend on the goals and objectives of the management strategy, although certain common practices, which are discussed below, are also applicable. These include air moni-

BOX 10-1 Monitoring Surface-Sediment PCB Concentrations

There are good scientific reasons for the common assumption that PCBs in surface sediments are an important source of most exposures of fish of PCBs via direct uptake from water in contact with sediments and via feeding on prey that are contaminated by interactions with the sediment and interstitial and overlying water (see Chapter 6). Thus, surface-sediment PCB concentrations have become a focus of monitoring and assessments. For this reason, there is a need for a much clearer understanding and reporting of what is meant by the term surface sediment.

In several assessments at specific sites, the upper 3 inches (7.6 cm) is defined as surface sediment. That might be acceptable in some cases where bioturbation or physical mixing rapidly mixes the upper 7.6 cm of the sediment. However, in other cases, surface sediment should be defined as 0.5 cm or less, because that is where the most immediate interaction is between the PCB-contaminated sediment, the overlying waters, and the biotic sectors of the ecosystems. Bioturbation and physical mixing are less intense in this area. Knowledge of the sediment dynamics in question and depth profiles in the sediments of PCB concentrations, coupled with modeling, can provide an estimate of the depth of surface sediment that should be sampled. Ideally, for monitoring assessments, cores of sediment should be cut horizontally into sections of 0.5 to 1 cm to larger intervals, depending on the biological and physical mixing rates of the sediments, followed by analysis for PCBs. At present, there are few such data in monitoring assessments.

Very fine-grained sediment and organic particles can be resuspended during dredging and resettle on areas that have been dredged. These particles have large surface area to volume (mass) and high sorption capacity for PCBs. Thus, the concentrations of PCBs can be high. A thin layer of high PCB concentrations can form on or in the surface sediments. In some circumstances, this layer can result in higher measured concentrations of PCBs in surface sediments after dredging compared with the desired and expected outcome of lower surface-sediment concentrations. Much depends on the individual site conditions and the depth of sediment sampled and analyzed as the surface sediment.

Given this situation, assessments of risks within various time frames are important. High concentrations within surface sediments can mean greater short-term risk from exposure of biota to PCBs. In the long term, however, combinations of bioturbation, physical mixing, and microbial degradation might significantly reduce these surface-sediment concentrations.

toring to protect human health during and after remediation and sediment monitoring to determine the concentrations of PCBs remaining in the sediments and the amount of sediment removed.

The committee details the assessment monitoring conducted at the General Motors Massena site because it has the most complete set of results. However, the committee also reviews some monitoring that was conducted at other PCB-contaminated sites around the United States, including New Bedford Harbor in Massachusetts and Manistique Harbor in upper Michigan.

General Motors, Massena, New York

The monitoring data were reported by BBL (1996), Scenic Hudson (2000), and General Electric (2000).

The General Motors Powertrain facility near Massena, New York, began operations in 1959. Between then and 1973, PCB-containing hydraulic fluids were used at the plant, and there were leaks and discharges that resulted in contamination of soils, groundwater, and nearby sediments of the St. Lawrence River. In 1984, the site was placed on the National Priorities List. Primary remediation goals at the site were as follows:

- Remove PCB-containing sediments within a specified geographic area (an embayment of the St. Lawrence River—about 11 acres of near-shore area of about 13,000 yd^3) to achieve a target cleanup goal of PCBs at no more than 1 ppm in sediments, if technically feasible.
- Dewater sediment sufficiently to pass EPA Method 9095—Paint Filter Liquids Test.
- Treat dredge-return water and other process water to remove solids and meet discharge limits for PCBs, oil and grease, and suspended solids (TSS) prior to discharge to the river.
- Prevent release of PCBs, suspended solids, and other constituents to the air and river to the maximum extent practicable.
- Substantially complete the work before December 1995.

The environmental monitoring plan for this effort consisted of the following:

- Monitor the St. Lawrence River sediment-removal project to verify that the project performed in a manner protective of human health and the environment.

- Provide procedures for sampling, analyzing, and measuring the effectiveness of removal operations.
- Specify the necessary steps to be taken should remediation activities cause exceedances in project action levels.

Environmental monitoring prior to sediment removal consisted of the following:

- Conduct a pre-dredge bathymetric survey.
- Determine optimal sampling depth and location for turbidity measurements and water-column samples.
- Collect nontransient, noncommunity potable water at the site and intake-water samples from the St. Regis Mohawk Tribe.
- Collect pre-dredging turbidity measurements.
- Collect air samples for ambient PCBs and dust.
- Take video documentation of bottom and connecting joints of sheet-pile wall.
- Conduct a turbidity study to determine adequacy of the sediment control system.

Monitoring sediment removal operations included the following:

- Visual inspection.
- Turbidity monitoring.
- Water-column testing.
- Demonstration and verification testing of temporary water-treatment system.
- Intake-water sampling at the General Motors site and from the St. Regis Mohawk Tribe.
- Air sampling for ambient PCBs and dust.

The monitoring action levels for the project are presented in Table 10-2. The sampling activities and results of analyses were reported on a daily basis to the U.S. Army Corps of Engineers onsite representative, the New York State Department of Environmental Conservation, and the St. Regis Mohawk Tribe Environmental Division in the *St. Lawrence River Environmental Monitoring Daily Report.* If there was an exceedance of an action level, the remediation contractor and the Corps of Engineers onsite representative were notified immediately. Prescribed modifications and procedures, which were agreed to before dredging began, were promptly executed.

TABLE 10-2 Action-Level Values for the St. Lawrence River Sediment Removal Project

Turbidity Monitoring:
Nephelometric turbidity units (NTUs): 28 NTUs at predetermined locations downstream greater than those at upstream location

Temporary Effluent Discharge Limitations:
PCBs: Nondetectable PCBs with a target detection limit of 0.065 μg/liter (L)[a]
TSS: 10 mg/L
Oil and Grease: 15 mg/L

PCB Water-Column Sampling:
PCBs: 2 μg/L at downstream monitoring stations

Airborne Particulate-Dust Monitoring:
Particulate Dust: 150 $\mu g/m^3$

Airborne PCB Monitoring:
PCBs: Daily average total = 0.1 $\mu g/m^3$

SLR Sediment-Sample Cleanup Goal:
PCBs: 1 mg/kg

[a] If PCB analyses of greater than 0.065 μg/L were indicated, the contractor was required to examine and resolve the treatment-system problems but could continue to discharge to the St. Lawrence River. If the PCB analyses of greater than 0.3 μg/L were indicated, the contractor was required to examine and resolve his treatment-system problems and also discontinue discharging to the St. Lawrence River until such levels dropped below 0.3 μg/L.

Turbidity Monitoring Results

There were a few exceedances of the turbidity action level due to turbid water escaping over the top of low sheet walls of the containment system. These exceedances were not substantial and dissipated within 5 to 15 minutes. However, a filter fabric was installed over the walls, and then additional sheet-pile was added to raise the walls. These actions corrected the problem, and only one exceedance of turbidity was measured for 1 day for the remainder of the project. This exceedance was due to a storm event.

Water-Column Monitoring Results

The action level was exceeded only one time because mechanical removal activities associated with a sheetpile wall without a protective silt screen in place. This situation was corrected the following day by installing a silt curtain inside along the sheetpile wall. Final water-column sampling after the sheetpile wall removal at the end of the remediation sediment removal indicated all PCB concentrations within the dredging area had approached detection limits for the sampling and analysis methods. These concentrations were definitely below the action level.

Temporary Water-Treatment Sampling

Three exceedances of the action level occurred on 3 days. Two were caused by high water-flow rates, causing shorter contact time through the activated carbon treatment. This problem was corrected by reducing the flow rate through the system and adding additional activated carbon in a unit to increase sorption contact time and sorption capacity for PCBs. The third exceedance was caused by a switch in the alignment and flow pattern through the activated carbon treatment system. The system was realigned, and the problem was corrected within 4 days. In the interim, the discharge of the treated effluent from the facility to the St. Lawrence River inside the sheetpile wall in the dredging area was recycled to the equalization basin of the dewatering treatment system until corrective realignment had been accomplished.

Drinking-Water-Intake Sampling

Intake samples were taken of the General Motors nontransient, noncommunity potable-water supply facilities and the St. Regis Mohawk Tribe potable-water treatment facilities. The General Motors intake showed action-level PCB concentrations in the early days of the monitoring, and these levels were evidently due to leaks in the system. Bottled water was provided to General Motors employees until the system was fixed. There were no action-level exceedances at the St. Regis Mohawk Tribe facility.

Air Monitoring

Baseline testing showed that 2 of 17 samples from one monitoring station prior to the dredging were above the action level, and the remainder of the samples and all 14 samples from another area were below the action level, in fact below the detection limits. Ninety-eight air samples were collected and analyzed from the area immediately adjacent to the sediment stockpile area, and 50 of these analyses exceeded the action level. After additional sampling stations were set up and results were received, it was determined that these exceedances were a result of grading activities of soils contaminated with low levels of PCBs adjacent to the sediment stockpile area. These activities were being conducted by a contractor for another site's remediation activities. Data from the more distant air monitor showed little or no impact on off-site areas associated with the sediment removal and stockpiling.

Monitoring After Sediment-Removal Operations

- *Bathymetric survey after sediment removal.* Bathymetric surveys were conducted along specified lines after each dredging event. At the project initiation, an estimated 16,800 yd^3 of sediment were targeted for removal from the designated dredge area. At project completion, the bathymetric survey estimated that an in situ volume of about 13,000 yd^3 had been removed.
- *Verification of sediment sampling and analyses of the river bottom.* The goal was to remove sediment so that residual PCB concentrations in river sediment were at or below 1 ppm. Final PCB concentrations in several of the individual sediment samples at the end of the project did not meet the 1-ppm goal. Further dredging removal was determined to be technically impractical, because there was only a thin layer of sediment remaining, and mechanical removal activities were removing more underlying clean materials than surface sediments with PCB concentrations above 1 ppm. A sediment cap for one of the areas of concentrations above 1 ppm was approved by EPA and installed. (See Chapters 6 and 9 for more details.)

New Bedford Harbor, Massachusetts

The monitoring data are from reports by the NRC (1997) and Bergen et al. (1998).

New Bedford Harbor is a marine Superfund site undergoing remediation. The general background on the site has been described in Chapter 9. Remedial actions are focused on the hot-spot areas, the upper and lower harbor, and the Buzzards Bay area. Dredging of the hot-spot sediment was completed in 1995 (see Chapter 9). The record of decision (ROD) for remediation of the remainder of the upper and lower New Bedford Harbor was issued in September 1998. The selected remedy calls for dredging and shoreline containment of approximately 450,000 yd^3 of contaminated sediments with PCB concentrations greater than 10 ppm in four confined disposal facilities constructed along the edge of the harbor.

The results of the monitoring of the air and water are summarized in Chapter 9. The possible effect of hot-spot remediation dredging on the PCB concentrations in the remainder of the upper and lower harbor were assessed in a sediment monitoring program (Bergen et al. 1998). The sampling matrix of 80 monitoring stations in the harbor and Buzzards Bay was sampled before (October 1983) and after (October 1995) hot-spot dredging and analyzed for 18 PCB congeners. "Although major re-distributions of contaminated sediments were confined to the immediate vicinity of remedial activities, there is evidence that low-molecular-weight PCBs were transported further." Those findings were not evident in the more traditional analyses of concentrations of total PCBs. The application of qualitative graphical analyses coupled with exploratory statistical techniques were required to notice the findings of transport from the immediate vicinity of the remedial action (Bergen et al. 1998). Simply stated, there was transport of PCB-contaminated sediments into other areas of the upper and lower harbor, but the amounts were not large enough to detect by analyzing for total concentrations of PCBs.

Manistique River and Harbor, Michigan

The monitoring data are from reports by Scenic Hudson (2000) and BBL (2000).

Manistique River is in the upper peninsula of Michigan and drains into the north shore of Lake Michigan. EPA estimated that there were about 16,000 pounds of PCBs in the Manistique harbor sediments and immediately upstream in the river. Remediation included dredging of contaminated sediments in several areas between 1996 through 2000. A recent report (BBL 2000) of PCB concentrations in cores of sediments details the concentrations of PCBs in the upper 0-3 inches of sediment from dredged and undredged areas. The report presents detailed comparisons of PCB concentrations in surface sediments (0-3 inches) for 1999 samples and for some 1997 and 1998 samples with 1993 pre-remediation dredging action. The emphasis in the report is again on surface sediments, and thus the data for deeper sections of the cores are presented only for the 1999 samples and not for the 1993 samples. Thus, it is difficult to discern from this report what happened regarding removal of the deeper sediments, perhaps some with higher PCB concentrations. This problem becomes important when considering the issue of mass removal of PCBs versus the issue of initial effects on surface-sediment PCB concentrations. Among the conclusions of this report are two that illustrate that point (BBL 2000):

- Data suggest that while a decline in PCB concentrations might be occurring naturally in undredged areas, no decline is apparent in dredged areas. This observation suggests that dredging might have slowed the rate of naturally occurring surface-sediment PCB reductions in dredged areas by re-exposing previously buried PCBs. Whereas the median PCB concentration in the undredged area declined significantly, the median values for 1993 and 1999 dredged area samples showed no statistically significant change in PCB concentrations.

- Data demonstrate that although PCB mass removal occurs during dredging, limitations exist with respect to the ability of large-scale environmental dredging to reduce risks associated with PCBs in aquatic sediments. Thus, dredging at Manistique to date has not resulted in appreciable reductions in surface-sediment PCB concentrations.

Once again, this example emphasizes the debate about mass removal versus surface-sediment concentrations and about balancing increased short-term risks against long-term risks.

The committee notes that remediation activity was still ongoing at Manistique after the report was prepared. Dredging was completed in August 2000. Further sediment data after that time are not available as this report goes to press. It is possible and feasible to use the existing monitoring data and new monitoring data to be collected to guide further action in terms of additional dredging if that is warranted after evaluating the data using the framework approach that the committee recommends.

The NRC (1995) review of the EPA Environmental Monitoring and Assessment Program (EMAP) provides a good discussion of the scientific issues associated with environmental monitoring that are also pertinent to assessment monitoring. EMAP was established to monitor ecological status and trends by estimating "changes in selected indicators of condition of the nation's ecological resources on a regional basis with known confidence."

The issue of human health concerns associated with the presence of PCBs in sediments and in connection with various management options has been discussed earlier in this report (see Chapters 2, 6, and 7 and Appendix G). A useful human health monitoring and epidemiological study is difficult to implement because of the many interactive and confounding factors and limited sample size (NRC 1991). Nevertheless, the need for such studies is clear. General guidance for the design and implementation of such studies is available (IOM 1999). Furthermore, even if the risks to human health are expected to be minimal, carrying out human-health monitoring is beneficial and is a safeguard against unexpected effects. It is reassuring to the affected communities to have this type of monitoring in place. The challenge is to have a human-health monitoring capability that provides meaningful results for reasonable cost. The risk-assessment framework we recommend provides a process for meeting such a challenge.

CONCLUSIONS AND RECOMMENDATIONS

Monitoring for a few sites and remediation actions have been completed or are ongoing. The committee's assessment of the monitoring plans and reported monitoring results is that the data are helpful and allow some short-term assessments of the efficacy of a few projects. However, the data and reviews are limited, because they are based on total PCBs and not on congener-specific analyses and risk calculations using congener-specific risk factors. Based on measurements of total PCBs and all the caveats associated with such measurements, monitoring and interpretation of the data protected public health from immediate risks associated with inhalation of airborne PCBs. In addition, the monitoring data helped minimize the release of suspended or dissolved PCBs from the dredging activities. In the future, additional monitoring will provide better information and a longer-term record of evaluation.

There are few instances where monitoring has assessed ecological health beyond the measurements of total PCBs in selected fish samples. Given the current capabilities for assessing PCBs in congener-specific measurements and coupling these with biological-effects responses, as noted in Chapter 6 and Appendix G, the lack of such monitoring data is disappointing.

In addition, there are few baseline data for concentrations of PCBs in fish at many of the sites undergoing remediation. The lack of baseline data precludes an assessment of the immediate effect of a remedial action when compared with measurements made after the remedial action. This data gap has to be acknowledged when interpreting monitoring data. It might be many years before any expected reductions in PCB concentrations can be measured because of the length of turnover times of PCBs in older and larger fish, even when inputs to their habitats and exposure levels are decreased (Chapter 6). For new sites and those still undergoing management actions, the committee strongly recommends that baseline monitoring data be collected before management actions are undertaken.

The monitoring of sediments for PCB concentrations is confounded by a lack of adjusting the sampling and analysis strategy to account for the dynamics of sediment systems and particles in aquatic systems. Furthermore, a much clearer understanding about the meaning of the term surface sediments is needed in many of the studies reviewed to avoid confusion. Often, in the evaluations cited above, the upper 3 inches (7.6 cm) of sediment is defined as surface sediments. However, sampling 0 to 3 inches of surface sediments from carefully obtained core samples is an unnecessary expenditure of effort if the samples are homogenized before analysis. Sampling of only the upper few centimeters would reveal more useful information about the PCB concentrations in surface sediments (see Box 10-1).

The result of the monitoring of surface sediments is the subject of controversy in terms of what the data mean for risk management. This issue is only partly a monitoring strategy issue as noted in the preceding paragraph and is more an issue of agreeing to the actual risk-management goals.

Information about monitoring economic and social aspects of remedial actions is missing. For example, there were careful plans to control the traffic of trucks with dewatered sediments leaving the New Bedford containment facility for an out-of-state disposal facility. However, the committee has no information regarding how this system worked in the neighborhoods. Were there adverse economic impacts on nearby businesses? Once again, the committee notes that there is a lack of data for this and other sites.

Although there were references in the monitoring reports about the reporting of data to a select set of affected parties, there seemed to be no process to involve affected communities beyond notifications through newsletters of the results of the monitoring. Each site should have a policy and management mechanism by which the affected parties can have rapid and easy access to monitoring data and a clear understanding of the implications of the data. That mechanism needs to be coupled with an agreed upon mechanism for interruption or modification of the remediation process if the monitoring data indicate agreed upon deviations from the expected results.

- Short-term and long-term assessments of the efficacy of management actions require carefully planned and adequately funded monitoring to inform all interested parties and affected communities of the outcomes of the management actions.
- Information from assessments of completed and ongoing remediation projects should be assessed within the recommended risk-management framework to inform decisions about management options at other sites and projects.
- The collection of baseline data for pre-remediation characterization is essential for new sites and ongoing sites where additional management actions are being contemplated. There is often a lack of adequate data for pre-remediation baseline assessment at sites undergoing remediation. Many of these sites will involve future management activities, at the very least monitored natural attenuation. Continued collection of monitoring data at these sites is needed. Such data are needed not only for assessment of completed management actions, but also for monitoring in anticipation of future remediation. The latter may require different types of monitoring than the former.
- PCB monitoring should utilize congener-specific analyses, allowing for congener-specific risk calculations.

- Assessment of remedial action requires social science, economic, cultural monitoring in addition to technological, human health, and ecological health monitoring.
- Each site requires a policy and management mechanism by which the affected parties can have rapid and easy access to monitoring data and a clear understanding of the implications of the data. This mechanism needs to be coupled with an agreed upon mechanism for interruption or modification of the risk-management process if the monitoring data indicate agreed upon deviations from the expected results.
- Contaminated sites can act as both sources and sinks for the global redistribution of PCBs. Monitoring of those sites can provide estimates of the net contribution of PCB-contaminated sites to the global burden of PCBs.

REFERENCES

BBL (Blasland, Bouck & Lee). 1996. St. Lawrence River Sediment Removal Project Remedial Action Completion Report. Prepared for General Motors Powertrain, Massena, New York. Blasland, Bouck & Lee, Syracuse, New York. June.

BBL (Blasland, Bouck & Lee). 2000. Dredging-Related Sampling of Manistique Harbor—1999 Field Study. Report prepared for Fox River Group. Syracuse, New York: Blasland, Bouck & Lee Engineers and Scientists. June.

Bergen, B.J., K.A. Rahn, and W.G. Nelson. 1998. Remediation at a marine superfund site- Surficial sediment PCB congener concentration, composition, and redistribution. Environ. Sci Technol. 32(22):3496-3501.

GE (General Electric Company). 2000. Environmental Dredging: An Evaluation of Its Effectiveness in Controlling Risks. General Electric Company, Blasland, Bouck & Lee, Inc., and Applied Environmental Management, Inc., Albany, New York. August.

IOM (Institute of Medicine). 1999. Gulf War Veterans: Measuring Health. Washington, DC: National Academy Press.

Niagara Mohawk. 2000. Annual Fish Tissue Sampling Program Data Report of the 1999 Results and Five Year Summary (1995-1999). Prepared for the Niagara Mohawk Power Corporation by Parsons Engineering Science, Inc. April 2000.

NRC (National Research Council). 1991. Human Exposure Assessment for Airborne Pollutants: Advances and Opportunities. Washington, DC: National Academy Press.

NRC (National Research Council). 1995. Review of EPA's Environmental Monitoring and Assessment Program: Overall Evaluation. Washington, DC: National Academy Press.

NRC (National Research Council). 1997. Contaminated Sediments in Ports and Waterways. Washington, DC: National Academy Press.

PCCRARM (Presidential/Congressional Commission on Risk Assessment and Risk Management). 1997. Framework for Environmental Health Risk Management: Final Report. Washington, DC: The Commission.

Scenic Hudson. 2000. Accomplishments at Contaminated Sediment Cleanup Sites Relevant to the Hudson River. Scenic Hudson. September.

Appendixes

Appendix A

Biographical Information on the Committee on Remediation of PCB-Contaminated Sediments

JOHN W. FARRINGTON (Chair) is associate director for education and dean of graduate studies at Woods Hole Oceanographic Institution. He earned his Ph.D. in oceanography from the University of Rhode Island. His research interests include biogeochemistry of natural organic and contaminant organic chemicals in the marine environment, marine environmental quality, science policy interactions and science education.

RAYMOND C. LOEHR (Vice Chair) is the H.M. Alharthy Centennial Chair and a professor of civil engineering at the University of Texas in Austin. He received a Ph.D. in sanitary engineering from the University of Wisconsin. Dr. Loehr's research interests include environmental health engineering, water and wastewater treatment, hazardous waste treatment, industrial waste management, and land treatment of wastes.

ELIZABETH L. ANDERSON is president of Sciences International, an Alexandria, VA consulting firm specializing in qualitative and quantitative risk assessment. She received her Ph.D. in organic chemistry from The American University. Dr. Anderson specializes in analysis of the relationship between environmental contaminants, carcinogenesis, and risk assessment for pesticides, industrial chemicals, and pharmaceuticals.

W. FRANK BOHLEN is a professor of physical oceanography at the University of Connecticut in Groton. He received a Ph.D. in oceanography from the Massachusetts Institute of Technology/Woods Hole Oceanographic Institution joint program. Dr. Bohlen's research focuses on turbulence and sediment transport processes, and the dispersal of contaminated sediments.

YORAM COHEN is a professor of chemical engineering and Director of the Center for Environmental Risk Reduction at the University of California at Los Angeles. He holds a Ph.D. in chemical engineering from the University of Delaware. Dr. Cohen's research program includes environmental multimedia analysis and membrane separations technologies. His multimedia analysis work includes modeling, contaminant transport in the soil matrix and chemical volatilization from shallow water bodies.

KEVIN J. FARLEY is an associate professor of environmental engineering at Manhattan College in New York. Dr. Farley received a Ph.D. in civil engineering from the Massachusetts Institute of Technology. He has worked on modeling the fate of PCBs in a contaminated river and has examined remediation processes including natural attenuation for contaminated sites for both surface waters and groundwaters.

JOHN P. GIESY is a professor of zoology in the College of Natural Science at Michigan State University. Dr. Giesy received his Ph.D. in limnology from Michigan State University. His research interests include cycling of heavy metals, uptake and availability of PCBs from soil and water and their toxicity to wildlife.

DIANE S. HENSHEL is an associate professor in the School of Public and Environmental Affairs at Indiana University and president of Henshel EnviroComm, engaged in technical consulting on environmental health considerations. She received her Ph.D. in neuroscience from Washington University. Dr. Henshel's applied research involves an evaluation of the effects of dioxins, PCBs and other pollutants on avian wildlife exposed in ovo in the wild, with model studies also being carried out in parallel in the laboratory.

STEPHEN U. LESTER is science director at the Center for Health, Environment and Justice in Falls Church, Virginia. He earned M.S. degrees in toxicology from Harvard University and environmental health from New York University. Mr. Lester's interests include the interaction between science, policy, and public health.

KONRAD J. LIEGEL is a partner with Preston Gates & Ellis, LL.P. in Seattle, Washington. Mr. Liegel received a J.D from Cornell Law School. His work focuses on environmental and land use issues for private and public clients. He participated in the development of state sediment standards and dredged material disposal guidelines.

PERRY L. McCARTY is a professor of environmental engineering and science at Stanford University and director of the Western Region Hazardous Substance Research Center. He earned his Sc.D from Massachusetts Institute of Technology. Dr. McCarty's research is focused on the control of hazardous substances in treatment systems and groundwater and biological processes for water quality control.

JOHN L. O'DONOGHUE is director of Health and Environment Laboratories at Eastman Kodak Company. He earned his V.M.D. and Ph.D. degrees from the University of Pennsylvania. He is a Diplomate of the American Board of Toxicology. His main research interest is neurotoxicology, including clinical and pathology methods development and the effect of solvents.

JAMES J. OPALUCH is a professor of environment and natural resource economics at the University of Rhode Island. He received his Ph.D. in Agricultural and Resource Economics from the University of California at Berkeley. Dr. Opaluch's research interests include uncertainty analysis and assessment of natural resource damages from spills of oil and hazardous substances, groundwater quality, wetlands and facility siting.

DANNY D. REIBLE is the Chevron Professor of Chemical Engineering and Director, Hazardous Substance Research Center at the Louisiana State University in Baton Rouge. Dr. Reible received a Ph.D. in chemical engineering from the California Institute of Technology. His research interests include remediation technologies such as capping of contaminated sediments and the transport, behavior, and fate of contaminants in sediments.

Appendix B

Participants at Public Sessions

June 7, 1999 - Washington, DC

Michael Powers, Office of Congressman John Sweeney (R-NY, 22nd District)
E. Timothy Oppelt, U.S. Environmental Protection Agency (EPA), National Risk Management Research Laboratory
Tudor Davies, EPA, Office of Water
Larry Reed, EPA, Office of Emergency and Remedial Response
Dick Jensen, Sediment Management Workgroup (ad hoc industry group)
John Haggard, General Electric Corporation
John Connolly, Quantitative Environmental Analysis
E. John List, Flow Science Inc.
Robert Engler, U.S. Army Corps of Engineers
Emily Green, Sierra Club

September 27, 1999 - Green Bay, WI

Rebecca Katers, Clean Water Action Council
David Ludwig, Exponent, Inc.
Mark Brown, Blasland, Blouck and Lee
Paul Doody, Blasland, Blouck and Lee
Tom Nelson, Oneida Nation
John Kennedy, Green Bay Metropolitan Sewage District
Hallet J. Harris, Professor Emeritus, University of Wisconsin-Green Bay
William Farin, Fox Valley Engineering, Inc., Neenah, WI
Gordon E. Doule, Imtech, WI

Charlotte Arendt, citizen
Bill Acker, citizen
Jim Olmstead, citizen
Mickey Marique, citizen
Mike Lenczuk, citizen

September 28, 1999 - Green Bay, WI

David Allen, U.S. Fish and Wildlife Service
Larry Reed, EPA, Office of Emergency and Remedial Response
Bill Muno, EPA Region 5
James Hahnenberg, EPA Region 5
Donald F. Hayes, Civil and Environmental Engineering, University of Utah

November 8, 1999 - Afternoon - Albany, NY

John Quensen, Michigan State University
Peter Moreau, Niagara Mohawk Corporation
Ken Jock, St. Regis Mohawk Tribe
Doug Tomchuk and Alison Hess, EPA Region 2
Ron Sloan, New York State Department of Environmental Conservation
Edward Horn, New York State Department of Health
John Smith, Alcoa Corporation
Merrilyn Pulver, Fort Edward Town Supervisor

November 8, 1999 - Evening - Albany, NY

David Adams, Saratoga County Environmental Management Council
John Haggard, General Electric
Jim Berg, Adirondack Regional Chamber of Commerce
Tom Borden, Agricultural Liaison Committee
Cara Lee, Scenic Hudson
Aaron Mair, Arbor Hill Environmental Justice Corporation
Tim Gray, Housatonic River Initiative
Judy Schmidt-Dean, Citizen's Liaison Group/Community Interaction Program
Andi Bartczak, Hudson River Sloop Clearwater

Appendix C

Public Access Materials

1. Bergen, B.J., Rahn, K.A., and W.G. Nelson. 1998. Remediation at a marine superfund site: Surficial sediment PCB congener concentration, composition, and redistribution. Environmental Science and Technology 32:3496-3501. (published citation)

2. Bremle, G., and P. Larsson. 1998. PCB concentration in fish in a river system after remediation of contaminated sediment. Environmental Science and Technology 32:3491-3495. (published citation)

3. Chen, P. Zhou, W., and Tavlarides, L.L. 1997. Remediation of polychlorinated biphenyl contaminated soils/sediments by supercritical fluid extraction. Environmental Progress 16(3):227-236. (published citation)

4. Chiarenzelli, J., Scrudato, R., Bush, B., et al. 1998. Do large-scale remedial and dredging events have the potential to release significant amounts of semivolatile compounds to the atmosphere? Environmental Health Perspectives 106(2):47-49. (published citation)

5. Cleland, J. 1997. Advances in dredging contaminated sediment: New technologies and experience relevant to the Hudson River PCBs site. Poughkeepsie, NY: Scenic Hudson, Inc. (available from Scenic Hudson)

6. Cogliano, V.J. 1998. Assessing the cancer risk from environmental PCBs. Environmental Health Perspectives 106(6):317-323. (published citation)

7. Eco Logic. 1998. PCB-contaminated soil and sediment treatment using Eco Logic's gas-phase chemical reduction process. Rockwood, Ontario: ELI Eco Logic International Inc. (NRC only)

8. EPA. 1998. EPA's Contaminated Sediment Management Strategy (foreword and executive summary). U.S. Environmental Protection Agency, Office of Water. EPA-823-F-98-001. April (also available at http://www.epa.gov/ost)

9. Froese, K.L., Verbrugge, D.A., Snyder, S.A., et al. 1997. PCBs in the Detroit River water column. J. Great Lakes Research 23(4):440-449. (published citation)

10. Froese, K.L., Verbrugge, D.A., Ankley, G.T., et al. 1998. Bioaccumulation of polychlorinated biphenyls from sediments to aquatic insects and tree swallow eggs and nestlings in Saginaw Bay, Michigan, USA. Environmental Toxicology and Chemistry 17(3):484-492. (published citation)

11. General Electric. 1997. An evaluation of the feasibility of environmental dredging in the upper Hudson River: Including an evaluation and critique of Scenic Hudson's dredging report. GE Corporate Environmental Programs. May. (NRC only)

12. Giesy, J.P. and K. Kannan. 1998. Dioxin-like and non-dioxin-like toxic effects of polychlorinated biphenyls (PCBs): Implications for risk assessment. Critical Reviews in Toxicology 28(6):511-569. (published citation)

13. Giesy, J.P., Jude, D.J., Tillitt, D.E., et al. 1997. Polychlorinated dibenzo-*p*-dioxins, dibenzofurans, biphenyls and 2,3,7,8-tetrachlorodibenzo-*p*-dioxin equivalents in fishes from Saginaw Bay, Michigan. Environmental Toxicology and Chemistry 16(4):713-724. (published citation)

14. Holton, W.C. 1998. No safe harbor. Environmental Health Perspectives 106(5):A228-A233. (published citation)

15. Kannan, K., Nakata, H., Stafford, R., et al. 1998. Bioaccumulation and toxic potential of extremely hydrophobic polychlorinated biphenyl congeners in biota collected at a Superfund site contaminated with Aroclor 1268. Environ. Sci. Technol. 32:1214-1221. (published citation)

16. Kimbrough, R.D., Doemland, M.L., LeVois, M.E. 1999. Mortality of male and female capacitor workers exposed to polychlorinated biphenyls. Journal of Occupational and Environmental Medicine 41(3):161-171. (published citation)

17. Klasson-Wehler, E., Bergman, A., Athanasiasou, M., et al. 1998. Hydroxylated and methylsulfonyl polychlorinated biphenyl metabolites in albatrosses from Midway Atoll, north Pacific Ocean. Environmental Toxicology and Chemistry 17(8):1620-1625. (published citation)

18. Korrick, S.A., Altshul, L. 1998. High breast milk levels of polychlorinated biphenyls (PCBs) among four women living adjacent to a PCB-contaminated waste site. Environmental Health Perspectives 106(8):513-518. (published citation)

19. Mousa, M.A., Ganey, P.E., Quensen, J.F., et al. 1998. Altered biologic activities of commercial polychlorinated biphenyl mixtures after microbial reductive dechlorination. Environmental Health Perspectives 106 (Suppl 6):1409-1417. (published citation)

20. Murk, A.J., Legler, J., Denison, M.S., et al. 1996. Chemical-activated luciferase gene expression (CALUX): A novel in vitro bioassay for Ah receptor active compounds in sediments and pore water. Fundamental and Applied Toxicology 33:149-160. (published citation)

21. National Research Council. 1997. Contaminated Sediments in Ports and Waterways: Cleanup Strategies and Technologies. Washington, DC: National Academies Press. (available from NAP)

22. Quensen, J.F., Mousa, M.A., Boyd, S.A., et al. 1998. Reduction of aryl hydrocarbon receptor-mediated activity of polychlorinated biphenyl mixtures due to anaerobic microbial dechlorination. Environmental Toxicology and Chemistry 17(5):806-813. (published citation)

23. Remediation Technologies Development Forum. 1999. Summary of the Remediation Technologies Development Forum Sediments Remediation Action Team Meeting. Available: http://www.rtdf.org/public/sediment/minutes/011399/Summary.htm.

24. Renner, R. 1998. "Natural" remediation of DDT, PCBs, debated. Environmental Science & Technology 360A-363A, August 1. (published citation)

25. Sanderson, J.T., Kennedy, S.W., Giesy, J.P. 1998. In vitro induction of ethoxyresorufin-*o*-deethylase and porphyrins by halogenated aromatic hydrocarbons in avian primary hepatocytes. Environmental Toxicology and Chemistry 17(10):2006-2018. (published citation)

26. Sanderson, J.T., Janz, D.M., Bellward, G.D., Giesy, J.P. 1997. Effects of embryonic and adult exposure to 2,3,7,8-tetrachlorodibenzo-p-dioxin on hepatic microsomal testosterone hydroxylase activities in great blue herons (Ardea herodias). Environmental Toxicology and Chemistry 16(6): 1304-1310. (published citation)

27. Sanderson, J.T., Aarts, J.M., Brouwer, A., et al. 1996. Comparison of Ah receptor-mediated luciferase and ethoxyresorufin-O-deethylase induction in H4IIE cells: Implications for their use as bioanalytical tools for the detection of polyhalogenated aromatic hydrocarbons. Toxicology and Applied Pharmacology 137:316-325. (published citation)

28. Sediments Remediation Action Team. 1999. Bibliography [36 records]. Available at http://www.rtfd.org:591/FMPro.

29. Starodub, M.E., Miller, P.A., Gerguson, G.M., et al. 1996. A risk-based protocol to develop acceptable concentrations of bioaccumulative organic chemicals in sediments for the protection of piscivorus wildlife. Toxicological and Environmental Chemistry 54:243-259. (published citation)

30. Van den Berg, M., Birnbaum, L., Bosveld, A.T.C., et al. 1998. Toxic equivalency factors (TEFs) for PCBs, PCDDs, PCDFs for humans and wildlife. Environmental Health Perspectives 106(12):775-792. (published citation)

31. Woodyard, J.P. 1998. Navigating the 1998 PCB Disposal Amendments. Environmental Management, pp. 13-27, November. (published citation)

32. Representative John Sweeney (R-NY, 22nd District). Remarks Before the Committee on Contaminated Sediments, National Academy of Sciences. June 7, 1999. (read by Michael Powers) (NRC only)

33. U.S. Environmental Protection Agency. U.S. EPA's Perspectives. By Tim Oppelt, Office of Research and Development/National Risk Management Research Laboratory; Bill Farland, Office of Research and

Development/National Center for Environmental Assessment; Tudor Davies, Office of Water; Larry Reed, Office of Emergency Response and Remediation. Presented to the NRC Committee, June 7, 1999. (NRC only)

34. Sediment Management Work Group. Effective Management of Contaminated Sediments: Risk-Based Approach. Presented by Dick Jensen, E.I. DuPont de Nemours to the NRC Committee, June 7, 1999. (NRC only)

35. General Electric. Major Contaminated Sediment Site Database. Presented by John Haggard to the NRC Committee, June 7, 1999. [Database is available on the Internet at http://www.hudsonwatch.com/mcss/]

36. Connolly, John. The Role of Models in Evaluating Contaminated Sediments. Quantitative Environmental Analysis, LLC. Presented to the NRC Committee, June 7, 1999. (NRC only)

37. List, John E. Remediation Risks: Palos Verdes Shelf, California. Flow Science, Inc. Presented to the NRC Committee, June 7, 1999. (NRC only)

38. Engler, Robert M. Dredging Overview. U.S. Army Corps of Engineers, Waterways Experiment Station. Presented to NRC Committee, June 7, 1999. (NRC only)

39. Green, Emily. Great Lakes Sediment Clean-up. Sierra Club. Presented to NRC Committee, June 7, 1999. (NRC only)

40. Savitz, Jacqueline. Coast Alliance Concerns. Coast Alliance, Washington, DC, Presented to NRC Committee, June 7, 1999. (NRC only)

41. Matta, Mary. NOAA's Concerns Regarding PCB Contaminated Sediments. National Oceanic and Atmospheric Administration, Seattle, WA. Submitted to the NRC Committee, June 8, 1999. (NRC only)

42. King, Patricia. 1995. Clean Lakes, Clean Sediments: Citizen's Guide and Action Plan. Sierra Club Great Lakes Ecoregion Program, Madison, Wisconsin. (available from Sierra Club)

43. Clean Water Action Council of Northeast Wisconsin, Inc. Letter to Mr.

James Hahnenberg, EPA Region 5 from Rebecca Katers, Executive Director, dated July 15, 1999. 6 pp. (NRC only)

44. Natural Attenuation in the Management of Contaminated Sediments. Presented by David Ludwig of Exponent, Inc., and Mark P. Brown of Blasland, Blouck and Lee, September 27, 1999. 18 pp. (NRC only)

45. Effectiveness of Sediment Removal: An Analysis of EPA Region V's Claims. Presented by Paul Doody of Blasland, Blouck and Lee, September 27, 1999. 34 pp. (NRC only)

46. Position Statement from the Science and Technical Advisory Committee of the Lower Fox River and Green Bay Remedial Action Plan Regarding Remediation of Contaminated Sediments in the Lower Fox River, September 27, 1999. 4 pp. (NRC only)

47. Green Bay Natural Resource Damage Assessment (NRDA) Presentations. Presented by David Allen, U.S. Fish and Wildlife Service, 1 p outline with reference to website at http://www.fws.gov/r3pao/nrda.

48. Remediation of Contaminated Sediments: A National Perspective. Presented by Larry Reed, Office of Solid Waste and Emergency Response, U.S. EPA, Washington, DC. September 28, 1999. 17 pp. (NRC only)

49. A Great Lakes Perspective of Sediment Contamination. Presented by William Muno, U.S. EPA Region 5. September 28, 1999. 10 pp. (NRC only)

50. Lower Fox River and Sediment Site Cleanups. Presented by James Hahenberg, U.S. EPA Region 5. September 28, 1999. 8 pp. (NRC only)

51. Dredging PCB Contaminated Sediments. Presented by Donald F. Hayes, Civil and Environmental Engineering, University of Utah. September 28, 1999. 39 pp. (NRC only)

52. Statement from Hallet J. Harris, Professor Emeritus, University of Wisconsin-Green Bay. September 27, 1999. 1 p. (NRC only)

53. Dredging of the Fox and PCBs Remediation by Thermal Oxidation. Presented by William Farin, Fox Valley Engineering, Inc. Neenah, WI. September 27, 1999. 5 pp. (NRC only)

54. Presentation from Gordon E. Doule, Imtech, WI. September 27, 1999. 3 pp. (NRC only)

55. Wisconsin Department of Health and Family Services. 1999. Human Health Impact of the Land Application of PCB-Contaminated Materials [draft]. October 5, 1999. 23 pp. (NRC only)

56. Wisconsin Department of Health and Family Services. 1999. Human Health Impact of the Land Application of PCB-Contaminated Materials: Summary of Comments. 7 pp (NRC only)

57. Wisconsin Department of Health and Family Services. 1999. Wildlife Soil Criterion for Polychlorinated Biphenyls (PCBs) [draft]. October 5, 1999. 23 pp. (NRC only)

58. Advisory Committee. 1999. Summary of AC Comments on Draft PCB Soil Criteria for Wildlife Protection. 8 pp (NRC only)

59. General Electric. 1999. Restoration Work Plan: Bay Road pond, Queensbury, New York. Albany, NY. August 1999; with cover letter dated August 10, 1999 from John Haggard to Mr. Michael O'Toole, New York State Department of Environmental Conservation. (NRC only)

60. EPA Region 2. Hudson River PCBs Site—Reassessment RI/FS: Schedule. June 10, 1999. 1 p. (available from EPA Region 2)

61. EPA Region 2. The Hudson River PCBs Site: Reassessment Reports Overview. 6 pp (available from EPA Region 2)

62. EPA Region 2.. Phase 2 Report - Review Copy. Further Site Characterization and Analysis, Database Report, Hudson River PCBs Reassessment RI/FS. Executive Summary. October 1995. 3 pp (available from EPA Region 2)

63. EPA Region 2. Phase 2 Report - Review Copy. Further Site Characterization and Analysis. Volume 2B - Preliminary Model Calibration Report, Hudson River PCBs Reassessment RI/FS. Executive Summary. October 1996. 7 pp (available from EPA Region 2)

64. EPA Region 2. Executive Summary: Data Evaluation and Interpretation Report. 8 pp (available from EPA Region 2)

65. EPA Region 2. Low Resolution Sediment Coring Report: Executive Summary. July 1998. 6 pp (available from EPA Region 2)

66. EPA Region 2. Baseline Modeling Report: Executive Summary. May 1999. 7 pp (available from EPA Region 2)

67. EPA Region 2. Ecological Risk Assessment: Executive Summary. August 1999. 9 pp (available from EPA Region 2)

68. EPA Region 2. Human Health Risk Assessment; Upper Hudson River: Executive Summary. August 1999. 6 pp (available from EPA Region 2)

69. EPA Region 2. Phase 1 Report - Review Copy. Interim Characterization and Evaluation: Hudson River PCB Reassessment RI/FS. Executive Summary. August 1991. 14 pp (available from EPA Region 2)

70. EPA Region 2. Report of the Hudson River PCBs Site: Modeling Approach Peer Review - Final Report. November 10, 1998. 70 pp (available from EPA Region 2)

71. EPA Region 2. Report on the Peer Review of the Data Evaluation and Interpretation Report and Low resolution Sediment Coring Report for the Hudson River PCB Superfund Site. Final Report. June 3, 1999. 49 pp (available from EPA Region 2)

72. EPA Region 2. Evaluation of Removal Action Alternatives, Thompson Island Pool: Early Action Assessment. March 1999. Approx 130 pp (available from EPA Region 2)

73. EPA Region 2.. Hudson River PCBs Reassessment RI/FS- Phase 3 Feasibility Study: Scope of Work. September 1998. 52 pp (available from EPA Region 2)

74. EPA Region 2. Hudson River PCBs Reassessment RI/FS: Responsiveness Summary for Phase 3 - Feasibility Study: Scope of Work. June 1999. 110 pp (available from EPA Region 2)

75. Wisconsin Department of Natural Resources. 1999. Model Evaluation Workgroup: Technical Memorandum 2g - Quantification of Lower Fox River Sediment Bed Elevation Dynamics through Direct Observations. July 23, 1999. 79 pp (available from EPA Region 2)

76. Velleux, Mark. 1999. Technical response to Fox River Group comments regarding Technical Memorandum 2g. Memorandum from Wisconsin Department of Natural Resources to Lower Fox River Administrative File. June 18, 1999. 10 pp (NRC only)

77. Velleux, Mark. 1999. Response to FRG comments on Technical Memorandum 2 g dated May 21, 1999. Letter from Wisconsin Department of Natural Resources to Mark Travers, de maximus, inc., dated June 18, 1999. (NRC only)

78. Akwesasne Task Force on the Environment. 1996. Protocol for Review of Environmental and Scientific Research Proposals. Research Advisory Committee. 14 pp (NRC only)

79. New York State Department of Health. 1999. Health Consultation: 1996 Survey of Hudson River Anglers - Hudson Falls to Tappan Zee Bridge at Tarrytown, New York. Public Review Draft, February. Cerclis No. NYD980763841. (available from NYS DOH)

80. Abramowicz, D.A. 1999? Aerobic PCB biodegradation and anaerobic PCB dechlorination in the environment. 10th Forum in Microbiology, pp 42-46. (published article)

81. Manchester-Neesvig, J.B., A.W. Andren, and D.N. Edgington. 1996. Patterns of mass sedimentation and of deposition of sediment contaminated by PCBs in Green Bay. J. Great Lakes Res. 22(2):444-462. (published article)

82. Johnson, B.L., H.E. Hicks, D.E. Jones, W. Cibulas, A. Wargo, and C.T. DeRosa. 1998. Public health implications of persistent toxic substances in the Great Lakes and St. Lawrence Basin. J. Great Lakes Res. 24(2): 698-722. (published article)

83. Grand River Conservation Authority. Undated. Guelph All-Weather Sequential Events Runoff Model. Cambridge, Ontario. http://www.grandriver.on.ca

84. Boyd, D.K. and H. Schroeter. Undated. Application of GAWSER surface water model to estimate water budget. Grand River Conservation Authority. 8 pp (NRC only)

85. U.S. Public Health Service, Agency for Toxic Substances and Disease Registry and U.S. Environmental Protection Agency. Undated. Public health implications of exposure to polychlorinated biphenyls (PCBs). 33 pp (NRC only)

86. Fox, J.C., B.L. Johnson, and L.R. Goldman. 1998. Letter addressed to "Dear Health Care Professional." Agency for Toxic Substances and Disease Registry. December. 3 pp (NRC only)

87. Perciasepe, R., W. Schultz, and L.R. Goldman. 1996. Letter addressed to "Dear Governor." U.S. Environmental Protection Agency and U.S. Department of Health and Human Services. June 26, 1996. 3 pp. (NRC only)

88. Fitzgerald, E.F., D.A. Deres, S.-A. Hwang, B. Bush, B.-Z. Yang, A. Tarbell, and A. Jacobs. 1999. Local fish consumption and serum PCB concentrations among Mohawk men at Akwesasne. Environmental Research Section A 80:S97-S103. (published article)

89. Carpenter, D.O. 1995. Communicating with the public on issues of science and public health. Environmental Risk Perspectives103(S6):127-130. (published article)

90. The Akwesasne Task Force on Environment Research Advisory Committee. 1997. Superfund clean-up at Akwesasne: A case study in environmental justice. International Journal of Contemporary Sociology 34(2):267-290. (published article)

91. Sediment Priority Action Committee, Great Lakes Water Quality Board. 1999. Sediment Management for Ecosystem Recovery. Fact sheet. 2 pp. (NRC only)

92. Zarull, M.A., J.H. Hartig, and L. Maynard. 1999. Ecological Benefits of Contaminated Sediment Remediation in the Great Lakes Basin. Report to the International Joint Commission. Sediment Priority Action Committee, Great Lakes Water Quality Board. August. 45 pp. (available from NRC or IJC)

93. Krantzberg, G., J. Hartig, L. Maynard, K. Burch, and C. Ancheta. 1999. Deciding When to Intervene: Data Interpretation Tools for Making Sediments Management Decisions Beyond Source Control. Based on a

workshop to evaluate data interpretation tools held at the Great Lakes Institute for Environmental Research, University of Windsor, December 1-2, 1998. Sediment Priority Action Committee, Great Lakes Water Quality Board. August. 87 pp. (available from NRC or IJC)

94. National Research Council. 1999. National Symposium on Contaminated Sediments: Coupling Risk Reduction with Sustainable Management and Reuse. Washington, DC: National Academy Press, Transportation Research Board. (available from National Academies Press)

95. Carpenter, D.O. 1998. Polychlorinated biphenyls and human health. International Journal of Occupational Medicine and Environmental Health 11(4):291-303. (published article)

96. Headrick, M.L., K. Hollinger, R.A. Lovell and J.C. Matheson. 1999. PBBs, PCBs and dioxins in food animals, their public health implications. Veterinary Clinics of North America - Food Animal Practice 15(1):109-31. (published article)

97. Sediment Management Work Group. 1999. Contaminated Sediment Management Technical Papers. 37 pp with 4 Appendices. Fall. (NRC only)

98. Akwesasne Task Force on the Environment. Undated. Background Information on Environmental Contamination at Akwesasne. 90 pp. (NRC only)

99. Negoita, S., and L. Swamp. 1997. Human Health & Disease Patterns at Akwesasne. Akwesasne Task Force on the Environment. 62 pp. (NRC only)

100. Akwesasne Research Advisory Committee. Undated. Superfund Cleanup Akwesasne: A case study in environmental justice. Akwesasne Task Force on the Environment. 38 pp. (NRC only)

101. The Fox River Group. 1999. Sensible Solutions for the Fox River: The Role of Nature and Technology in Restoring the Fox River While Preserving the Vitality of Our Communities - A Strategy for Restoring the River. June. 90 pp. (NRC only)

102. Wisconsin Department of Natural Resources. 1999. Baseline Human Health and Ecological Risk Assessment: Lower Fox River, Wisconsin.

Prepared by ThermoRetec Consulting Corporation. February 24, 1999. 1017 pp. With cover letter from James Hahnenberg, U.S. EPA, Region 5, dated November 9, 1999. (available from EPA Region 5, Chicago)

103. Clark, M., J. Hahnenberg, B. Jones, K. Klewin, and M. Sprenger. 1999. Comments on the Draft Remedial Investigation Lower Fox River Wisconsin; Baseline Human Health and Ecological Risk Assessment Lower Fox River Wisconsin and Draft Feasibility Study Lower Fox River Wisconsin - Prepared by ThermoRetec, February 24, 1999. U.S. Environmental Protection Agency, Region 5. Dated April 12, 1999. 38 pp. (available from NRC)

104. U.S. Army Corps of Engineers. 1999. Analytical Results and Quality Assurance Report; Appleton Seepage Barrier. DLZ Laboratories, Inc. Cover letter from Robert Davis, COE, to William Muno, U.S. EPA. 423 pp. (available from EPA Region 5, Chicago)

105. Kemble, N.E., D.G. Hardesty, C.G. Ingersoll, B.T. Johnson, F.J. Dwyer, and D. MacDonald. 1999. Evaluation of the toxicity and bioaccumulation of contaminants in sediment samples from Waukegan Harbor, Illinois: U.S. Geological Survey Final Report for the U.S. Environmental Protection Agency Great Lakes National Program Office. U.S. EPA, Great Lakes National Program Office, Chicago, IL. September 22, 1999. 104 pp. (NRC only)

106. Saratoga County Environmental Management Council. 1999. Package of 13 letters and comments to U.S. Environmental Protection Agency on the Hudson River Superfund PCB Reassessment from July 21, 1992 to October 19, 1999. Cover letter from David Adams to Robbie Wedge, NRC. 69 pp. (NRC only)

107. Donohoe, A.M. 1999. GIS Site Evaluation of Polychlorinated Biphenyls (PCBs) Contamination of Fish in Lake Hartwell, South Carolina. South Carolina Department of Health and Environmental Control. 4 pp. (NRC only)

108. U.S. EPA. 1999. Housatonic River Overview. November 1, 1999. 9 pp. (NRC only)

109. U.S. EPA. 1999. Aluminum Company of America, New York. EPA ID# NYD980506232. U.S. EPA Region 2. St. Lawrence County, Massena. November 4, 1999. 2 pp. (NRC only)

110. Tomchuk, D. 1999. Reassessment of the Hudson River PCBs Site: Presentation for the National Research Council Panel on Remediation of PCB-Contaminated Sediment. U.S. EPA, Region 2. November 8, 1999. 15 pp. (NRC only)

111. Hess, A.A. 1999. Hudson River PCBs Reassessment: Risk Assessment and Remediation Alternatives. U.S. EPA, Region 2. November 8, 1999. 13 pp. (NRC only)

112. Sloan, R.J. 1999. Hudson River Fish and the PCB Perspective: A Presentation to the National Research Council. New York State Department of Environmental Conservation. November 8, 1999. 33 pp with two attachments - Hudson River Basin 1998 Summary Table - Hg - Spring, 16 pp; Striped Bass PCB Decline - Commercial Reopening Consideration, 31 pp. (NRC only)

113. Sloan, R., B. Young, and K. Hattala. 1995. PCB Paradigms for striped bass in New York State. Technical Report 95-1 (BEP). New York State Department of Environmental Conservation. October. 116 pp. (NRC only)

114. Smith, J.R. 1999. Non-time-critical removal action (NTCRA) pilot dredging in Grasse River. Pittsburgh, PA: Alcoa Inc. November 8, 1999. 13 pp. (NRC only)

115. Pulver, M. 1999. Fort Edward Town Councilwoman Merrilyn Pulver: Written Comments to National Academy of Sciences. November 8, 1999. 5 pp. (NRC only)

116. Sweeney, J.E. 1999. Remarks by Congressman John E. Sweeney (R/C-Clifton Park) presented to the National Academy of Sciences. November 8, 1999. 3 pp. (NRC only)

117. Adams, D.D. 1999. Statement by David D. Adams, PE, member Saratoga County Environmental Management Commission (SCEMC) to the National Research Council Committee on the Remediation of PCB-Contaminated Sediments. November 8, 1999. 5 pp with attachment, Adams, D.D., Hudson River PCBs Reassessment, Presentation to Steering Committee, July 14, 1993, 2 pp. (NRC only)

118. Haggard, J. 1999. Supplemental Technical Information in support of

the presentation "Hudson River Framework for Evaluation of Remedial Options" to the National Academy of Sciences Committee on Remediation of PCB-Contaminated Sediments. November 8, 1999. General Electric Company, Albany, NY. Folder contains the following documents: Copy of presentation; 1984 Record of Decision (EPA); listing of communities and groups who have passed resolutions opposing dredging and sample resolution; Hudson River Program Overview; summary charts of river data; summary of GE/EPA scientific investigations in support of remedial decision making; review of contaminated sediment sites remediated in the U.S.; dechlorination of PCBs - an overview; and mortality in male and female capacitor workers exposed to PCBs. (NRC only)

119. Borden, T. 1999. Testimony to NRC on PCB Contaminated Sediments. Washington County Farm Bureau, NY. November 8, 1999. 4 pp with fax cover sheet. (NRC only)

120. Lee, C. 1999. Materials presented to National Research Council, Committee on the Assessment of Risks of Remediating PCB-Contaminated Sediment. Scenic Hudson, Inc. November 8, 1999. Folder contains the following materials: 6 fact sheets; 2 sets of comments on US EPA's Human Health Risk Assessment Reassessment RI/FS; 12 articles and reports. (NRC only)

121. Mair, A. 1999. Two documents submitted by Aaron Mair, Arbor Hill Environmental Justice Corporation: Hudson River Fish Advisories Outreach and Education Project: The Sign Development Process (April 1999; 8 pp) and The Hudson River Chronicles: Feeder canal detour an unforseen delight by Fred LeBrun in the Albany Times Union (September 25, 1998; 2 pp). November 8, 1999. (NRC only)

122. Hirschhorn, H.S. 1994. Technology Assessment - Treatment Technologies for PCB Contaminated Sediment: A White Paper on the Hudson River PCB Cleanup. Hudson River Sloop Clearwater, Inc. November 1994. 40 pp. (NRC only)

123. Schmidt-Dean, J. 1999. Comments from Judy Schmidt-Dean, Chair of the Citizen's Liaison Group of the community Interaction Program. Schuyler Yacht Basin, Schuylerville, NY. November 8, 1999. 5 pp. (NRC only)

124. Barclay, B. 1993. Hudson River Angler Survey- A report on the adherence to fish consumption health advisories among Hudson River anglers. Hudson River Sloop Clearwater. March 1993. 56 pp. (NRC only)

125. Farrington, J.W., R.W. Tripp, A.C. Davis, and J. Sulanowski. 1985. One view of the role of scientific information in the solution of enviro-economic problems. Proceedings of the International Symposium on Utilization of Coastal Ecosystems: Planning, Pollution and Productivity, 21-27 Nov. 1982. Volume 1: 73-102. (published article)

126. Farley, K.J. and R.V. Thomann. 1998. Fate and Bioaccumulation of PCBs in Aquatic Environments. Chapter 119 in Environmental and Occupational Medicine, Third Edition, William N. Rom, editor. Philadelphia, Lippincott-Raven Publishers. pps 1581-1593. (published article)

127. U.S. EPA. 1999. Use of Monitored Natural Attenuation at Superfund, RCRA Corrective Action, and Underground Storage Tank Sites. Office of Solid Waste and Emergency Response Directive No. 9200.4-17P. 40 pp. (available from EPA)

128. U.S. EPA. 1993. Guidance on Conducting Non-Time-Critical Removal Actions Under CERCLA. Office of Solid Waste and Emergency Response. EPA 540-R-93-057. August 1993. 57 pp. (available from EPA)

129. Canter, D.A., M. Lorber, C. Braverman, R.O. Warwick, J.F. Walsh, and S.T. Washburn. 1998. Determining the "Margin of Incremental Exposure": An Approach to Assessing Non-Cancer Health Effects of Dioxins. (NRC only)

130. U.S. EPA. 1999. Volume 2F - Human Health Risk Assessment: Hudson River PCBs Reassessment RI/FS. Phase 2 Report - Review Copy, Further Site Characterization and Analysis. U.S. Environmental Protection Agency, Region 2 and U.S. Army Corps of Engineers, Kansas City District. August 1999. Also available at: www.epa.gov/region02/superfnd/hudson/hhra-fulltext/pdf. (available from EPA)

131. U.S. EPA. 1999. Volume 2E - Baseline Ecological Risk Assessment: Hudson River PCBs Reassessment RI/FS. Phase 2 Report - Review Copy, Further Site Characterization and Analysis. 3 volumes. U.S.

Environmental Protection Agency, Region 2 and U.S. Army Corps of Engineers, Kansas City District. August 1999. www.epa.gov/region02/-superfnd/hudson/bera-book1-full.pdf. (available from EPA)

132. Berg, J.A. 1999. The economic and social consequences of an upper Hudson River dredge and dump project. Adirondack Regional Chambers of Commerce. November 8, 1999. 8 pp. (NRC only)

133. Fox J.M. 1999. Letter to Peter Balet, Saratoga County Environmental Management Council from Jeanne M. Fox, EPA Region 2 Administrator. November 8, 1999. 3 pp. (NRC only)

134. McCabe, W. 1999. Letter to Merrilyn Pulver, Fort Edward Town Councilwoman from William McCabe, EPA Region 2, Emergency and Remedial Response Division. October 21, 1999. 2 pp. (NRC only)

135. New York State Department of Environmental Conservation. 1999. Technical Guidance of Screening Contaminated Sediments. Division of Fish, Wildlife and Marine Resources. January 1999. 39 pp. (available from NYSDEC)

136. Newell, A.J., D.W. Johnson, and L.K. Allen. 1987. Niagara River Biota Contamination Project: Fish Flesh Criteria for Piscivorous Wildlife. New York State Department of Environmental Conservation. Technical Report 87-3. July 1987. 182 pp. (available from NYSDEC)

137. Niagara Mohawk Power Corporation. 1999. Annual Fish Tissue Sampling Program Data Report of 1998 Results: Queensbury Site, Town of Queensbury, Warren County, New York. February 1999. 67 pp. (NRC only)

138. Quensen, J. 1999. Microbial dechlorination of PCBs: Evaluation & Implications. Presentation made to Committee on Remediation of PCB Contaminated Sediments. November 8, 1999. 41 pp. (NRC only)

139. Adams, D.D. 1999. Memorandum from David D. Adams, Saratoga County Environmental Management Council, to NRC Committee. Dated December 13, 1999. 3 pp. (NRC only)

140. Svirsky, S. 1999. Housatonic River NAS Response. U.S. EPA Region 1. Dated December 22, 1999. 5 pp. (NRC only)

141. Skinner, L.C. 1993. Dioxins and furans in fish below Love Canal, New York: Concentration reduction following remediation. New York State Department of Environmental Conservation, Division of Fish and Wildlife. August 1993. 52 pp. (available from NYSDEC)

142. U.S. Army Corps of Engineers. 1992. Field notes: Saginaw Bay/River & Augres Harbor. July 1992. 93 pp. With attachments from Detroit COE, Saginaw Bay, Delivery Order #0027, July 12, 1989, Approach Channel Bay, 18 pp, and U.S. Army Corps of Engineers, Saginaw River and Saginaw Bay Channel, Sediment Sampling Methodology, undated, 87 pp. (available from COE)

143. Skip Fox. 2000. Surface sediments of the Lower Duwamish Waterway have been sampled intensively for PCBs. Handout from Skip Fox, Boeing Corporation. Dated February 2, 2000. 6 pp. (NRC only)

144. Houghton, Jon. 2000. Environmental Design Considerations for Waterfront Remediation Projects. Dated February 1, 2000. 5 pp. (NRC only)

145. Verduin, John. 2000. John Verduin of Anchor Environmental submitted the following materials with cover letter: (a) PCB Cleanup Operations: Duwamish Waterway, Seattle District, U.S. Army Corps of Engineers, 19 pp; (b) Closed bucket pilot study feasibility and evaluation for east waterway dredging, prepared by Anchor Environmental for Department of the Army, Corps of Engineers, Northwest Division, Seattle District, dated September 24, 1999, 49 pp; (c) Draft treatability dredge study for the Alcoa (Point Comfort/Lavaca Bay Superfund Site, dated January 2000, 84 pp. (NRC only)

146. Wakeman, John. 2000. The following materials were submitted by John Wakeman, U.S. Army Corps of Engineers on February 2, 2000: 9a) Old Navy Dump, Manchester, Washington, [documents available online at http://www.nws.usace.army.mil/geotech/manchest/manchest.htm] 2 pp; (b) Wyckoff/Eagle harbor, East Harbor Marine Operable Unit 1, 2 pp; (c) 1999 Environmental Monitoring Report: Wyckoff/Eagle Harbor Superfund Site, East Harbor Operable Unit, Bainbridge Island, Washington, draft dated December 3, 1999, 6 pp; (d) Operations, maintenance and monitoring plan, East Harbor Operable Unit, Wyckoff/Eagle Harbor Superfund Site, Work Plan, dated July 17, 1995, 90 pp (available from COE)

147. Verduin, J., et al. undated. Eagle Harbor West Harbor Operable Unit

Case Study: The successful implementation of a contaminated sediment remedial action. Cover page from John Verduin, Anchor Environmental, dated January 17, 2000. 16 pp. (NRC only)

148. Sumeri, Alex. 2000. Email from Alex Sumeri, U.S. Army Corps of Engineers, Seattle District, to Robbie Wedge, NRC with attachments. Attachments are: (1) Information Paper - East Waterway Channel Deepening, dated 10/01/99, 1 p; (2) Information Paper - Blair Waterway Channel Deepening, dated 10/01/99, 2 pp; (3) 1.5.1 Problem, undated, 3 pp; (4) Memorandum for Record: Supplemental determination on the stability of selected dredged material tested under the East Waterway Stage II project, September 7, 1999, 5 pp; (5) Memorandum for Record: Determination on the suitability of dredged material tested under the East Waterway Stage II project, November 2, 1999, 10 pp; (6) Appendix 2. East Waterway dredging project: PSDDA Evaluation Summary, undated, 8 pp; (7) Appendix 9: Proposed approach for developing an interim target tissue level for total PCBs based on risk to human health, December 21, 1999, 5 pp; (8) Table 1 - Calculation of allowable bottom fish tissue concentration based on risk to tribal and Asian/Pacific Islander consumers, undated, 2 pp. (available from COE)

149. Sumeri, Alex. 2000. Fax cover sheet from Alex Sumeri, U.S. Army Corps of Engineers, Seattle District, to Roberta Wedge, NRC. With attachments of: (1) Stage I Project Report: East Waterway Channel Deepening, Seattle Harbor, Washington, dated July 1998, 26 pp; and (2) JRB Associates. May 4, 1984. Removal and mitigation of sediments contaminated with hazardous substances, prepared for U.S. EPA, 11 pp. (available from COE)

150. Sumeri, Alex. Undated. Dredged material is not spoil. Status on the use of dredged material in Puget Sound to isolate contaminated sediments. U.S. Army Corps of Engineers, Seattle District. 38 pp. (NRC only)

151. Wakeman, John. 2000. Email from John Wakeman, U.S. Corps of Engineers to Alex Sumeri regarding “Advice on seismic design criteria for a confined aquatic disposal site.” January 27, 2000, 2 pp. (NRC only)

152. Monitoring Plan for Denny Way Sediment Capping: Attachment B. Submitted by Alex Sumeri, U.S. Army Corps of Engineers. February 2, 2000. 12 pp. (NRC only)

153. Monitoring Plan for Pier 53: Sediment capping site and enhanced natural recovery area. 1992. Submitted by Alex Sumeri, U.S. Army Corps of Engineers. Dated September 1992. 7 pp. (available from COE)

154. Summary and Overview (St. Paul Waterway). 1992. Submitted by Alex Sumeri, U.S. Army Corps of Engineers. February 1992. 12 pp. (available from COE)

155. Truitt, Clifford. 1986. The Duwamish Waterway capping demonstration project: Engineering analysis and results of physical monitoring. Technical Report D-86-2. U.S. Army Engineering District, Seattle. March 1986. 54 pp. (available from COE)

156. Phillips, K. and J. Malek. Undated. Dredging as a remedial method for a Superfund site. Submitted by John Malek, EPA Region 10, Seattle. February 2, 2000. 13 pp. (NRC only)

157. Monitoring Objectives (Section 1) and Approach (Section 2). Undated. Submitted by John Malek, EPA Region 10, Seattle. February 2, 2000. 7 pp (NRC only)

158. Havis, Robert. Undated. Sediment resuspension at the point of dredging. Submitted by John Malek, EPA Region 10, Seattle. February 2, 2000. 8 pp. (NRC only)

159. Marcy, K., et al. Undated. Lessons learned while monitoring a sediment cap for PAH's: Wyckoff/Eagle Harbor Superfund Site. Submitted by John Malek, EPA Region 10, Seattle. February 2, 2000. 32 pp. (NRC only)

160. Reed, Larry. 2000. Cover letter with attachments from Larry Reed, EPA Office of Emergency and Remedial Response to Robbie Wedge, NRC, dated January 28, 2000. Attachments are

    a. A summary of information supporting the decision to dredge for Manistique Harbor. A video tape is also included. Materials include "Questions and Answers on the Manistique River/Harbor time-critical removal site," 4 pp; Final Report of the Manistique Technical Review Team, 21 pp; Assessment of the Remediation Technologies, Manistique River and Harbor Area of Concern, Interagency Review Team, Final Report, 28 pp; Assessment of Technologies for Manistique River and Harbor Area of Concern, Report of U.S. Army

Corps of Engineers to the Interagency Review Team, Michael Palermo, Jan Miller. March 30, 1995, 88 pp. (NRC only)

b. Superfund Guidance Documents for Monitoring, undated, 5 pp. (NRC only)
c. Superfund Site Cleanups and Natural Resource Damages, undated, 3 pp. (NRC only)
d. PCB sites recommended for evaluation, undated, 1 p. (NRC only)
e. Sangamo Weston/Twelve Mile Creek/Lake Hartwell Site, Operable Unit 2- Pickens, SC - PCB levels, monitoring & cost information, undated, 2 pp. (available from EPA)
f. Feasiblity study report for the Sangamo Weston, Inc./Twelve Mile Creek/Hartwell Lake Site - Pickens, Pickens County, South Carolina, prepared by Bechtel Environmental, March 1994, 258 pp. (available from EPA)
g. Final Record of Decision for the Sangamo Weston/Twelvemile Creek/ Lake Hartwell PCB Contamination Superfund Site - Operable Unit Two, Pickens, Pickens County, South Carolina. June 1994. 123 pp. (available from EPA)
h. Lake Hartwell 1999 Fish and Sediment Study: Operable Unit 2 ROD Monitoring Program, October 1999. Environmental Resources Management. 54 pp. (available from EPA)
i. Final Operable Unit (OU) 13 Long-Term Monitoring Plan (LTMP), Loring Air Force Base, November 1998. 62 pp with large maps. (available from EPA)
j. Final Remedial Action Report, Source Control Remedial Action, Re-Solve, Inc., Superfund Site, North Dartmouth, Massachusetts. February 1996. 87 pp. (available from EPA)
k. St. Lawrence River Sediment Removal Project Environmental Monitoring Plan - General Motors Powertrain Site, Massena, New York. May 1995. 73 pp. (available from EPA)

161. Reed, Larry. 2000. Cover letter with attachments from Larry Reed, EPA Office of Emergency and Remedial Response to Robbie Wedge, NRC, dated March 30, 2000. Attachments are

a. Table of PCB levels and costs for four sites, with supporting summaries for each site.
b. Additional information on Sangamo Weston, Inc., site.
c. Remediation of Basewide Surface Water/Sediment (OU13), Final Loring Air Force Base Remedial Action Report, August 1999, Volumes 1 & 2.
d. 1999 Wetlands Monitoring Annual Report, January 2000 draft, Wet-

lands Management Program, Loring Air Force Base (documents monitoring of wetlands restoration).

e. Pre-and Post-Excavation Sampling, East Wetland Remedial Area (map with tables), Re-Solve Site, January 16, 1996.

162. Fox River Group. 2000. Dredging-related Sampling of Manistique Harbor - 1999 Field Study. Prepared by Blasland, Bouck & Lee, Inc. June. 45 pp.

163. Fox River Group. 2000. Fox River Dredging Demonstration Projects at Sediment Deposit N and Sediment Management Unit 56/57: Environmental Monitoring Report. Prepared by Blasland, Bouck, & Lee, Inc. June.

164. General Electric Company. 2000. Environmental Dredging: An Evaluation of its Effectiveness in Controlling Risks. Prepared with assistance from Blasland, Bouck & Lee, Inc. August. 98 pp.

165. Scenic Hudson. 2000. Accomplishments at Contaminated Sediment Cleanup Sites Relevant to the Hudson River: An Update to Scenic Hudson's Report *Advances in Dredging Contaminated Sediment*. Draft report prepared by Joshua Cleland. September 2000.

166. Wisconsin Department of Natural Resources. 2000. Fox River Deposit N Report: Evaluation of the Effectiveness of Remediation Dredging, November 1998-January 1999. Prepared by the Fox River Remediation Advisory Team, Madison, WI. Available at www.dnr.state.wi.us/org/water/wm/lowerfox/sediment/frratdepositnreport.

Appendix D

Case Studies

ST. PAUL WATERWAY PROBLEM AREA
TACOMA, WASHINGTON

The St. Paul Waterway Area Remedial Action and Habitat Restoration Project is an example of how sediment pollution control may be successfully coupled with natural-resource restoration, allowing for a healthy maritime economy and marine ecology. The project is also an example of how early and open agency, tribal, and public consultation can result in community consensus and support for the environmental cleanup and restoration plan that emerges from these discussions.

Background

In 1983, the Commencement Bay Nearshore Tideflats were placed on the priority list of sites requiring investigation and cleanup under the EPA's Superfund authorities. In 1985, the Simpson Tacoma Kraft Company purchased a pulp and paper mill located at the mouth of the Puyallup River and within the federal Superfund site. There was extensive sediment contamination just off-shore from the plant site, resulting from 6 decades of mill operations that poured untreated effluent into Commencement Bay directly in front of the plant and created a 17-acre "hot spot" of toxicants (mostly phenols)—the St. Paul Waterway Problem Area (one of eight "problem areas" in the Superfund site requiring remediation).

Upon acquiring the mill in 1985, Simpson assumed a National Pollution

Discharge Elimination System (NPDES) permit and an administrative order issued by the Washington State Department of Ecology (Ecology), which required the construction of a new outfall. Encouraged by Ecology to address several environmental problems at the same time, Simpson began in 1986 to investigate and implement better control of sources of pollution at the mill and, together with Champion International Corporation (who acquired the mill as a result of a merger with the St. Regis Corporation shortly before its sale to Simpson), to plan remedial action for the contaminated sediments. This remedial action planning for the St. Paul Waterway Problem Area proceeded in parallel with the federal remedial-investigation/feasibility-study (RI/FS) process for the Commencement Bay Nearshore/Tideflats Superfund site.

The project was the first completed Superfund cleanup in U.S. marine waters and the first natural-resource damages settlement in the United States without litigation and with all federal, state, and tribal trustees (EPA 1991). It addressed tribal fishing rights, Section 404, Endangered Species Act, Coastal Zone Management, and Growth Management issues, as well as sediment cleanup.

Strategy Chosen

Before proposing any actions, Simpson and Champion consulted with the Puyallup Tribe; environmental groups and interested citizens; federal, state, and local officials; and agency staff beginning in January 1987. Out of these discussions and after analysis of many remedial technologies, as well as their effectiveness in—and impact on—marine waters, a comprehensive environmental cleanup and restoration approach emerged (Weiner 1991). The approach included

- A new outfall for the secondary treatment plant.
- Permanent isolation of the contaminated sediments from marine life by capping the area with clean sediments from the nearby Puyallup River.
- Habitat restoration and enhancement of nearshore and intertidal areas, including a long-term monitoring and adaptive management plan.
- Preventive measures against future sediment contamination from the cap and the mill, including source control within the mill, monitoring, and contingency plans.

After site preparation and source-control actions initiated in December 1987, the 17-acre area was capped with clean sediment in July and August of 1988. The cleanup action was integrated with natural-resource restoration to produce new intertidal and shallow-water habitat in Commencement Bay, an area that had lost about 90% of such habitat over the last 100 years. This new habitat was achieved through a cap that ranged in thickness from approximately 4 to 20 feet, depending on the area being capped and the desired tide-flat habitat elevations (Weiner 1991; Sumeri 1989, Sumeri et al 1994). The habitat restoration was based on an assessment of landscape-scale needs in the estuary and led to a more comprehensive restoration planning effort for Commencement Bay and the lower Puyallup River watershed in the 1990s.

Overall Assessment

As described above, the project is designed to provide (1) permanent isolation from the environment of chemical contamination found in marine sediments; and (2) restoration of intertidal and shallow-water habitat. Project monitoring, which included physical, chemical, and biological studies, occurred before, during, and after project construction. Monitoring before construction helped not only with project design but also with establishing baseline conditions for evaluating future monitoring results. Monitoring during and immediately after construction ensured pollution control and verified that the remedial work conformed to the remedial design.

In general, the tenth year confirmational monitoring results indicate that the project and new habitat are both functioning as planned (Parametrix 1999). The project provides habitat for diverse biological communities of benthic and epibenthic organisms, as well as algae. Shorebirds and salmon use the site for feeding and rearing, and tide pools observed at low tide are abundant with invertebrates. Productive shoreline habitat now exists at the project site where essentially no productive habitat existed prior to project construction. The cap-sediment elevations had minimal changes, with only minor redistribution of materials at higher intertidal levels. The project is now in the contingency monitoring phase of its adaptive management program. The project continues to be one of the lowest cost cleanups in Commencement Bay and one of the most cost-effective sediment cleanups undertaken in Puget Sound.

Lessons Learned

- Sediment remediation projects can present opportunities for integrating natural-resource protection and economic redevelopment with sediment pollution control.
- Community support is critical to obtaining permits and other regulatory approvals in a timely manner.

- Community support is earned through early and open consultation and a role in shaping the cleanup strategy that will be chosen for the site.
- Site-specific factors play an important role in determining appropriate and effective technologies for a given site.
- In situ capping might be an effective technology for the sediment remediation and habitat restoration.
- Long-term confirmational monitoring and an adaptive management program are necessary to ensure that a cleanup performs as designed.
- Sediment cleanup can be successfully integrated with water quality, natural-resource damages, tribal fishing, brownfields, and growth-management issues to produce cost-effective, environmentally sustainable solutions in existing industrial harbor areas.

ASARCO TACOMA SMELTER

The Asarco Tacoma Smelter is located along Commencement Bay in Ruston, Washington, a small town surrounded by the metropolitan city of Tacoma. Owned by ASARCO, Inc., the smelter processed lead and copper for close to a century. The plant released sulfur dioxide gases and dust particles (containing arsenic and other metals) into the air. Much of the dust settled nearby contaminating the soil and the waters of Commencement Bay. ASARCO poured hot slag, a waste product of the smelting process (containing lead, arsenic, copper, and other metals), into Commencement Bay to cool and harden, creating an artificial shoreline. Some slag was cooled on land, resulting in a black, rock-like material that was sold to residents and businesses in the community for landscaping purposes, driveways, sandblast grit, fill, and other purposes. The cleanup involved smelter demolition, site and marine cleanup, and residential soil cleanup.

An expedited action to clean up 11 of the most seriously contaminated properties showed Region 10 a hint of the problem it faced. The property owners refused access because they did not think any response was necessary. Residents were distrustful of government and its warnings of health risks with no irrefutable causal link. And they were loyal to ASARCO, the town's sole tax source, which had employed the residents for generations. When the company made the business decision to close and move overseas, the Ruston community blamed EPA for the job losses.

Cleanup of hundreds of residential yards still needed to be done. Region 10 management realized that if it did not change and find a way to break through the resistance, the community would block the work. According to Community Involvement Coordinator Michelle Pirzadeh, "regardless of how good a recommended solution is, it cannot be implemented without getting people on board. To prove ourselves we needed to visibly address concerns, follow through on promises and build trust—with no surprises."

The Ruston/North Tacoma Community Workgroup formed in 1990 when the planning for the residential cleanup began. The workgroup included community members and met on a monthly basis until the start of cleanup activities in 1993. Initially, the workgroup was somewhat negative and reactive, according to Mary Kay Voytilla, remedial project manager for the residential area. Workgroup members would not actively engage in discussions about how the cleanup could best meet the needs of the community. Instead they questioned the need for the cleanup. Charlene Hagan, Town Council member, described the relationship between EPA and the community as adversarial. "EPA was heavy-handed and the community operated like a lynch mob."

The members of the site team recognized it would have to redouble its efforts to reach out and include the community in cleanup. They became more personally involved and began knocking on doors, interviewing residents on the process and the type of involvement desired. They held workshops and created opportunities for dialogue, such as one-on-one availability sessions. For the first time, residents began to feel that EPA really was listening to and interested in what they had to say. The site team took other steps to reach out and get involved in the community.

- Clayton Johnson, a Ruston resident, was asked to be the community liaison and serve as EPA's eyes and ears at key meetings. He answered residents' questions and help to facilitate issues even though some community members still saw him as an EPA spy.
- EPA formed the Ruston/North Tacoma Coordinating Forum to facilitate discussion and coordination among the various agencies and organizations involved in or affected by the residential cleanup. The forum assisted in the development and selection of a remedy that would be implementable in the community.
- The site team stopped holding large meetings. Instead they used smaller public meetings in residents' homes.
- Fact sheets, mailers and a Residential Soils Bulletin on the progress of the cleanup were routinely distributed to residents, property owners, businesses, and schools. Under a cooperative agreement, the Tacoma-Pierce County Health Department developed two brochures related to the handling and disposal of contaminated soil.
- The region provided informational brochures and bankers' seminars for professionals involved in property transactions. They wrote up typical real estate questions and answers related to property values and set up a database for property transactions.

- The site team showed its commitment to the community by volunteering at a local fund raising event to be dumped in a water tank.

Lessons Learned

- Community buy-in is critical. Regardless of how good a technical solution might be, it will not be successful without community support.
- Resistance is not the end of the road. Persist in building relationships and proactively reaching out to break through suspicion and opposition.
- Create a community liaison. A resident who is willing to serve as a go-between and facilitator is an invaluable resource.

KODAK COMPANY FACILITY

Kodak Park, the largest industrial complex in the northeast United States, is located on more than 1,300 acres and stretches for nearly 4 miles through the city of Rochester and the town of Greece, New York. Much of its 22-mile perimeter borders residential neighborhoods. Approximately 13,000 households and 550 businesses are located close enough to the site to be considered plant neighbors.

Until 1988, the site operations relied on the corporation's reputation and general community networks to address most community relations issues. Early that year, groundwater contamination at the site brought on a tidal wave of negative press and neighborhood concerns. Local neighborhood groups demanded that action be taken by local government, regulatory agencies, and of course, the company.

Initial Actions Taken

A special Neighborhood Relations Team was established at the site. Its mission was to establish a direct communication path between Kodak Park and its neighbors. Reporting to senior site management, the team had access to and support from company community relations, communications, and health and environmental resources.

A Neighborhood Information Center was opened in the neighborhood. The center was easily accessible to the public and, as the name implies, provided health, safety, and environmental information to neighbors. It also became a resource center for subsequent community programs addressing issues associated with recovery.

KP management and the Neighborhood Relations Team conducted a series of community meetings to share information with neighbors and get direct input regarding their concerns.

A survey was conducted to ascertain the issues and concerns of the neighbors. The survey results were evaluated in detail and incorporated into an action plan to address neighborhood concerns. Neighbors were invited to participate in open forums to discuss the plan and provide additional input into it.

A Neighborhood Leaders Council was formed with representatives of all local home associations, special interest groups, and Kodak Park. The council tracked progress on resolving environmental and related issues.

As in any issue dealing with contamination, questions about possible health effects arose. Kodak worked with local public-health agencies to collect exposure data that could be used to address these concerns and provided funds to enable the community to hire an independent, third-party scientist to validate health and environmental data and conclusions. Thus, the neighbors were not dependent on the company or public officials to interpret the study results.

A second major issue that surfaced was concern about property values. A Value Protection Plan was put into place that guaranteed the value of housing in the neighborhood for 10 years and allowed homeowners to stay in the neighborhood or move without financial penalty. This plan was accomplished through low interest loans for home financing, home improvement grants and loans, and reimbursement of relocation and real-estate transactions costs.

To maintain communications with neighbors, a newsletter was mailed out on a monthly basis to give a status report on progress.

12 Years Later

The Neighborhood Relations Team has developed into the Neighborhood Relations Office, a permanent group responsible for the management of liaison and issues associated with all neighborhoods bordering the company's three Rochester plant sites.

Each plant site has developed and maintains strategic neighborhood relations plans, which outline four primary goals:

- Build public confidence.
- Generate external support to facilitate plant-site success.
- Gain internal commitment to minimize adverse operational impact on the community.
- Reinforce the plant site as a favorable community asset.

The Neighborhood Relations Office's annual operating plans, which support the remediation strategies, are in-place and tracked for status and results.

The Kodak Park Neighborhood Leaders Council continues to meet twice monthly to maintain an ongoing dialogue and relationship with plant and community leaders.

In addition, a Community Advisory Council (CAC) has been established at each of the three plants. Their membership includes representatives from local government, schools, community associations, and other special interest groups near the plants. The primary purpose is the exchange of information and opinions regarding community and company interests.

Kodak Park continues to publish a neighborhood newsletter (*The Update*) to keep neighbors, local businesses, and employees informed of environmental, manufacturing, and local community issues associated with the plant. Five issues are mailed to 13,000 nearby households, 550 businesses, and 15,000 employees.

Each year, Kodak Park conducts two local neighborhood opinion surveys, one written (received by every neighbor and Kodak Park employee) and the other a statistically random telephone survey. The surveys ask for opinions and ratings on the site's performance in communication, accessibility to the public, minimizing operational impact, and environmental responsibility. The results of the survey, combined with input from other community forums, are used to assist with Neighborhood Relations Office planning.

The Bottom Line

Involvement in this situation has shown Kodak that becoming a "neighbor of choice" is not just a lofty goal for business and industry. In today's environment of high operating costs, extreme competitive pressure, and public participation, a company cannot operate effectively without the consent of those who live near a plant's borders.

Earning public trust and sustaining a positive relationship with neighbors can only be realized through comprehensive planning and a commitment at all levels to progress. Like any relationship, it's hard work but the rewards far outweigh the effort.

Kodak Park and its neighbors have benefitted from the mutual willingness to communicate and address operational issues before they become a crisis. In the process, Kodak Park management and employees have become much more aware of the issues and interests of neighboring communities. In many ways, Kodak Park and its neighbors regard each other as partners, supporting each other's plans, and benefitting from each other's successes.

The Prince William Sound Regional Citizens' Advisory Council (RCAC)

It is generally accepted that environmental protection measures are more likely to succeed if the affected parties and in particular the affected community are involved in the decision-making process. Too often, community involvement is a token involvement limited to having the public informed and allowed to submit comments during specified periods. A truly pro-active approach to community involvement in environmental protection is exemplified by The Prince William Sound Regional Citizens' Advisory Council (RCAC).

The RCAC is an independent nonprofit corporation formed as a result of the 1990 Oil Pollution Act. The act established two demonstration projects in Alaska, one in Prince William Sound and the other in Cook Inlet, both designed to promote partnership and cooperation among local citizens, industry, and government. Another important objective of the act was building trust and providing citizen oversight of environmental compliance by oil terminals and tankers. The RCAC's 18 member organizations are communities in the region affected by the 1989 Exxon Valdez oil spill. These include groups involved in aquaculture, commercial fishing, environment, recreation, and tourism.

The mission of the RCAC is to promote environmentally safe operation of both the Alyeska Pipeline Service Company terminal in Valdez and the oil tankers that use it. The RCAC's structure and responsibilities are derived from a contract with Alyeska, which operates the trans-Alaska pipeline as well as the Valdez terminal. Under this contract, the council receives funding from Alyeska to service the public. The second guiding document, enacted after the council was created, was the federal Oil Pollution Act of 1990, which required citizen oversight councils for Prince William Sound and Cook Inlet.

Each year, the U.S. Coast Guard assesses whether the RCAC fosters the general goals and purposes of the Oil Pollution Act and whether the council is broadly representative of the community's interests as envisioned in the act. As the council for Prince William Sound pursuant to the act, the RCAC advises and makes recommendations on policies, permits, and regulations relating to the oil terminal and tankers. The RCAC monitors, comments, and makes recommendations regarding Alyeska's oil-spill-response and prevention plans and associated operations, prevention and response capabilities, and environmental protection capabilities. The council also reviews and monitors the actual and potential environmental impacts of terminal and tanker operations in Prince William Sound. The RCAC also recommends standards and modifications for terminal and tanker operations to minimize the risk of oil spills and other environmental impacts and to enhance prevention and response capabilities. As part of its activities, the RCAC participates in the monitoring and assessment of environmental, social, and economic consequences of oil-related accidents. The council also comments on and participates in the selection of research and development projects.

The RCAC was initially funded at $2 million per year. This funding is renegotiated every 3 years and is at a level of $2.5 million per year. Although the council works closely with Alyeska, which funds the council, its independence is specified as a condition of its contract with Alyeska.

The RCAC has been in operation for about a decade. Its positive impact on environmental protection in Alaska is now well-recognized. It has aided in ensuring that health risk assessment and the monitoring of studies in Port Valdez are conducted at the highest level of scientific scrutiny. It has conducted a pro-active program of public participation and information dissemination to ensure that the sensitive region is utilized in an environmentally responsible manner by both industry and the region's citizens.

REFERENCES

EPA(U.S. Environmental Protection Agency). 1991. News Release regarding lodging of consent degree for St. Paul Waterway Area Remedial Action and Habitat Restoration Project. June 24, 1991

Parametrix. 1999. St. Paul Waterway Area Remedial Action and Habitat Restoration Project. 1998 Monitoring Report. Unpublished report to Simpson Tacoma Kraft Company, Tacoma, Washington, and Champion International, Stanford, Connecticut for the Washington State Department of Ecology and U.S. Environmental Protection Agency. Parametrix, Seattle, Washington.

Sumeri, A. 1989. Confined Aquatic Disposal and Capping of Contaminated Bottom Sediments in Puget Sound. Proceedings of the WODCON XII, Dredging: Technology, Environmental, Mining, World Dredging Congress, Orlando, FL, 2-5 May, 1989.

Sumeri, A., T.J. Fredette, P.G. Kullberg, J.D. Geermano, D.A. Carey and P. Pechko. 1994. Sediment Chemistry Profiles of Capped Dredged Sediment Deposits Taken 3 to 11 Years After Capping. Dredging Research Technical Note. DRP-5-09. Vicksburg, MS: U.S. Army Engineer Waterways Experiment Station. May.

Weiner, K.S. 1991. Commencement Bay Nearshore/ Tideflats Superfund Completion Report for St. Paul Waterway Sediment Remedial Action. Submitted to the U.S. Environmental Protection Agency for Simpson Tacoma Kraft Company and Champion International Corporation. January 1991.

Appendix E

PCB Biodegradation

PCBs are not a single compound. They are a class of compounds, all of which contain the biphenyl two-ring structure with 1 to 10 chlorine atoms attached to each molecule. The chlorine atoms replace some or all of the 10 hydrogen atoms on the biphenyl molecule. As such, there are 209 possible PCB congeners, each having its own physical and chemical properties and potential for biodegradation. In general, those with fewer chlorine atoms tend to be more readily biotransformed under aerobic conditions, and the higher chlorinated congeners are more readily biotransformed under anaerobic conditions. The potential for biodegradation is a function not only of the number of chlorine atoms on a given PCB molecule but also of their placement. PCB congeners with the chlorine atom on the ortho carbon (that ring position closest to the bond connecting the two rings) tend to be more difficult to biotransform than those with the chlorine atom in the meta or para positions, the ones farther away from the connecting bond. Various reviews have been written on PCB biodegradation (Brown et al. 1987; Abramowicz 1990; Boyle et al. 1992; Higson 1992; Mohn and Tiedje 1992; Haluska et al. 1993; Tiedje et al. 1993).

Aerobic Biotransformation

Aerobic biotransformation of PCBs is similar to aerobic biodegradation of the biphenyl molecule itself. The first step in this process is oxidation of the ring by a 2,3-dioxygenase, which substitutes two hydrogens with two hydroxyl groups at adjacent ortho and meta positions on the molecule. The ring is then cleaved through the assistance of another dioxygenase at either of two locations. The opened ring is then further metabolized by well-known pathways. PCB congeners containing several chlorine groups, especially in the 2 (ortho) or 3 (meta) positions, can block the first oxygenation of a PCB. Other researchers (Bedard and Haberl 1990) have shown that a 3,4-dioxygenase might also be effective, thus permitting somewhat broader susceptibility of PCB congeners to aerobic biodegradation. Because chlorine atoms on the PCB ring effectively block the action of the oxygenating enzymes, only PCBs with relatively few chlorine atoms are readily susceptible to aerobic biodegradation.

Anaerobic Biotransformation

Anaerobic biotransformation of PCBs is significantly different from aerobic biodegradation (Tiedje et al. 1993) and is most effective with more highly chlorinated PCBs. Under anaerobic conditions, PCBs are transformed by reductive dehalogenation. Here, a chlorine atom is removed from the molecule and substituted with hydrogen. Reductive dehalogenation of organic molecules has become recognized in recent years as a general process that is effective for dehalogenating a variety of halogenated organic compounds, from pesticides and many aromatic compounds, such as PCBs, to aliphatic compounds, such as chlorinated solvents (Holliger et al. 1998; Tiedje et al. 1993). In such reductions, the haloorganic serves as an electron acceptor, the role that oxygen serves under aerobic conditions. Such dehalogenation might be a fortuitous event brought about by enzymes that are designed for other purposes. In this case, the process is called a cometabolic one. However, in some cases, organisms can use the electron-accepting potential of haloorganics in energy metabolism, in which case the process is called dehalorespiration (Holliger et al. 1998) or simply halorespiration. Dehalorespiration has been demonstrated in the case of biotransformation of chlorobenzoates (Dolfing and Tiedje 1987) and tetrachloroethene (Holliger et al. 1993). The electron donor for these transformations frequently is molecular hydrogen (H_2), which is commonly formed as an intermediate in normal anaerobic fermentation of complex organic materials. In other cases, simple organic compounds, such as acetate or lactate, might serve as the electron donor. Dehalorespiration tends to be a more efficient dehalogenating process because the organisms can grow catalytically by the process; reaction rates tend to be higher, and the use of available electron donors is more efficient. Whether cometabolic or through dehalorespiration, reductive dehalogenation generally requires the presence of other organic matter, such as decaying vegetable matter, to provide the electron donor required for the process to occur.

Reductive dehalogenation of PCBs was first recognized as an important process by Brown et al. (1987). Whether the reductive dehalogenation that occurs is by cometabolism or dehalorespiration, or a mixture of the two, is not well known. Dehalogenation of meta chlorines is generally favored over para chlorines, which in turn are favored over ortho chlorines (Tiedje et al. 1993). There are reports of ortho dechlorination (Natarajan et al. 1996). However, there is general doubt that reductive dehalogenation will lead to complete chlorine removal from congeners with chlorine atoms in the ortho position. However, congeners with chlorine atoms only in the meta or para positions can be completely dehalogenated under proper conditions to biphenyl.

Natarajan et al. (1999) studied anaerobic biodegradation of biphenyl by a methanogenic consortium. They reported complete conversion of the biphenyl molecule to carbon dioxide and methane. This conversion occurred whether biphenyl was added as the sole carbon source or as a cosubstrate along with other glucose and methanol.

Complete dehalogenation of all congeners in typical PCB formulations has not yet been demonstrated. Different PCB dehalogenating organisms have different abilities to dehalogenate, and thus it is perhaps not surprising that PCB dehalogenating patterns have been found to be different with different sediments (Tiedje et al. 1993). Some were found primarily to remove meta chlorines, others para chlorines, and some both meta and para chlorines. Combinations of sediments with their different organism types would thus appear to be capable of more complete dehalogenation. However, mixing two sediments with different dehalogenating abilities did not prove more efficient except when they were used sequentially (Tiedje et al. 1993).

Factors Affecting Transformation Rates

Tiedje et al. (1993) suggested various factors that might affect transformation rates: PCB concentration, bioavailability, inhibitors, temperature, and nutrients. Of course, the potential and rate also depend on the microorganisms that happen to be present at a given site. Bioavailability refers partly to the fact that PCBs are highly insoluble and sorb strongly to sediments, making them less available to bacteria for dehalogenation. At times, PCBs are mixed with mineral oils that also make them less bioavailable. More highly chlorinated congeners tend to partition more strongly onto soils so that solution concentrations are less than those for less highly chlorinated species. This partitioning process also means that the more highly chlorinated species have less tendency to move from sediments to overlying waters. Transformation rates are generally directly proportional to solution concentrations rather than total sediment concentrations. Solution concentrations are generally in the low nanogram per liter concentration for most individual congeners. For this reason, transformation rates can be very slow. Quensen et al. (1988) indicated that optimal rates of reductive dehalogenation of PCB occurred for concentrations in the range of several hundred to thousands of parts per million in sediments. Below 50 ppm, rates often are very slow or negligible. A similar lower threshold concentration of 35 to 45 ppm was reported by Sokol et al. (1998). The latter indicated that at higher concentrations, removal rates of complex PCB mixtures, such as Arochlor 1248, were linear initially, but eventually a plateau was reached at which no further dechlorination resulted. This plateau appeared to occur once the meta chlorines had been removed, leaving the more resistant ortho chlorines. It is unclear whether there is an absolute threshold for PCBs. If cometabolism were involved and sufficient food for growth of dehalogenating organisms were available, dehalogenation might continue. It would continue, however, at rates that are so slow that they are difficult to measure, but they might still be significant over periods of years. In any event, complete anaerobic dehalogenation of PCBs in sediments has not yet been reported in the peer-reviewed literature.

Combined Anaerobic and Aerobic Biodegradation

Because anaerobic conditions are best for decomposition of more-chlorinated PCBs and aerobic conditions for less-chlorinated PCBs, it would appear that sequential anaerobic and aerobic treatment could result in complete PCB degradation. For example, anaerobic degradation in sediments would lead to the formation of less-chlorinated congeners that are more mobile and would diffuse into overlying aerobic waters in a river, where aerobic degradation could occur (Tiedje et al. 1993). Because of the slowness of the process, this route of complete decomposition has not been adequately demonstrated at given sites. Generally, when PCB concentrations are high in surface sediments, they are also high in fish. Although complete destruction of some congeners might result from this combination of processes, success in protecting fish from PCB contamination has not been demonstrated adequately.

References

Abramowicz, D.A. 1990. Aerobic and anaerobic biodegradation of PCBs: a review. Crit. Rev. Biotechnol. 10(3):241-251.

Bedard, D.L., and M.L. Haberl. 1990. Influence of chlorine substitution pattern on the degradation of polychlorinated-biphenyls by 8 bacterial strains. Microb. Ecol. 20(2):87-102.

Boyle, A.W., C.J. Silvin, J.P. Hassett, J.P. Nakas, and S.W. Tanenbaum. 1992. Bacterial PCB biodegradation. Biodegradation 3(2/3):285-298.

Brown, J.F., Jr., R.E. Wagner, H. Feng, D.L. Bedard, M.J. Brennan, J.C. Carnahan, and R.J. May. 1987. Environmental dechlorination of PCBs. Environ. Toxicol. Chem. 6(8):579-594.

Dolfing, J., and J.M. Tiedje. 1987. Growth yield increase linked to reductive dechlorination in a defined 3-chlorobenzoate degrading methanogenic coculture. Arch. Microbiol. 149(2):102-105.

Haluska, L., S. Balaz and K. Dercova. 1993. Microbial degradation of polychlorinated-biphenyls. Chemicke Listy 87(10):697-708.

Higson, F.K. 1992. Microbial degradation of biphenyl and its derivatives. Adv. Appl. Microbiol. 37:135-164.

Holliger, C., G. Schraa, A.J Stams, and A.J. Zehnder. 1993. A highly purified enrichment culture couples the reductive dechlorination of tetrachloroethene to growth. Appl. Environ. Microbiol. 59(9):2991-2997.

Holliger, C., G. Wohlfarth, and G. Diekert. 1998. Reductive dechlorination in the energy metabolism of anaerobic bacteria. FEMS Microbiol. Rev. 22(5):383-398.

Mohn, W.W., and J.M. Tiedje. 1992. Microbial reductive dehalogenation. Microbiol. Rev. 56(3):482-507.

Natarajan, M.R., W.-M. Wu, J. Nye, H. Wang, L. Bhatnagar, and M.K. Jain. 1996. Dechlorination of polychlorinated biphenyl congeners by an anaerobic microbial consortium. Appl. Microbiol. Biotechnol. 46(5-6):673-677.

Natarajan, M.R., W.-M. Wu, R. Sanford, and M.K. Jain. 1999. Degradation of biphenyl by methanogenic microbial consortium. Biotechnol. Lett. 21(9):741-745.

Quensen, J.F. III, J.M. Tiedje, and S.A. Boyd. 1988. Reductive dechlorination of polychlorinated-biphenyls by anaerobic microorganisms from sediments. Science 242(4879):752-754.

Sokol, R.C., C.M. Bethoney, and G.Y. Rhee. 1998. Effect of Aroclor-1248 concentration on the rate and extent of polychlorinated biphenyl dechlorination. Environ. Toxicol. Chem. 17(10):1922-1926.

Tiedje, J.M., J.F. Quensen III, J. Chee-Sanford, J.P. Schimel, and S.A. Boyd. 1993. Microbial reductive dechlorination of PCBs. Biodegradation 4(4):231-240.

Appendix F

Methods of Analysis of PCBs in Sediments, Water, and Biota

METHODS OF QUANTIFICATION

A number of methods are available to quantify concentrations of PCBs in sediments and tissues. Each technique has advantages and disadvantages. The resources and equipment required vary greatly among methods. The method selected for use in quantifying PCBs, either in the exposure quantification phase or in the subsequent monitoring phases, should be matched to the requirements of the analysis in which the data will be used. Some of the methods are sufficiently robust to allow congener-specific analyses while allowing the generation of total concentrations of PCBs that can be compared with historical information. Below is an overview of the available methods and their utility for use in risk assessment and monitoring efforts.

Gas-Chromatography-Based Methods

The gas chromatography schemes that are used in the quantification of PCBs can generally be divided into two types: PCB quantification based on analogy to Aroclor technical mixtures is referred to as the Aroclor method and that based on quantification of individual PCB congeners present in samples is referred to as the congener-specific method. The typical Aroclor analysis is based on estimating residue PCB concentrations by measuring a subset of PCB congeners or by approximating the profile as an appropriate set of technical mixtures (Newman et al. 1998). In the United States, PCB water-quality criteria established by the Clean Water Act (Title 40 CFR part 125.58) to protect humans and wildlife are based on Aroclor values (EPA 1995). Unfortunately, the environmental weathering of Aroclors modulates mixture toxicity (Quensen et al. 1998). As such, carcinogenic risk-assessment guidelines recommend the calculation of congener-specific or total PCB data when available (EPA 1994c). Congener-specific analyses utilize the direct quantification of each unique PCB congener. The result is a precise description of PCB profiles, which can highlight physiological, spatial, and temporal changes that might not be apparent in Aroclor values. Both the Aroclor and the congener-specific methods of PCB analysis rely on gas chromatography.

Gas-chromatography methods require the extraction of PCB from the environmental matrix and usually require the cleanup of the PCB extract by one or more methods before analysis. A gas chromatograph (GC) is used to separate individual PCB congeners or combinations of congeners based on physical properties, such as volatility and polarity. Over the past 20 years, open tubular capillary GC columns have replaced older packed GC columns for routine laboratory work. The capillary GC columns offer improved resolution, better selectivity, and increased sensitivity compared with packed GC columns. For that reason, many standard EPA methods have been rewritten to allow the use of capillary GC columns for the analysis of environmental samples. The use of open tubular capillary GC columns in a GC system is termed high-resolution gas chromatography (HRGC). All the analytical methods presented below for PCB analysis use HRGC.

Within a GC system, PCBs are then detected with electron capture detection (ECD), electrolytic conductivity detection (ELCD), or mass spectrometry (MS). For MS detection of PCBs, two systems are considered: low- or medium-resolution MS and high-resolution MS. The data output of a GC is called a chromatogram and individual PCBs are identified as peaks on the chromatogram. For GC-ECD and GC-ELCD, PCBs are identified by order of elution from the GC, also called retention time. For MS systems, PCBs are also identified by molecular mass. Certain coplanar PCBs, which are known or suspected carcinogens, are sometimes measured with high-resolution mass spectrometry (HRMS) techniques originally developed for the analysis of dioxins and furans in environmental samples. The HRMS technique provides lower limits of detection for PCB congeners but requires additional sample cleanup beyond that normally required and requires special instrumentation, the high-resolution mass spectrometer.

The utility of the two GC-based methods of PCB analysis (Aroclor and congener-specific), based on costs and benefits, is summarized in Table F-1. These two methods are discussed in detail in the following sections.

TABLE F-1 Utility of Aroclor and Congener-Specific Methods of PCBs Analysis

Analysis Method	Cost	High Utility	Low Utility
Aroclor analysis	Low	Identification of neat mixture Preliminary site screening; soil Preliminary site screening; sediment PCB quantification Soils and sediments a. single PCB mixture b. low organic carbon c. short residence time	Biological samples Weathered PCBs
Congener-specific analysis	High	Biological samples Weathered PCBs Data utilized for toxicology analysis	

Aroclor Method of PCB Quantification

For many years, PCB concentrations in environmental media have been quantified by comparison to Aroclor standards. Early gas chromatography–electron capture detection (GC-ECD) methods, such as the EPA's solid waste 846 (SW 846) methods 8080 and 8081 (EPA 1994a,b), relied on comparing the chromatographic pattern of peaks in the environmental sample with the pattern or number of peaks in a series of pure Aroclor standards. However, due to biological processes, such as biodegradation and metabolism, or environmental processes, such as volatilization, congener profiles of PCBs in environmental samples are different from technical Aroclor standards. The alteration of the congener pattern in environmental samples can be influenced by several factors, including source and type of PCBs, type of environmental media (air, water, soil, sediment, and biota), physical-chemical properties of the media (temperature, pH, organic carbon content), the congeners present in technical mixtures, and the type and abundance of microfauna and flora. Therefore, the methods might not be applicable to environmental samples containing weathered Aroclors or complex mixtures of Aroclors.

Aroclor analyses are estimations that are prone to error. Two important sources of variance in the methodology as defined by EPA are the subjective assignment of Aroclor speciation and response factors (Draper et al. 1991) and the assumption that Aroclor response factors are representative of weathered profiles. As an example, determination of the concentrations of PCBs based on the COMSTAR algorithm, a statistical procedure to determine total PCBs based on marker congeners and peak ratios (Burkhard and Weininger 1987), overestimated the concentrations determined by summing the concentrations of individual congeners. This artifact occurs because COMSTAR estimates total concentrations of unweathered PCB congeners that would have been present in the original technical Aroclor mixtures (Giesy et al. 1997). Similarly, several other computer programs incorporating multivariate analysis have been developed to normalize PCB concentrations in samples to those in Aroclor standards (Stalling et al. 1985). Despite that, the Aroclor method does not adequately represent the concentrations found in weathered environmental samples. The discrepancies in the congener composition between the commercial mixture and real-world environmental exposures imply that the predictive value of studies based on commercial mixtures might be limited with respect to estimating risks from environmental exposure. Determination of all PCB isomers and congeners present in environmental matrices using a mixture of technical PCB preparations (such as an equivalent mixture of Aroclors 1016, 1242, 1254 and 1260) as standards might provide a better estimate of PCB concentrations for use in risk assessment.

The advantage of the Aroclor method is its simplicity, cost, and comparability to historical and contemporary PCB concentrations.

Detection Limits and Method Performance

The method detection limits (MDLs) for Aroclors vary in the range of 0.054 to 0.9 mg/L in water and 57 to 70 ng/g in soils, with higher MDLs for the more heavily chlorinated Aroclors. Estimated quantification limits for PCBs as congeners vary by congener in the range of 5 to 25 ng/L and 160 to 800 pg/g in soils, the higher values being for the more heavily chlorinated congeners.

Cost

Aroclor analysis is relatively inexpensive in comparison to congener-specific methodologies. The cost on a per sample basis ranges from $50 to $500, depending on the sample matrix, number of samples to be analyzed, and extent of quality-assurance–quality-control protocols.

Implications of Aroclor Analysis in Risk Assessment

The differences in the composition of PCB residues in environmental matrices have implications not only for quantification but also in hazard evaluation, particularly when considering the differences in the biological activity, both qualitatively and quantitatively, among isomers as well as congeners. Several studies have demonstrated the differences in both mechanisms and toxic potentials of individual PCB congeners (see Giesy and Kannan 1998). Thus, the impacts of PCBs on the environment and biota are due to the individual components of these mixtures and the additive and/or nonadditive (synergistic and antagonistic) interactions among them and other chemical classes of pollutants. Therefore, the development of scientifically based regulations for the risk assessment of PCBs requires analytical and toxicological data on the individual PCB congeners present in any technical mixture and information regarding interactive effects. Although developments in high-resolution isomer-specific PCB analysis have enabled identification and quantification of individual PCB congeners present in commercial mixtures and environmental samples, there are important challenges associated with risk assessment of PCBs due to their different mechanisms of biological activity and toxicity. Most of the earlier in vivo animal exposure—and in vitro bioassay—studies have exposed animals to commercially available technical PCB mixtures. Only in the past 10 years has the toxicity of individual congeners been studied. Due to the differences in metabolism and/or biodegradation rates of individual congeners, the compositions of the original commercial technical mixtures are different from the compositions of the mixtures to which humans or wildlife are exposed (McFarland and Clarke 1989).

Due to changes in the relative proportions of individual congeners in PCB mixture, reference doses (RfDs) derived from laboratory studies for technical Aroclor mixtures might not be appropriate for the PCB mixture found in environmental samples. The environmental weathering of PCB mixtures can result in a reduction or enrichment of toxic potency over time (Quensen et al. 1998, Corsolini et al. 1995a,b; Williams et al. 1992). For example, microbially mediated anaerobic reductive dechlorination is common in sediments in which chlorines are removed from the PCB biphenyl rings (Bedard and Quensen 1995). Dechlorination of PCBs can eliminate the most toxic coplanar congeners (Zwiernik et al. 1998). As a result, the toxic potency of the resulting PCB mixture can be reduced by as much as 98% (Quensen et al. 1998). Moreover, Aroclor-based analysis will not detect a change in either the quantity or the quality of the above PCB mixture, including toxic coplanar congeners. In this case, estimation of hazard based on RfDs from laboratory exposure to technical PCB mixtures would overestimate the risk. In contrast, certain aquatic animals selectively enrich arylhydrocarbon receptor (AhR)-active congeners, an enrichment that can result in greater relative proportion of these toxic congeners in tissues than in technical mixtures (Williams et al.

1992; Corsolini et al. 1995a,b). In this case, estimation of hazard based on RfDs from laboratory exposure to technical PCB mixtures would underestimate the risk.

Aroclor-based analysis and risk assessment offer minimal insight into toxicokinetics, because animals exhibit interspecies differences in their abilities to metabolize specific PCB congeners. Toxicity of PCB congeners to different species might be modulated by species-specific differences in lipid metabolism, quantitative differences in binding of PCBs to receptors in target organs, enzyme induction, or other differences in toxicokinetics.

Another uncertainty associated with estimates of toxicity based on exposure to commercial PCB mixtures (i.e., Aroclors) is related to the relative amounts of polychlorinated dibenzofurans (PCDFs) and polychlorinated naphthalenes (PCNs) identified as contaminants in technical PCB preparations or as covariates in complex environmental mixtures. Concentrations of total PCDFs and PCNs in Aroclor preparations were in the ranges of 0.6 to 7.5 μg/g and 2.6 to 170 μg/g, respectively. Certain PCN congeners are bioaccumulative and exhibit toxic effects similar to those reported for PCBs. In most studies, the PCDF and PCN contents were not quantified, and their contribution to technical PCBs-induced toxicity is unknown.

Accepted Methods for Aroclor-Based Analysis

The recent EPA method 8082 is a performance-based procedure that describes a dual capillary -column gas-chromatographic analysis of PCBs as Aroclors or as individual PCB congeners in extracts from solid or aqueous matrices (EPA 1996a). This procedure also relies on Aroclor pattern recognition techniques. This method is an improvement of earlier iterations of the procedure and provides greater laboratory flexibility. Unfortunately, the method has procedural recommendations that might be counterproductive. The use of decachlorobiphenyl (PCB 209) as an internal standard will result in negative bias if Aroclor 1268 is present in environmental samples (Aroclor 1268 contains 4.8% of PCB 209) (Kannan et al. 1997). More important, method 8082 suggests instrument calibration with appropriate Aroclors when a PCB contaminant source mixture is known but relies on 1:1 mixture of Aroclors 1016 and 1260 to calibrate for samples containing PCBs from an uncharacterized source. Single concentrations of subjectively identified Aroclors are then analyzed, and the 1016 and 1260 calibration is transformed based on the results. This practice has two underlying problems. First, the use of single-point extrapolations assumes linear-response factors throughout the range of PCB concentrations to be analyzed, when in fact that assumption is rarely the case. Second, the Aroclor mixture has a distinct congener percent composition; therefore, the assumption that 1016:1260 congener co-elution domain response factors represent the response factors for individual Aroclors is fundamentally unsound (Newman et al. 1998)

Despite the concerns discussed above, Aroclor determination might be required when compliance measurements specify Aroclors. The details of one Aroclor-based method are given below and summarized in Table F-2.

Method: EPA SW 846 Method 8082/PCBs by Gas Chromatography for Aroclor (EPA 1996a)

Analytes Measured: PCBs are measured as Aroclor formulations or congener. Aroclor 1016, Aroclor 1221, Aroclor 1232, Aroclor 1242, Aroclor 1254, Aroclor 1248, Aroclor 1254, and Aroclor 1260 have been tested.

Instrumentation: High-resolution gas chromatography with electron capture detection (HRGC-ECD) or electrolytic conductivity detection (ELCD).

Quantification Method: Uses the external standard method to quantify Aroclor. A solution of Aroclor 1016 and Aroclor 1260 is used for a 5-point calibration, which is then used for all Aroclors. All Aroclors are also analyzed separately for pattern identification and single-point calibration. Decachlorobiphenyl (PCB209) is used as a surrogate to determine recovery.

MDL: The MDL ranges from 57 to 70 μg/kg of soil and from 0.054 to 0.090 μg/L of water.

Discussion: EPA method 8082 was released in December 1996, replacing EPA method 8080. PCBs can be measured as Aroclors or congeners by this

TABLE F-2 Summary of One Aroclor-Based Analytical Method Used for the Examination of PCB in the Environment

Method	Analytes	Instrumentation	Quantification method	MDL	Typical cost per sample
EPA SW 846 method 8082/ PCBs by gas chromatography	PCBs as Aroclor formulation	HRGC-ECD or HRGC-ELCD	External standard method, single surrogate, 5-point calibration of Aroclor solution	57 to 70 μg/kg, soil/sed.	$250-300

method. The surrogate compound is spiked into the sample, the sample is then extracted by an appropriate method, and the extract is concentrated. Typically for PCBs, the extracts are cleaned up using concentrated sulfuric acid cleanup. The internal standard is added to the clean extract, and the sample is analyzed by HRGC-ECD. Individual PCB peaks are identified by retention time. The peaks should be confirmed by dual-column analysis. The method recommends the analysis of reference material and samples spiked with compounds of interest following SW 846 guidance.

The method warns users that acceptable qualitative and quantitative analysis of Aroclor in environmental samples using this method or other methods is difficult to obtain because of weathering. Aroclor analysis is recommended only for compliance measurements where Aroclor concentrations need to be specified.

Congener-Specific Analysis of PCBs

As analytical capabilities have increased over the past 30 years, the standard approach applied to the analysis of PCBs has slowly shifted from technical grade approximation to component-based analyses. Component-based analysis or congener-specific analysis has improved the quality and the toxicological relevance of the resulting data. Many research applications, particularly those investigating ecological effects require comprehensive, quantitative, congener-specific analysis of PCBs (Frame 1997). By this method, congener profiles resulting from combinations of technical mixtures can be easily identified. Likewise, congener profiles differing from the original technical mixtures in profile and toxicity because of weathering can now be addressed effectively. Congener-specific analysis can be performed by both GC-MS and GC-ECD techniques. Both techniques use analyte separation via a split valve injector and capillary-column stationary phase. Mass-spectrometric detection of ions selective for individual congener groups provides confirmation. Selective mass analysis is used to individually quantify sets of co-eluting congeners. By this method, interfering compounds can be identified easily and therefore false-positives can be eliminated.

Detection Limits and Method Performance:

The GC-MS method provides better precision and resolution than the GC-ECD techniques. Detection limits of individual congeners can be achieved in the lower parts per billion to parts per trillion. The non-ortho coplanar PCBs can be more reliably measured by GC-MS because of their low concentration in most environmental media compared with other congeners. Non-ortho coplanar PCBs can be measured in biological tissues (20 g matrix) at 1 to 10 ppt range, depending upon the nature of the matrix (egg, liver, or fat).

Data Summing

The manner of summing and reporting PCB congener data is a data-reporting issue, not an analytical one, and the approach varies depending on the application and data need. Typical approaches for handling PCB congener data include

- Reporting and using only individual congener data.
- Summing the identified congeners to arrive at a total observed PCB concentration.
- Summing the identified congeners and using a regression analysis based on environmental data to report levels assumed present.
- Summing the individual congeners by level of chlorination; (e.g., summing all congeners containing three chlorine atoms and reporting Σ Cl_3-PCB).
- Summing the individual congeners identified as toxic or potentially toxic and calculating a TODD-equivalent concentration.

Individual congener data provides the most flexibility for supporting environmental management decisions, because the congeners provide the raw data that can be analyzed numerically or statistically by the environmental manager, case by case, as needed. For example, advanced numerical analysis using hierarchical cluster analysis (H.A.) and principal component analysis (PCA) of PCB congeners may be used to evaluate the linkage between the samples and likely point or nonpoint sources—PCB "fingerprinting" analogous to techniques used to identify fugitive petroleum products (Stout et al 1998).

Costs

Congener-specific analysis is costlier than Aroclor-based methodologies. The cost on a per sample basis for GC-MS analysis ranges from $800 to $2,000, depending on the sample matrix, number of samples to be analyzed, and the extent of QA/QC required. GC-ECD congener-specific methods are slightly less costly at $500 to $1,200 per sample. These higher costs are in part related to the necessity of highly trained professionals to perform rigorous cleanup techniques to remove interferences from coplanar PCBs. In addition, GC-ECD congener-specific methods require second column confirmation.

Implications of Congener-Specific Analysis in Risk Assessments:

Congener-specific analysis is recommended for risk assessment because of the differences in the toxic potentials of individual congeners in technical mixtures. Not all 209 PCB congeners are toxic. Several studies have confirmed a correlation between the structure-activity relationships for PCBs. Non-ortho coplanar PCBs 77 (3,3′,4,4′-T4CB), 126 (3,3′,4,4′,5-P5CB), and 169 (3,3′,4,4′,5,5′-H6CB), which are substituted in both para, at least 2 meta. and no ortho positions, are clearly the most toxic congeners of PCBs (Giesy and Kannan 1998). Similar to non-ortho PCBs, mono-ortho congeners, which have one chlorine atom substituted in the ortho position can also elicit AhR binding activity (Safe 1994). However, mono-ortho congeners are relatively less toxic than non-ortho congeners. These studies suggested a need for measuring PCBs at a congener level for application in risk assessment. Toxicological studies conducted in the past 10 years have focused on individual congeners rather than technical mixtures. There is considerable toxicological information for individual congeners from which risk assessment can be performed more reliably than that based on Aroclor measurements (Van den Berg et al. 1998).

Total PCB concentrations can also be measured by summing concentrations of all the individual PCB congeners identified in samples. The homologue and congener profile might reveal the source or origin of the contamination. Evaluation of these profiles might indicate whether mixed Aroclors were present and the degree of weathering.

Accepted Methods for Congener-Based Analysis

This section presents a technical summary of several of the accepted congener-based methods for the analysis of PCBs in environmental samples, including relevant instrumentation, limits of detection, and quality-assurance issues. The technical summary is not intended to be comprehensive. Interested readers are requested to review the original methods. All of the GC analytical methods below are appropriate for the analysis of PCBs in sediments, soils, water, or biological tissue, provided appropriate extraction and cleanup methods are used.

The methods below identify between 18 and 78 congeners, which are reported individually or as two or three co-eluting congeners. The 18 to 78 congeners represent a subset of the total 209 possible congeners but include those most commonly found in environmental samples. Collectively, about 18 to 22 congeners usually represent about one-half of the total PCB that might be found in an environmental sample, if an exhaustive analysis of all congeners is performed. The EPA method 1668 measures 13 toxic or suspected toxic congeners that usually constitute a small fraction (<5%) of the PCBs found in environmental samples.

Although the standard HRGC-ECD methods presented below identify and report between 18 and 22 PCB congeners, it is important to note that each method can be expanded to include additional PCB congener analytes. All PCB congeners are chemically similar and behave similarly in terms of their extraction and analysis. The addition of new congeners to an existing method is a simple matter of adding the new congener to the calibration standards and performing appropriate validation steps to demonstrate the method. PCB congener standards of documented purity are available for all 209 PCB congeners.

The methods presented for congener-specific PCB analyses are summarized in Table F-3. Detailed descriptions of the methods follow the table.

Method: EPA SW 846 Method 8082/ PCBs by Gas Chromatography for Congener (EPA 1996a)

Analytes Measured: The 19 PCB congeners tested by the method include PCB 1, PCB 5, PCB 18, PCB 31, PCB 44, PCB 52, PCB 66, PCB 87*, PCB 101*, PCB 110, PCB 138*, PCB 141, PCB 151, PCB 153, PCB 170*, PCB 180, PCB 183, PCB 187*, PCB 206. (Reported in the EPA congener list: *PCB 87 comprises three co-eluting congeners PCBs 87/115/81, PCB 101 is the co-eluting pair PCBs 101/90, PCB 138 is PCBs 138/160/163; PCB 170 is PCBs 170/190, and PCB 187 is PCBs 187/182.)

Instrumentation: High-resolution gas chromatography with electron capture detection (HRGC-ECD) or electrolytic conductivity detection (ELCD).

Quantification Method: Uses the internal standard method with a decachlorobiphenyl (PCB 209) as the single internal standard is used. A solution of the 19 target congeners is used for a 5-point calibration. Tetrachloro-*m*-xylene is used as a surrogate to monitor recovery.

MDL: The MDLs for congeners are not provided in the method but are assumed similar to the National Oceanic and Atmospheric Administration (NOAA) method and the U.S. Army Corps of Engineers (USACE) New York

TABLE F-3 Summary of Congener-Specific Analytical Methods Used for the Examination of PCBs in the Environment

Method	Analytes	Instrumentation	Quantification Method	MDL	Typical Cost per Sample
EPA SW 846 method 8082/ PCBs by GC	19 PCB congeners specified, does not include all individual coplanar PCB congeners	HRGC-ECD or HRGC-ELCD	Single internal standard, single surrogate, 5-pt calibration.	Not published for individual congeners, MDL of total PCB assumed similar to MDL range for Aroclor	$300-400
NOAA NS&T/EPA-EMAP method (NOAA 1998)	18 PCB congeners, does not include all individual coplanar PCB congeners	HRGC-ECD	Multiple internal standards, 4-pt calibration; multiple surrogate standards	0.1 to 0.4 µg/kg, soil; 0.5 to 3.6 µg/kg biological tissue dry weight	$300-400
EPA draft method 1668	13 coplanar PCB congeners known or suspected to be mammalian carcinogens specified	HRGC-HRMS	Multiple internal standards (labeled PCB), multiple surrogate standards (labeled PCB)	Ultra-low level detection 0.02 to 0.05 µg/kg soil; 0.05 to 0.1 µg/kg tissue dry weight (estimated)	$1,100-1,500
Battelle method	78 congeners (plus 28 co-eluting groups of 2 or 3 congeners)	HRGC-MS	Multiple internal standards (labeled PCB), multiple surrogate standards (labeled PCB)	0.1µg/kg soil, 0.4 µg/kg tissue dry weight	$500-600
Michigan State University - ATL method	69 congeners (plus 14 co-eluting groups of 2-3 congeners) includes all major non-ortho and mono-ortho, PCB congeners	HRGC-ECD GC-MSD	Multiple internal standards (labeled PCB), multiple surrogate standards (labeled PCB)	Ultra-low level detection, 0.1µg/kg for most congeners; 0.001µg/kg for coplanar congeners; dry wt for soil and wet wt for tissue	$1,000-1,200

District, EPA Region 2 method for soils and sediments provided the same amounts of sample and similar cleanup are used.

Discussion: The surrogate compound is spiked into the sample, the sample is then extracted by an appropriate method, and the extract is concentrated. Typically for PCBs, the extracts are cleaned up using concentrated sulfuric acid cleanup. The internal standard is added to the clean extract, and the sample is analyzed by HRGC-ECD. Individual PCB peaks are identified by retention time against authentic congener standards. Peak identification is confirmed by dual-column analysis. The method recommends analysis of reference material and samples spiked with compounds of interest following SW 846 guidance.

This method is similar to the NOAA National Status and Trends (NS&T) method but is less rigorous than the NOAA method in several ways. The use of decachlorobiphenyl (PCB 209) as an internal standard will result in a negative bias if Aroclor 1268 is present in the environmental samples. (Aroclor 1268 contains 4.8% PCB 209.) PCB 209 has the highest molecular weight and elutes from the HRGC well beyond the other congeners. Thus, the recovery and GC response of PCB 209 might not be representative, and its use as an internal standard might lead to a higher degree of imprecision compared with the NOAA method. Otherwise, method 8082 is acceptable for measuring the 19 PCB congeners in environmental samples. Adding congeners to the calibration standards and demonstrating acceptable MDLs for those standards might expand the list of 19 PCB analytes.

Method: Quantitative Determination of Chlorinated Hydrocarbons (NOAA NS&T 1998)

Analytes Measured: The 18 PCB congeners tested by the method include PCB 8*, PCB 18, PCB 28, PCB 44, PCB 52, PCB 66, PCB 101*, PCB 105, PCB 118, PCB 128, PCB 138*, PCB 153, PCB 170*, PCB 180, PCB 187*, PCB 195, PCB 206, PCB 209. (Reported in the NOAA NS&T: * PCB 8 is the co-eluting pair PCBs 8/5, PCB 101 is PCBs 101/90, PCB 138 is PCBs 138/160/163, PCB 170 is PCBs 170/190, and PCB 187 is PCBs 187/182.)

Instrumentation: High-resolution gas chromatography with electron capture detection (HRGC-ECD).

Quantification Method: Uses the internal standard method with tetrachloro-*m*-xylene (TCMX) as the single internal standard. A solution of the 19 target congeners is used for a 4-point calibration. Three surrogates,

4,4′-dibromooctcafluorobiphenyl (DBOFB), PCB 103, and PCB 198, are used to determine sample recovery.

MDL: The MDLs for the target congeners range from 0.1 to 0.4 μg/kg of sediment, 0.5 to 3.6 μg/kg of biological tissue dry weight.

Discussion: The NOAA NS&T PCB congener method is a performance-based method that allows for the documentation of methodology and laboratory performance through time as analytical procedures change. The basic method has been used in the NOAA NS&T program since 1984 for the routine analysis of PCB congeners in sediments and biological tissue samples. The surrogate compounds are spiked into the sample, the sample then is extracted by an appropriate method, and the extract is concentrated. Typically for tissue samples, sample extracts are cleaned with alumina and size-exclusion high-pressure liquid chromatography (HPLC). For sediments and soils, sample extracts are cleaned with copper and alumina. The internal standards are added to the clean extract, and the sample is analyzed by HRGC-ECD.

The original NOAA NS&T congener list included several coplanar PCB that are now analyzed separately for coplanar PCBs using an HRGC-HRMS method similar to that presented below. The NS&T method has found general acceptance nationally because of the strong NS&T QA program that is co-administered by NOAA and National Institutes of Standards and Technology (NOAA 1993). All NS&T contractors and many other environmental laboratories participate in this laboratory intercomparison program. The NOAA NS&T PCB congener method is acceptable for measuring the 18 PCB congeners in environmental samples.

The NOAA NS&T PCB congener method has found wide acceptance. Examples include USACE New York District, EPA Region 2 testing procedures to fulfill Green Book testing requirement for the testing of dredged material for ocean dumping (USACE 1992), EPA Region 9 San Francisco Bay, St. Johns (Florida) Water Management District, Puget Sound Dredge Disposal Activity (PSDDA), and EPA EMAP marine assessment program (Heitmuller and Valente. 1993).

Method: EPA Method 1668, Toxic PCBs by Isotope Dilution High-Resolution Gas Chromatography–High-Resolution Mass Spectrometry (EPA 1997)

Analytes Measured: The 13 congeners tested by the method include PCB 77, PCB 105, PCB 114, PCB 118, PCB 123, PCB 126, PCB 156, PCB 157, PCB 167, PCB 169, PCB 170, PCB 180, and PCB 189.

Instrumentation: High-resolution gas chromatography with high-resolution mass-spectrometry (more than 10,000) detection (HRGC-HRMS).

Quantification Method: Uses the four labeled PCBs, one for each level of chlorination measured (Cl_4-PCB, Cl_5-PCB, Cl_6-PCB, and C17-PCB). Stable isotopically labeled analogues of the analytes serve as surrogates.

MDL: The MDLs are ultra-low. For PCB 126, the MDL is 40 pg/L (40×10^{-12} g/L). MDLs for congeners in sediment are not provided, but they are assumed to be 10 times lower than achievable by EPA method 8082 or NOAA NS&T methods.

Discussion: The method is based on previously written EPA methods for the analysis of chlorinated dioxins and furans in environmental samples (EPA 1994). The labeled compounds are spiked into the sample, the sample is then extracted by an appropriate method, and the extract is concentrated. The extracts are cleaned up using back extraction with sulfuric acid, followed by gel permeation, silica gel, Florisil, and activated carbon chromatography. Further, HPLC can be used to isolate specific fractions for analysis. Prior to the above cleanup, extracts of biological tissue pre-cleaned with a separate size-exclusion procedure to remove fats. The elaborate cleanup and use of HRMS makes method 1668 the most costly of PCB analytical procedures. However, the method provides ultra-low detection limits that might be needed for some research studies. Method 1668 is currently being revised to expand the full list of 209 PCB congeners.

Method: PCB Congener and Homologue Analysis By High-Resolution Gas Chromatography–Low-Resolution Mass Spectrometry (Durell and Seavey 2000)

Analytes Measured: The PCB congeners tested by the method include 78 individual PCB congeners and 28 co-eluting groups of two to three congeners as listed: PCB 1, PCB 3, PCBs 4/10, PCB 6, PCBs 7/9, PCBs 8/5, PCBs12/13, PCBs 16/32, PCBs 17/15, PCBs 18, PCBs 19, PCBs 21, PCB 22, PCBs 24/27, PCB 25, PCB 26, PCB 28, PCB 29, PCB 31, PCBs 33/20, PCB 40, PCBs 41/64/71, PCBs 42/37, PCB 43, PCB 44, PCB 45, PCB 46, PCBs 47/75, PCB 48, PCB 49, PCB 51, PCB 52, PCB 53, PCBs 56/60, PCB 59 PCB 63, PCB 66, PCBs 70/76, PCB 74, PCB 82, PCB 83, PCB 84, PCB 85, PCBs 87/115/81, PCB 89, PCB 91, PCB 92, PCB 95, PCB 97, PCB 99, PCB 100, PCBs 101/90, PCB 105, PCBs 107/147, PCBs 110/77, PCB 114, PCB 118,

PCB 119, PCB 124, PCB 128, PCBs 129/126, PCB 130, PCB 131, PCB 132, PCB 134, PCBs 135/144, PCB 136, PCB 137, PCBs 138/160/163, PCBs 141/1709, PCB 146, PCBs 149/123, PCB 151, PCB 153, PCB 156, PCB 158, PCB 167, PCB 169, PCBs 170/190, PCBs 171/202, PCB 172, PCB 173, PCB 174, PCB 175, PCB 176, PCB 177, PCB 178, PCB 180, PCB 183, PCB 184, PCB 185, PCBs 187/182, PCB 189, PCB 191, PCB 193, PCB 194, PCBs 195/208, PCB 197, PCB 198, PCB 199, PCB 200, PCBs 201/157, PCBs 203/196, PCB 205, PCB 206, PCB 207, PCB 209 (PCB congeners presented as PCB/PCB represent co-eluting PCB peaks under most analytical conditions. The most abundant single congener is presented first.)

Instrumentation: High-resolution gas chromatography with low-resolution mass spectrometry (HRGC-MS).

Quantification Method: Uses the internal standard method with PCB 114, PCB 34, PCB 104, and PCB 112, those not found in environmental samples, as the internal standards. A solution of the 19 target congeners is used for a 4-point calibration. Three surrogates, PCB 96, PCB 103, and PCB 166, also not found in environmental samples, are used to determine sample recovery.

MDL: The MDLs for the target congeners range from 0.1 µg/kg of sediment, 0.4 µg/kg of biological tissue dry weight.

Discussion: The Battelle PCB congener method by low-resolution mass spectrometry is a performance-based method based on EPA method 8270b (EPA 1996b), a method that has been validated for PCB analysis in hazardous waste and environmental samples. Changes in methodology include the use of PCB-specific congeners for internal standards and surrogate materials and use of extraction and cleanup methods developed for the NOAA NS&T program. The surrogate compounds are spiked into the sample, the sample is then extracted by an appropriate method, and the extract is concentrated. Typically for tissue samples, sample extracts are cleaned with alumina and size-exclusion HPLC. For sediments and soils, sample extracts are cleaned with copper and alumina. Sulfuric acid is used for samples containing interfering chlorinated pesticides. The internal standards are added to the clean extract, and the sample is analyzed by HRGC-MS.

The congener list has been developed to include most PCB congeners found in environmental samples and recovers between 95% and 99% of total PCB in all samples analyzed (G. Durell, personal communication). The analyte list includes most, but not all the suspected toxic PCB measured with EPA method 1668. However, two of the most toxic and prevalent coplanar

PCB congeners, PCB77 and PCB126, are reported as minor components of co-eluting pairs with PCB110 and PCB129, respectively. Thus, this method might not be suitable for calculating TODD-equivalent concentrations.

Method: Extraction and Analysis of PCB and Non-ortho-Coplanar PCBs in Biological Matrices (Cannan and Giesy 1998)

Analytes Measured: The PCB congeners tested by the method include 69 individual PCB congeners, plus 14 co-eluting groups of two to three congeners, and all major non-ortho and mono-ortho PCB congeners as listed: PCB 4, PCBs 5/8, PCB 6, PCB 9, PCB 10, PCB 15, PCB 16, PCB 17, PCB 18, PCB 19, PCBs 53/33/20, PCB 22, PCB 25, PCB 26, PCB 27, PCBs 31/28, PCB 32, PCB 37, PCB 40, PCB 41, PCB 42, PCB 44, PCB 45, PCBs 75/47, PCB 49, PCB 52, PCB 56, PCBs 95/66, PCB 70, PCB 74, PCB 77, PCB 81, PCBs 151/82, PCB 83, PCB 84, PCB 85, PCB 87, PCB 90, PCB 91, PCB 92, PCB 97, PCB 99, PCB 101, PCB 105, PCB 107, PCB 110, PCB 113, PCB 117, PCB 118, PCB 119, PCBs 120/136, PCB 126, PCB 128, PCBs 178/129, PCBs 153/132, PCB 133, PCB 134, PCBs 135/144, PCB 137, PCBs 138/158, PCBs 179/141, PCB 156, PCB 157, PCBs 167/185, PCB 169, PCB 170, PCB 171, PCB 174, PCB 176, PCB 177, PCB 180, PCB 183, PCB 187, PCB 194, PCBs 195/208, PCB 199, PCB 200, PCB 201, PCB 202, PCB 205, PCB 206, PCB 207, and PCB 209. (PCB congeners presented as PCB/PCB represent co-eluting PCB peaks under most analytical conditions. The most abundant single congener is presented first).

Instrumentation: High-resolution gas chromatography with low-resolution mass spectrometry (HRGC-MS)

Quantification Method : Uses the external standard method with all the above listed PCB congeners. A solution of all the target congeners is used for a calibration. A surrogate, PCB 30, not found in environmental samples, is used to determine sample recovery.

MDL: The MDLs for the target congeners are 0.1 μg/kg for sediment (dry weight) and biological tissue (wet weight). Additional cleanup and instrumental analysis of non-ortho coplanar PCB congeners provide an MDL of 1-10 pg/g for sediment (dry weight) and biological tissue (wet weight).

Discussion: This PCB congener method by low-resolution mass spectrometry is a performance-based method based on EPA method 8270b (EPA

1996b), a method that has been validated for PCB analysis in hazardous waste and environmental samples. Changes in methodology include the use of PCB-specific congeners for internal standards and surrogate materials and use of extraction and cleanup methods developed for the NOAA NS&T program. The surrogate compounds are spiked into the sample, the sample is then extracted by an appropriate method, and the extract is concentrated. Typically for tissue samples, sample extracts are cleaned with multilayer silica gel and acidic silica columns. For sediments and soils, sample extracts are cleaned with copper and silica gel. Sulfuric acid is used for samples containing interfering chlorinated pesticides. The internal standards are added to the clean extract, and the sample is analyzed by HRGC-MS.

The congener list has been developed to include most PCB congeners found in environmental samples and recovers between 95% and 99% of total PCBs in all samples analyzed. The analyte list includes most of the toxic PCBs measured with EPA method 1668. Thus, this method is suitable for calculating TODD-equivalent concentrations.

BIOLOGICALLY BASED METHODS

Immunoassay Methods

PCBs can be quantified by immunoassay methods, such as enzyme-linked immunosorbent assay (ELISA). The ELISA method is based on the specific binding of PCB to antibodies and requires little effort for sample extraction or cleanup. Thus, ELISA is rapid and less expensive than traditional methods. This technique is a useful screening tool for detecting concentrations of PCBs in environmental samples in the low parts-per-million (0.5-1.5 ppm) range. This level of sensitivity is not sufficient for monitoring PCBs in environmental samples but may be applied for highly contaminated sediments and for evaluating the effectiveness of sediments dredging operations (Ritcher et al. 1994).

Immunoassay techniques for screening PCBs in soils have been developed and tested over the past 10 years (see Table F-4). Several of the approved methods are presented in EPA method 4020, Screening for Polychlorinated Biphenyls by Immunoassay (EPA 1996). The type of method is called competitive ELISA, and test kits are available for PCBs as well as other classes of compounds. The method is selective; that is, few other classes of compounds interfere with the test. However, the method has several drawbacks that limit its application to screening. PCBs are identified as total PCBs. Further identification as Aroclor or congener is not possible. The method is

TABLE F-4 Immunoassay Method for PCB Screening

Method	Typical Sample Size	Summary of Method
EPA method 4020, screening for PCBs by immunoassay	5 g of soil or sediment, or nonaqueous waste liquids	Soil samples are extracted with methanol and test kit reagents. A sample extract and an enzyme conjugate reagent are added to immobilized antibody that binds both. A second enzyme and dye are added, and the color intensity is measured with a spectrophotometer. The signal is indirectly proportional to the amount of PCB. The results are compared against three calibrators of 5, 10, and 50 ppm. PCB concentrations are semiquantitatively classified as below 5 ppm, between 5 and 10 ppm, between 10 and 50 ppm, and greater than 50 ppm. Sensitivity can be optimized to detect samples in the low (0.5-1.5 ppm) range

semiquantitative not quantitative. Different commercial PCB formulations show a wide range of sensitivity to the test method (factor of 8 between Aroclor 1254 and Aroclor 1268). However, because of the ease of use and rapidity, the method is often used for field screening of PCBs.

Cell Bioassays—H4IIE Cells

The H4IIE-luciferase induction assay is an in vitro technique for the identification of AhR-active compounds (Hilscherova et al. 2000). The technique uses rat hepatoma cells (H4IIE-luc) stably transfected with an AhR-controlled luciferase reporter gene construct (Sanderson et al. 1996). The assay is also referred to as the chemical-activated luciferase gene expression (CALUX) system (Murk et al. 1996). These cells express firefly luciferase in response to AhR agonists. Luciferase activity is measured conveniently and with high sensitivity as light emission using a plate-scanning luminometer. Luciferase induction potential is assessed by comparison of the response to that of 2,3,7,8-tetrachlorodibenzo-*p*-dioxin (TODD), the most potent agonist for the mammalian AhR.

This cell bioassay has been utilized for the screening and monitoring of Ah-active components in environmental extracts of sediment, water, air, and tissue (invertebrates, fish, birds, and mammals). The H4IIE rat hepatoma cell bioassay (Tillitt et al. 1991) is widely used for this purpose. In this assay, ethoxyresorufin-*O*-deethylase (EROD)-inducing potencies (ED_{50} values) of single compounds and environmental samples are determined from complete dose-response curves and compared with EC_{50} values of TODD to express the biological potency of the tested samples in TODD-equivalents. The bioassay integrates potential nonadditive interactions among AhR agonists and other compounds by measuring a final receptor-mediated response (Giesy et al. 1994). One of the primary purposes of this bioassay is to prioritize samples for more extensive quantification by instrumental analysis. It can also be used to direct fractionation steps in a toxicity identification and evaluation approach and to detect novel compounds that have biological activity similar to that of TODD. Excellent correlation has been observed between the TODD-equivalent concentrations determined from this bioassay and instrumental analysis when these methods have been applied to the same sample (Tillitt et al. 1996, Quensen et al. 1998). For a sample size of 20 g of tissue or soil and a final extract volume of 0.25 mL, the H4IIE-luc assay will detect 1 part per trillion (ppt; pg/g of wet weight) TODD-equivalents.

Sample Extraction Methods

The above analytical methods must be used with appropriate extraction and extract cleanup techniques to be effective. Broadly, PCBs are extracted from environmental matrices along with a potentially wide range of hydrophobic organic compounds of similar polarity and volatility. For example, in EPA method 1625, PCBs are extracted from water along with 175 other listed compounds extracted from water. In EPA method 8270, PCBs are listed with 243 other analytes. The selectivity of the extraction and cleanup procedures to isolate PCBs from other extracted compounds (so-called matrix interferences) usually determines the MDL—that is, the greater the level of interfering compounds in individual sample extracts, the greater the level of PCB required to overcome the interference. Ecological risk assessments currently require PCB MDLs in the low parts per billion for sediments and biological tissue. Appropriate and rugged sample extraction and extraction cleanup procedures are required to deliver those low MDLs routinely.

All of these methods use surrogate materials added at the beginning of the process to monitor performance.

Fortunately, environmental samples have been extracted for PCBs for over 25 years. During this time, sample extraction and cleanup procedures have been refined, and procedures delivering MDLs in the low parts per billion are well recognized. Quality-assurance programs have been established supporting low-level detection methods. Multiple standard reference materials, interim reference materials, and other quality-assurance samples are available to support PCB analytical programs.

This section is intended to provide a brief overview of the available approaches to sample extraction and extract cleanup and compares several of the methods. There are literally hundreds of combinations of methods currently in use for PCB extraction and cleanup, and no effort has been made for this review to be all inclusive. The methods presented are those in use by EPA, EPA regions, state agencies, and NOAA for determination of PCBs at environmental levels. An experienced laboratory will have little difficulty using the methods to determine PCBs at concentrations ranging from low parts per billion to high parts per million.

Note that many PCB extraction and cleanup methods, as well as the analytical methods, are performance based. That is, the methods are not prescriptive but rather present procedures that may be followed directly or used as guidelines for analysis at laboratory discretion. Instead of prescriptive methods, the federal and state agencies using PCB data establish quality-assurance requirements, including matrix-specific performance data. Examples of this method include the EPA method 8082 (including extraction method 3510 and 3520 for water and method 3540 and method 3541 for solid samples, and multiple cleanup options) and the NOAA NS&T method.

A summary of accepted extraction methods for sediments and soils, water, and biological tissue is presented below.

Table F-5 presents and compares several extraction procedures suitable for soils and sediments with a percent solids content of greater than 20%. Methods suitable for biological tissue extraction are presented in Table F-6. In general, many more methods exist for the extraction of PCBs from soils and sediments than for extraction from biological tissue. The EPA Office of Solid Waste and Office of Water have published method for soil and sediment. In addition, the NOAA NS&T method is presented. EPA has no published methods for extraction of PCB from tissue. However, the FDA has published methods for tissue, as well as NOAA NS&T. Used by an experienced laboratory, any of the listed methods will provide suitable results.

Cleanup Procedures

Cleanup methods—that is, laboratory techniques that remove interfering materials from extracts of environmental samples before analysis—are also presented.

Table F-7 presents and compares several cleanup procedures suitable for the isolation of PCB from other classes of organic compounds in extracts of environmental samples. All the above analytical methods presented in Tables F-3 and F-4 call for at least one cleanup method to remove chemical or biological interferences with the PCB analysis.

TABLE F-5 Sediment and Soil Extraction Methods

Method	Typical Sample Size	Summary of Method
EPA 3540C, soxhlet extraction	10 g of soil or sediment	Water layer, if any, is discarded. Any foreign objects, such as sticks, leaves, and rocks, are removed. Sample is mixed thoroughly and blended with equal amounts drying agent (sodium sulfate). PCB is extracted into acetone/hexane or acetone/methylene chloride using soxhlet extraction technique. Solvent is exchanged to hexane or other solvent for cleanup and analysis. Solvent concentration by Kuderna-Danish concentrator, nitrogen evaporation, or equivalent method.
EPA 3541, automated soxhlet extraction	10 g soil or sediment	Sample preparation as above. Sample may be air dried with no loss of PCB. Sample is mixed thoroughly and blended with equal amounts drying agent (sodium sulfate). PCB is extracted into acetone/hexane using an automated soxhlet extraction apparatus. Solvent is exchanged to hexane or other solvent for cleanup and analysis. Solvent concentration by Kuderna-Danish concentrator, nitrogen evaporation, or equivalent method.
EPA method 3550, ultrasonic extraction	30 g of soil or sediment	Sample preparation as above. Sample is mixed thoroughly and blended with equal amounts anhydrous sodium sulfate. The sample is extracted 3× with acetone/hexane or acetone/methylene chloride using an ultrasonic disrupter. Solvent is exchanged to hexane or other solvent for cleanup and analysis. Solvent concentration by Kuderna-Danish concentrator, nitrogen evaporation, or equivalent method.
EPA methods 3560 and 3561, super-critical extraction	3 g soil or sediment	Sample preparation as above. Wet samples may be mixed with drying agent or diatomaceous earth to enhance porosity. The sample is extracted with supercritical CO_2. PCBs that are extracted with the CO_2 are trapped in an organic solvent. Solvent is exchanged to hexane or other solvent for cleanup and analysis. Solvent concentration by Kuderna-Danish concentrator, nitrogen evaporation, or equivalent method.
NOAA NS&T method (MacLeod et al 1993)	10 g soil or sediment	Sample preparation as above. The sample is mixed thoroughly and blended with equal amounts anhydrous sodium sulfate. The sample is extracted 3× with dichloromethane using shaker or tumbler apparatus. Solvent is exchanged to hexane or other solvent for cleanup and analysis. Solvent concentration by Kuderna-Danish concentrator, nitrogen evaporation, or equivalent method.

TABLE F-6 Tissue Extraction Procedures

Method	Typical Sample Size	Summary of Method
Organo-chlorine residues, general methods for fatty foods (FDA 1999)	Ca. 30 g of wet tissue	Wet tissue is placed in a Teflon or glass extraction container along with equal amounts of anhydrous sodium sulfate and extracted with petroleum ether using a commercial high-speed blender. The solvent is filtered and dried before further clean up. Solvent concentration is by Kuderna-Danish concentrator, nitrogen evaporation, or equivalent method.
NOAA NS&T method (MacLeod et al 1993)	Ca. 30 g of wet tissue	This method is similar to the FDA method, from which it was originally taken. Wet tissue is placed in a Teflon extraction container along with equal amounts of anhydrous sodium sulfate and extracted with methylene dichloride using a commercial high-speed homogenizer. The solvent is exchanged to hexane or other solvent for cleanup and analysis. Solvent concentration is by Kuderna-Danish concentrator, nitrogen evaporation, or equivalent method.
NOAA NS&T soxhlet method (NOAA 1998	Ca. 30g of wet tissue	This method is similar to the soxhlet method presented above for sediment or soil. The sample is mixed thoroughly and blended with equal amounts drying agent (sodium sulfate). PCB is extracted into methylene chloride using soxhlet extraction technique. Solvent is exchanged to hexane or other solvent for cleanup and analysis. Solvent concentration by Kuderna-Danish concentrator, nitrogen evaporation, or equivalent method.

TABLE F-7 Methods Appropriate for the Cleanup of Environmental Extracts for PCB Analysis

Method	Summary of Method
EPA method 3610B, alumina cleanup	Alumina, a porous and granular form of aluminum oxide, is adsorptive and is used to separate classes of organic compounds based on their chemical polarity. A sample extract is added to an alumina column, which is eluted with a series of solvents of increasing polarity to isolate PCB. The method 3610B, which includes the use of commercial solid-phase extraction cartridges, is one of many written methods of the use of alumina PCB cleanup. Alumina is one of several adsorbents commercially available for this purpose. Others include silica gel and florisil, and accepted methods for using these also can be found in the literature. The method for extracts from water, waste water, soil, sediments, biological tissue, and other environmental samples upon method validation.
EPA method 3665A sulfuric acid/ permanganate cleanup	The sample extract is reacted with concentrated sulfuric acid alone or acid and potassium permanganate. This procedure destroys most organic chemicals, including most pesticides and other compounds that interfere with the PCB HRGC-ECD analysis. For this reason, the method cannot be used if non-PCB analytes are to be measured in the same extract. Acid cleanup is generally followed by alumina or other adsorption column cleanup to further isolate PCB.
Method 3640A, Gel-permeation (size-exclusion) chromatography	The method separates classes of compounds based on molecular size. The sample extract is passed through a porous gel or porous solid bead of uniform pore size, which is chosen to exclude molecules of a certain size or larger. Excluded molecules elute before smaller molecules that pass through the pores in the beads. The method is particularly effective at removing common interferences from environmental extracts, such as sulfur and plant or animal fats. Krahn et al (1988) presents an automated HPLC cleanup method based on the size exclusion principle that has been used extensively in the NOAA NS&T program.
NOAA NS&T copper method (MacLeod et al 1993)	Elemental sulfur, often present in freshwater or marine sediments, is extracted under most conditions presented in the above methods tables and must be removed. Activated copper is one of several reagents that can be used for this purpose. The activated copper is commonly added to concentrated extracts, but it also may be added in the thimbles of soxhlet extractors and extraction vessels of supercritical fluid extractors.

REFERENCES

Bedard, D.L., and J.F. Quensen, 3rd. 1995. Microbial reductive dechlorination of polychlorinated biphenyls. Pp 127-216 in Microbial Transformation and Degradation of Toxic Organic Chemicals, L.Y Young, and C.E. Cerniglia, eds. New York: Wiley-Liss.

Burkhard, L.P., and D. Weininger. 1987. Determination of polychlorinated biphenyls using multiple regression with outlier detection and elimination. Anal. Chem. 59(8):1187-1190.

Cannan, K. and J.P. Giesy. 1998. Extraction and Analysis of PCBs and non-ortho Coplanar PCBs in Biological Matrices. SOP No. 211, revision 1. Michigan State University, Aquatic Toxicity Laboratory.

Corsolini, S., S. Focardi, K. Kannan, S. Tanabe, A.Borrell, and R. Tatsukawa. 1995a. Congener profile and toxicity assessment of polychlorinated biphenyls in dolphins, sharks and tuna collected from Italian coastal waters. Mar. Environ. Res. 40(1):33-54.

Corsolini, S., S. Focardi, K. Kannan, S. Tanabe, and R. Tatsukawa. 1995b. Isomer-specific analysis of polychlorinated biphenyls and 2,3,7,8-tetrachloro-dibenzo-*p*-dioxin equivalents (TEQs) in red fox and human adipose tissue from central Italy. Arch. Environ. Contam. Toxicol. 29(1):61-68.

Durell, G., and J. Seavey Fredriksson. 2000. PCB Congener and PCB Homologue Analysis by HRGC/LRMS: A Low MDL, Cost Effective Analytical Alternative for Tomorrow's Quality Sensitive PCB Assessments. Paper presented on 23rd Conference on Analysis of Pollutants in the Environment, Pittsburgh PA, May 14-16, 2000.

Draper, W.M., D. Wijekoon, and R.D. Stephens. 1991. Speciation and quantitation of aroclors in hazardous wastes based on PCB congener data. Chemosphere 22(1-2):147-164.

EPA (U.S. Environmental Protection Agency). 1994. Method 8290, Polychlorinated Dibenzodioxins (PCDDs) and Polychlorinated Dibenzofurans (PCDFs) by High Resolution Gas Chromatography/High Resolution Mass Spectrometry (HRGC/HRMS), Revision 0, Test Methods for Evaluating Solid Wastes, Physical/ Chemical Methods, SW-846. Environmental Protection Agency, Office of Solid Waste, Washington DC. [Online]. Available: http://www.epa.gov/epaoswer/ hazwaste/test/sw846.htm

EPA (U.S. Environmental Protection Agency). 1994a. Method 8080A, Organochlorine Pesticides and Polychlorinated biphenyls by Gas Chromatography, Revision 1,Test Methods for Evaluating Solid Waste: Physical/ Chemical Methods, SW-846. 3rd Ed, Final Update II. EPA 530/SW-846. Environmental Protection Agency, Office of Solid Waste and Emergency Response, Washington, DC. September.

EPA (U.S. Environmental Protection Agency). 1994b. Method 8081, Organochlorine Pesticides and PCBs as Aroclors by Gas chromatography: Capillary Column Technique, Revision 0. Test Methods for Evaluating Solid Waste: Physical/

Chemical Methods, SW-846. 3rd Ed, Final Update II. EPA 530/SW-846. Environmental Protection Agency, Office of Solid Waste and Emergency Response, Washington, DC. September.

EPA (U.S. Environmental Protection Agency). 1994c. Method 1613, Tetra- Through Octa-Chlorinated Dioxins and Furans by Isotope Dilution HRGS/HRMS, Revision B, Environmental Protection Agency, Office of Water, Washington DC. [Online]. Available: http://www.epa.gov/ost/methods/1613.html. October.

EPA (U.S. Environmental Protection Agency). 1995. Great Lakes Water Quality Initiative Technical Support Document for Wildlife Criteria. EPA-820-B-95-009. Office of Water. Environmental Protection Agency, Washington, DC. NTIS PB95-187332.

EPA (U.S. Environmental Protection Agency). 1996. Method 4020, Screening for Polychlorinated Biphenyls by Immunoassay, Revision 0, Test Methods for Evaluating Solid Wastes, Physical/ Chemical Methods, SW-846. Office of Solid Waste, U.S. Environmental Protection Agency, Washington DC. [Online]. Available: http://www.epa.gov/epaoswer/hazwaste/test/sw846.htm. December.

EPA (U.S. Environmental Protection Agency). 1996a. Method 8082, Polychlorinated Biphenyls by Gas Chromatography, Revision 0, Test Methods for Evaluating Solid Wastes, Physical/ Chemical Methods, SW-846. Office of Solid Waste, U.S. Environmental Protection Agency, Washington DC. [Online]. Available: http://www.epa.gov/epaoswer/hazwaste/test/sw846.htm. December.

EPA (U.S. Environmental Protection Agency). 1996b. Method 8270c, Semivolatile Organic Compounds by Gas Chromatography /Mass Spectrometry (GC/MS). Revision 3, Test Methods for Evaluating Solid Wastes, Physical/ Chemical Methods, SW-846. Office of Solid Waste, Environmental Protection Agency, Washington DC. [Online]. Available: http://www.epa.gov/epaoswer/hazwaste/test/sw846.htm. December.

EPA (U.S. Environmental Protection Agency). 1997. Method 1668, Toxic Polychlorinated Biphenyls by Isotope Dilution High Resolution Gas Chromatography/High Resolution Mass Spectrometry. Draft. EPA 821/R-97-001. Office of Water, Environmental Protection Agency, Washington DC. March.

FDA (U.S. Food and Drug Administration) 1999. Pesticide Analytical Manual Vol. 1. (PAM), 3rd Ed. Center for Food Safety and Applied Nutrition, U.S. Food and Drug Administration. [Online]. Available: http://vm.cfsan.fda.gov/~frf/pami3.html [Jun 20, 2000].

Frame, G.M. 1997. A collaborative study of 209 PCB congeners and 6 Aroclors on 20 different HRGC columns. 1. Retention and coelution database. Fresenius J. Anal. Chem. 357(6):701-713.

Giesy, J.P., and K. Kannan. 1998. Dioxin-like and non-dioxin-like toxic effects of polychlorinated biphenyls (PCBs): implications for risk assessment. Crit. Rev. Toxicol. 28(6):511-569.

Giesy, J.P., J.P. Ludwig, and D.E. Tillitt. 1994. Dioxins, dibenzofurans, PCBs, and colonial fish-eating water birds. Pp. 249-307 in Dioxins and Health, A. Schecter, ed.. New York: Plenum Press.

Giesy, J.P., D.J. Jude, D.E. Tillitt, R.W. Gale, J.C. Meadows, J.L. Zajieck, P.H. Peterman, D.A. Verbrugge, J.T. Sanderson, T.R. Schwartz, and M.L. Tuchman. 1997. Polychlorinated dibenzo-p-dioxins, dibenzofurans, biphenyls and 2,3,7,8-tetrachlorodibenzo-p-dioxin equivalents in fishes from Saginaw Bay, Michigan. Environ. Toxicol. Chem. 16(4):713-724.

Hilscherova, K., M. Machala, K. Kannan, A.L. Blankenship, and J.P. Giesy. 2000. Cell bioassays for detection of aryl hydrocarbon (AhR) and estrogen receptor (ER) mediated activity in environmental samples. Environ. Sci. Pollut. Res.7(3)159-171.

Heitmuller, P.T., and R. Valente. 1993. Environmental Monitoring and Assessment Program EMAP-Estuaries Lousianian Province 1993 Quality Assurance Project Plan. EPA Contract No. 68-C3—309. Environmental Protection Agency, Office of Research and Development, Gulf Breeze, Florida. June 1, 1993.

Kannan, K., K.A. Maruya, and S. Tanabe. 1997. Distribution and characterization of polychlorinated biphenyl congeners in soil and sediments from a Superfund site contaminated with Aroclor 1268. Environ. Sci. Technol. 31(5):1483-1488.

Krahn, M.M., L.K. Moore, R.G. Bogar, C.A. Wigren, S.L. Chan, and D.W. Brown. 1988. High- Performance Liquid Chromatographic Method for isolating organic contaminants from tissue and sediment extracts. J. Chromatogr. 437(1):161-175.

McFarland, V.A., and J.U. Clarke. 1989. Environmental occurrence, abundance, and potential toxicity of polychlorinated biphenyl congeners: considerations for a congener-specific analysis. Environ. Health Perspect. 81:225-239.

MacLeod, W.D., J.W. Brown, A.J. Friedman, D.G. Burrows, O. Maynes, R.W. Pearce, C.A. Wigren, and R.G. Bogar. 1993. Standard analytical procedures of the NOAA National analytical facility, 1985 – 1986(Revised). Extractable toxic organic compounds. Pp. 1-51 in Sampling and Analytical Methods of the National Status and Trends Program, National Benthic Surveillance and Mussel Watch Projects 1984-1992, Vol. IV. Comprehensive Descriptions of Trace Organic Analytical Methods, G.G. Lauenstein and Y. Cantillo, eds. NOAA Tech. Memo. NOS ORCA 71. Silver Spring MD: U.S. Department of Commerce.

Murk, A.J., J. Legler, M.S. Denison, J.P. Giesy, C. van de Guchte, and A. Brouwer. 1996. Chemical-activated luciferase gene expression (CALUX): a novel in vitro bio-assay for Ah receptor active compounds in sediments and pore water. Fundam. Appl. Toxicol. 33(1):149-160.

Newman, J.W., J.S. Becker, G. Blondina, and R.S. Tjeerdema. 1998. Quantitation of aroclors using congener-specific results. Environ. Toxicol. Chem. 17(11):2159-2167.

NOAA (National Oceanic and Atmospheric Administration). 1993. Sampling and Analytical Methods of the National Status and Trends Program National Benthic Surveillance and Mussel Watch Projects: 1984 – 1992, Vol 1. Overview and Summary of Methods, G.G. Lauenstein and A.Y. Cantillo, eds. National Oceanic and Atmospheric Administration Technical Memorandum NOS ORCA 71. Silver Spring, MD: U.S. Department of Commerce.

NOAA (National Oceanic and Atmospheric Administration). 1998. Sampling and analytical methods of the National Status and Trends Program Mussel Watch Project: 1993 – 1996 Update, G.G. Lauenstein and A.Y. Cantillo, eds. National Oceanic and Atmospheric Administration Technical Memorandum NOS ORCA 130. Silver Spring MD: U.S. Department of Commerce.

Quensen, J.F, 3rd, M.A. Mousa, S.A. Boyd, J.T. Sanderson, K.L. Froese, and J.P. Giesy. 1998. Reduction of aryl hydrocarbon receptor-mediated activity of polychlorinated biphenyl mixtures due to anaerobic microbial dechlorination. Environ. Toxicol. Chem. 17(5):806-813.

Richter, C.A., J.B. Drake, and J.P. Giesy. 1994. Immunoassay monitoring of polychlorinated biphenyls (PCBs) in the Great Lakes. Part 1. General assessment principles. Environ. Sci. Pollut. Res. 1(2):69-74.

Safe, S.H. 1994. Polychlorinated Biphenyls (PCBs): Environmental impact, biochemical and toxic responses, and implications for risk assessment. Crit. Rev. Toxicol. 24(2):87-149.

Sanderson, J.T., J.M. Aarts, A. Brouwer, K.L. Froese, M.S. Denison, and J.P. Giesy. 1996. Comparison of ah receptor-mediated luciferase and ethoxyresorufin-o-deethylase induction in H4IIE cells: implications for their use as bioanalytical tools for the detection of polyhalogenated aromatic hydrocarbons. Toxicol. Appl. Pharmacol. 137(2):316-325.

Stalling, D.L., W.J. Dunn, T.R. Schwartz, J.W. Hogan, J.D. Petty, E. Johansson, and S. Wold. 1985. Application of soft independent method of class analogy (SIMCA) in isomer specific analysis of polychlorinated biphenyls. Pp. 195-234 in Trace Residue Analysis: Chemometric Estimations of Sampling, Amount and Error, D.A. Kurtz, ed. Symposium Series 284. Washington, DC: American Chemical Society.

Stout, S.A., A.D. Uhler, T.G. Naymik, and K.J. McCarthy. 1998. Environmental forensics: unraveling site liability. Environ. Sci. Technol. 32(11):260A-264A.

Tillitt, D.E., J.P. Giesy, and G.T. Ankley. 1991. Characterization of the H4IIE rat hepatoma cell bioassay as a tool for assessing toxic potency of planar halogenated hydrocarbons in environmental samples. Environ. Sci. Technol. 25(1):87-92.

Tillitt, D.E., R.W. Gale, J.C. Meadows, J.L. Zajicek, P.H. Peterman, S.N. Heaton, P.D. Jones, S.J. Bursian, T.J. Kubiak, J.P. Giesy, and R.J. Aulerich. 1996. Dietary exposure of mink to carp from Saginaw Bay. 3. Characterization of dietary exposure to planar halogenated hydrocarbons, dioxin equivalents, and biomagnification. Environ. Sci. Technol. 30(1):283-291.

USACE (U.S. Army Corps of Engineers). 1992. Guidance for Performanceing Tests on Dredged Material Proposed for Ocean Disposal. Draft. U.S. Army Corps of Engineers. New York District./ U.S. Environmental Protection Agency Region 2, New York, NY. December 18.

Van den Berg, M., L. Birnbaum, A.T.C. Bosveld, B. Brunström, P. Cook, M. Feeley, J.P. Giesy, A. Hanberg, R. Hasegawa, S.W. Kennedy, S.W., T. Kubiak, J.C. Larsen, F.X. van Leeuwen, A.K. Liem, C. Nolt, R.E. Peterson, L. Poellinger, S. Safe, D. Schrenk, D. Tillitt, M. Tysklind, M. Younes, F. Waern, and T. Zacharewski. 1998. Toxic equivalency factors (TEFs) for PCBs, PCDDs, PCDFs for humans and wildlife. Environ. Health Perspect. 106(12):775-792.

Williams, L.L., J.P. Giesy, N. DeGalan, D.A. Verbrugge, D.E. Tillitt, G.T. Ankley, and R.A. Welch. 1992. Prediction of concentrations of 2,3,7,8-TCDD equivalents (TCDD-EQ) from total concentrations of PCBs in fish fillets. Environ. Sci. Technol. 26(6):1151-1159.

Zwiernik, M.J., J.F. Quenson 3rd , and S.A. Boyd. 1998. FeSO4 amendments stimulate extensive anaerobic PCB dechlorination. Environ. Sci. Technol. 32(21):3360-3365.

Appendix G

Toxicity of PCBs

This appendix summarizes available data on PCB toxicity that are potentially useful for a risk assessment; it is not meant to provide a comprehensive review of PCB toxicity data. Data related to humans and aquatic organisms, such as freshwater and marine invertebrates, fish, birds, and marine mammals, that could be exposed to PCB-contaminated sediments are evaluated. The limitations of these toxicity data are also discussed, followed by a discussion of how this information is used in risk assessments, including toxic equivalency factors.

TOXIC EFFECTS OF PCBs

The toxicity of PCBs is well established from laboratory and field studies (Giesy et al. 1994a,b). Chronic toxicity has been observed in fish, birds, and mammals. Several studies demonstrate, however, that individual PCB congeners can act through different mechanisms and have different toxic potentials (Safe 1984, 1994; Strang et al. 1984; Seegal 1996). The overall impacts of PCBs on the environment and biota are due to not only the individual components of the mixtures but also the interactions (additive, synergistic, and antagonistic) among the PCB congeners present and between the PCBs and the other chemicals. Risk assessments of PCBs, therefore, require information on the levels of individual PCB congeners present in the PCB mixture and data on their interactions. Developments in high-resolution isomer-specific PCB analysis have made identification and quantitation of individual PCB congeners feasible, but challenges remain in the risk-assessment process because of the differing toxicity of individual PCBs.

Most in vivo animal studies and in vitro bioassays use commercially available, technical-grade mixtures of PCBs or individual congeners. PCBs present in the environment differ from commercially available mixtures, because different congeners are metabolized and biodegraded at different rates. Few studies have investigated the effects of environmentally altered mixtures of PCBs. The studies that have investigated those effects include field and controlled laboratory feeding experiments, but the co-occurrence of other toxicants, such as DDT, Toxaphene, and dieldrin, complicate their interpretation.

Commercial PCB mixtures elicit a broad spectrum of toxic responses that depend on several factors, including chlorine content, purity, dose, species, strain, age, and sex of animal, and route and duration of exposure. Immunotoxicity, carcinogenicity, neurotoxicity, and developmental toxicity, as well as the biochemical effects of commercial PCB mixtures, have been extensively investigated in various laboratory animals, fish, and wildlife species. The mechanisms and endpoints of PCB toxicity have been reviewed (Poland and Knutson 1982; Safe 1984; Barrett 1995; Silberhorn et al. 1990). Two main categories of PCBs have been designated based on mechanism of action: those that act through the arylhydrocarbon receptor (AhR) and those that do not.

AhR-Mediated Effects

The non- and mono-ortho-substituted PCBs are of particular concern, because these congeners can assume a planar or nearly planar conformation similar to that of 2,3,7,8-tetrachlorodibenzo-*p*-dioxin (TCDD) (Safe 1990; Giesy et al. 1994a; Metcalfe and Haffner 1995) and have toxic effects qualitatively similar to TCDD. These compounds act by the same mechanism as TCDD, that is, by binding to and activating the AhR, a cytosolic, ligand-activated transcription factor (Poland and Knutson 1982; Gasiewicz 1997; Blankenship et al. 2000). Each polychlorinated dibenzo hydrocarbon (PCDH) binds with different affinity to the AhR and, therefore, has different potency for biological effects (Safe 1990; Ahlborg et al. 1994; van den Berg et al. 1998).

A great deal of work has been conducted on the toxicity of dioxin, and there is a large amount of data on dioxin-like effects. The data come from experimental and epidemiological studies. Epidemiological studies have been conducted in individuals exposed to dioxin occupationally, in the chemical and the agricultural industries, individuals exposed environmentally (e.g., following an accident in Seveso, Italy, people living in farming communities were exposed to high levels of dioxin), and individuals exposed to Agent Orange, a herbicide that contained dioxin as a contaminant, during the conflict in Vietnam. Dioxin-like effects have been reviewed in great detail in other reports. In addition to numerous published review articles, a review was published by the Institute of Medicine (IOM) on the toxicity and epidemiological data on AhR-mediated effects in *Veterans and Agent Orange: Health Effects of Herbicides Used in Vietnam* and its subsequent updates (IOM 1994, 1996, 1999, 2000, 2001). EPA has also issued a draft health risk assessment of TCDD that summarizes and reviews the literature on the toxicity and epidemiology of dioxin and dioxin-like compounds (EPA 2000). Therefore, AhR-mediated health effects are only mentioned briefly here, and the reader is referred to other sources for a summary of the numerous studies.

As summarized by IOM (2001), dioxin-like effects comprise a diverse spectrum of sex-, strain-, age-, and species-specific effects, including carcinogenicity, immunotoxicity, reproductive or developmental toxicity, hepatotoxicity, neurotoxicity, chloracne, and loss of body weight. The wasting syndrome occurs at high concentrations of TCDD and is characterized by a loss of body weight and fatty tissue. Dioxin causes toxicity in the liver where lethal doses of TCDD cause necrosis (i.e., cell death). Effects on the morphology and the function of the liver are seen at lower doses. Dioxin affects the endocrine system of animals. Some experiments indicate that thyroid hormone levels are altered by activation of the AhR. Some neurodevelopmental effects have been seen in rats and monkeys after in utero TCDD exposure; some of those effects are not mediated by the AhR. In animals, one of the most sensitive systems to AhR-mediated toxicity is the immune system. Recent studies have demonstrated that dioxin can alter the levels of immune cells, the measured activity of those cells, and the ability of animals to fight off infection. Effects on the immune system, however, appear to be dependent on the species and strain of animal studied. Reproductive and developmental effects have been seen in animals exposed to dioxin. Effects on sperm counts, sperm production, and seminal vesicle weights have been seen in male offspring of rats treated with TCDD during pregnancy. Effects have also been seen on the female reproductive system following developmental exposure to TCDD. In some recent studies, however, the effects on the male and female reproductive systems were not accompanied by effects on reproductive outcomes. TCDD is a potent tumor promoter in laboratory rats. Evidence for an association between dioxin and some cancers (soft-tissue sarcoma, non-Hodgkin's lymphoma, Hodgkin's disease, and respiratory disease) is seen in epidemiological studies. Recent epidemiological studies also suggest an association between TCDD exposure and an increased incidence of diabetes.

Non-AhR-Mediated Effects

PCBs with two or more ortho chlorines do not interact with the AhR and elicit a different pattern of toxicity. These PCBs have been shown to elicit a diverse spectrum of non-Ah-receptor-mediated toxic responses in experimental animals, including neurobehavioral (Bowman et al. 1978, 1981; Schantz et al. 1991), neurotoxic (Fingerman and Russel 1980; Seegal et al. 1990; Seegal 1996; Kodavanti and Tilson 1997), carcinogenic (Barrett 1995; Ahlborg et al. 1995), and endocrine changes (Brouwer 1991; Van Birgelen et al. 1995). In addition, some of the metabolites of PCBs have antiestrogenic properties (Kramer et al. 1997) and cause hypothyroidism and decreased plasma vitamin A levels (Brouwer et al. 1995; Li and Hansen 1996; Brouwer 1991). These alterations in vitamin A and thyroid hormone concentrations might significantly modulate tumor promotion and developmental and adult neurobehavioral changes (Ahlborg et al. 1992). Although AhR-mediated toxicity is peliotropic, effects due to nondioxin-like PCBs might involve multiple unrelated mechanisms of action.

Developmental and cognitive dysfunctions observed in children born to mothers who consumed PCB-contaminated rice oil in Japan (Yusho) and Taiwan (Yu-Cheng) have been associated with exposure to halogenated aromatic hydrocarbons (HAHs) (Chen and Hsu 1994; Chen et al. 1994; Masuda 1985). The rice oil in both those incidents was contaminated with a mixture of HAHs, including PCBs, polychlorinated dibenzofurans (PCDFs), and PCQs (polychlorinated quarterphenyls) (Masuda 1985), and it has been difficult to determine which contaminants in the rice oil are responsible for the persistent behavioral and cognitive developmental alterations in the exposed children. Laboratory studies of ortho-substituted PCBs indicate that the developing nervous system is sensitive to PCBs and show effects similar to those seen following the accidental PCB poisonings and described in epidemiological reports (Kuratsune et al. 1972; Hsu et al. 1988; Jacobson et al. 1990; Seegal and Schantz 1994; Huisman et al. 1995; Seegal 1996). No or a poor correlation between the presence of AhR-mediated effects, such as chloracne and hyperpigmentation, and the observed cognitive dysfunctions suggest that the alterations in neurological function in the Yusho and Yu-Cheng children might be due to the nonplanar ortho-substituted PCB congeners present in many commercial mixtures of PCBs rather than to the coplanar contaminants that interact at the AhR (Yu et al. 1991; Rogan and Gladen 1992). The mechanisms of neurotoxic effects of ortho-substituted PCBs and the behavioral effects reported in epidemiological studies have been reviewed (Maier et al. 1994; Chu et al. 1996; Chishti et al. 1996; Eriksson and Fredriksson 1996a,b; Morse et al. 1996a,b; Gasiewicz 1997; Jacobson and Jacobson 1997; Kodavanti and Tilson 1997; Wong et al. 1997). The neurotoxicological effects of technical PCB mixtures in experimental studies are summarized in Table G-1. Although the mechanism of PCB-induced neurotoxicity has not been determined, various biochemical effects have been investigated.[1]

Aroclor 1254 (Greene and Rein 1977; Kittner et al. 1987; Seegal et al. 1989)[2] reduced dopamine (DA) content in pheochromocytoma (PC12) cells, a continuous cell line that synthesizes, stores, releases, and metabolizes DA in a manner similar to that of the mammalian central nervous system. Although the decrease in DA in one of those studies (Angus and Contreras 1996) was attributed to the cytotoxicity, it implies that the non-ortho PCBs could be lethal to cells at concentrations that are neurotoxic for certain ortho-substituted PCBs. Studies in a mouse neuroblastoma cell line (NIE-N115) also showed effects on the dopaminergic system (Seegal et al. 1991a,b). Of about 50 individual PCB congeners tested in PC12 cells, di-ortho- through tetra-ortho-substituted congeners were the most potent at affecting DA content, whereas coplanar PCB congeners were ineffective (Shain et al. 1991). In addition, chlorination in a meta-position decreased the potency of ortho-substituted congeners, but meta-substitution had little effect on congeners with both ortho- and para-substitutions. Changes in the content of neurotransmitters, such as DA, have been seen following exposure to PCBs in mice and nonhuman primates (Seegal et al. 1985, 1986a,b, 1991a,b,c, 1994). Those results suggest that some of the neurotoxicity associated with PCBs might be due to a mechanism independent of AhR activation. Rats appear to be less sensitive to the effects of PCBs on DA content (Morse et al. 1996a,b), suggesting that the potency of ortho-substituted PCBs in reducing brain DA might be species-specific. Other studies also suggest that the non-ortho coplanar congener 3,3′,4,4′-(PCB 77) alters DA concentrations in a species-, age- and dose-dependent manner (Agarwal et al. 1981; Chishti and Seegal 1992). Similarly, effects of PCBs on learning behavior appear to be sex-specific, females being more sensitive than males (Schantz et al. 1992), and age-specific, effects being prominent after prenatal exposure but less so after adult exposure (Seegal 1996). Cholinergic function was affected by early postnatal exposure of mice to non-ortho coplanar congeners (Eriksson et al. 1991; Eriksson and Fredriksson 1998).

Some PCB congeners also affect Ca^{2+} homeostasis and protein kinase C (PKC) translocation in cerebellar granule cells (Kodavanti et al. 1993a,b, 1994, 1996). That activity is congener-specific; ortho-substituted PCBs, but not AhR-activated congeners, affect Ca^{2+} homeostasis (Kodavanti et al. 1995).

---

[1]Some evidence suggests that effects on the dopaminergic system might be involved in the neurotoxicity of PCBs.

[2]And 3,3′,4,4′-5-(PCB 12b) (Angus and Contreras 1996).

TABLE G-1 Summary of Effects of Peri- and Postnatal Exposures to PCBs on Neurotoxic Effects in Animals

PCB Congener/ Mixture	Species, Sex, Age	Dose and Exposure	Effects and Effective Doses	Reference
In vivo studies				
3,3′,4,4′- (PCB 77)	CD-1 mice, pregnant female	32 mg/kg of birth weight (bw), oral, prenatal exposure, 10-16 d of gestation	Hyperactivity in offspring, neuromuscular dysfunction, learning and performance deficits, 'spinning' syndrome	Tilson et al. 1979
3,3′,4,4′- (PCB 77)	CD-1 mice, pregnant female	32 mg/kg bw, oral, prenatal exposure, 10-16 d of gestation	Hyperactivity in offspring, reduction in brain dopamine, behavioral alterations	Agarwal et al. 1981
3,3′,4,4′- (PCB 77)	NMRI mice, male, 10 d	0.41-41 mg/kg bw, oral, single postnatal exposure	Cholinergic system affected at 0.41 mg/kg bw, disturbed behavior	Eriksson et al. 1991
2,4,4′- (PCB 28)	NMRI mice, male, 10 d	0.18, 0.36, 3.6 mg/kg bw, oral, single postnatal exposure	After 4 months aberrations in spontaneous behavior, lack of effect on memory and learning and on nicotinic receptors, no effect on dopamine or serotonin, 0.36 mg/kg bw reduced total activity	Eriksson and Fredriksson 1996b
2,2′,5,5′- (PCB 52)	NMRI mice, male, 10 d	0.2, 0.41, 4.1 mg/kg bw, oral, single postnatal exposure	After 4 months aberrations in spontaneous behavior, deficits in memory and learning function, cholinergic nicotinic receptors affected, no effect on dopamine or serotonin, 4.1 mg/kg bw reduced total activity	Eriksson and Fredriksson 1996b
2,3′,4,4′,5- (PCB 118)	NMRI mice, male, 10 d	0.23, 0.46, 4.6 mg/kg bw, oral, single postnatal exposure	No significant changes in spontaneous and swim-maze behavior up to the dose of 4.6 mg/kg bw	Eriksson and Fredriksson 1996b
2,3,3′,4, 4′,5- (PCB 156)	NMRI mice, male, 10 d	0.25, 0.51, 5.1 mg/kg bw, oral, single postnatal exposure	No significant change in spontaneous and swim-maze behavior up to the dose of 4.6 mg/kg bw	Eriksson and Fredriksson 1996b

2,2′,5,5′-(PCB 52)	NMRI mice, male, 10 d	4.1 mg/kg bw, oral, single post-natal exposure	At 4 months decrease in rearing, locomotion and total activity	Eriksson and Fredriksson 1996a
3,3′,4,4′,5-(PCB 126)	Sprague-Dawley rats, both sexes, 5-7 wk (weanling)	0.1-100 ng/g in diet for 13 wk, oral, postnatal	Growth suppression, thymic atrophy, increased liver weight, anemia, no significant alterations in biogenic amines, NOAEL = 0.1 ng/g in diet or 0.01 mg/kg bw/d	Chu et al. 1994
3,3′,4,4′-(PCB 77)	Sprague-Dawley rats, both sexes, 5-7 wk (weanling)	10-10,000 ng/g in diet for 13 wk, oral, postnatal	Increased EROD activity, decreased vitamin A, altered dopamine and homovanillic acid in brain, histopathological changes in thyroid and liver, NOAEL = 100 ng/g in diet or 8.7 mg/kg bw/d	Chu et al. 1995
2,3′,4,4′,5-(PCB 118)	Sprague-Dawley rats, both sexes, 5-7 wk (weanling)	10-10,000 ng/g in diet for 13 wk for males, 2-2000 ng/g for females, oral, postnatal	Increased EROD activity, reduced dopamine and homovanillic acid in brain, histopathological changes in thyroid and liver, brain residues at the highest dose 0.36-1 mg/g, NOAEL = 200 ng/g in diet or 17 mg/kg bw/d	Chu et al. 1995
2,2′,4,4′,5,5′-(PCB 153)	Sprague-Dawley rats, both sexes, 5-7 wk (weanling)	50-50000 ng/g in diet for 13 wk, oral, postnatal	Increased EROD activity, reduction in hepatic vitamin A, decreased dopamine and its metabolites, females more sensitive, histological changes in thyroid and liver, highest dose brain residues 16-29 μg/g, NOAEL = 500 ng/g in diet or 34 μg/kg bw/d	Chu et al. 1996
2,2′,3,3′,4,4′-(PCB 128)	Sprague-Dawley rats, both sexes, 5-7 wk (weanling)	50-50000 ng/g in diet for 13 wk, oral, postnatal	Increased EROD activity, reduction in hepatic vitamin A, decreased dopamine and its metabolites, females more sensitive, histological changes in thyroid and liver, highest dose brain residues 5-10 μg/g, NOAEL = 500 ng/g in diet or 42 μg/kg bw/d	Lecavalier et al. 1997

TABLE G-1 *(Continued)*

PCB Congener/ Mixture	Species, Sex, Age	Dose and Exposure	Effects and Effective Doses	Reference
3,3′,4,4′,5- (PCB 126)	Lewis rats, adult female	10 and 20 μg/kg bw on days 9,11,13,15,17 and 19 days of gestation, oral, prenatal	Fetotoxicity, delayed physical maturation, reduced body weight in offspring, increased liver weight and EROD activity, no effect on learning or neuro-behavioral performance, no residues in brain, exhibited sex differences in neurotoxicity	Bernhoft et al. 1994
3,3′,4,4′,5- (PCB 126)	Lewis rats, adult female	2 μg/kg bw on days 10,12,14,16,18 and 20 days of gestation, oral, prenatal	Neurotoxic effects in offspring, no fetotoxicity, behavioral alterations, hyperactivity, impaired discrimination learning, no brain residues	Holene et al. 1995
2,3′,4,4′,5- (PCB 118)	Lewis rats, adult female	1 and 5 mg/kg bw on days 10,12,14,16,18 and 20 days of gestation, oral, prenatal	Neurotoxic effects in offspring, no fetotoxicity, behavioral alterations, hyperactivity, impaired discrimination learning, brain residues 6-982 ng/g	Holene et al. 1995
3,3′,4,4′- (PCB 77)	Wistar rats, adult female	1 mg/kg bw, days 7 to 18 of gestation, subcutaneous injection, prenatal	Behavioral effects in offspring, PCB concentrations in brain 0.15 μg/g	Weinhand-Härer et al. 1997
2,2′,4,4′- (PCB 28)	Wistar rats, adult female	1 mg/kg bw, days 7 to 18 of gestation, subcutaneous injection, prenatal	Behavioral effects in offspring, PCB concentrations in brain 0.61 μg/g	Weinhand-Härer et al. 1997
Fenclor 42	Fischer rats, adult female	5-10 mg/kg bw/d intake or 25-50 mg/kg, i.p., five injections daily, 2 wk prior to mating, prenatal	Neurotoxicity and behavioral alterations, 40 mg/kg resulted in significant postweaning behavioral effects, LOAEL = 10 mg/kg bw/d	Pantaleoni et al. 1988
Aroclor 1254	Wistar rats, adult female	0.2-26 μg/g in diet, preweaning, perinatal exposure	Impaired neurological development, LOAEL = 2.5 μg/g	Overmann et al. 1987

Aroclors 1254 and 1260	Wistar rats, adult male	500-1000 mg/kg bw, single oral exposure, postnatal	Decrease in dopamine, norepinephrine and serotonin concentrations in specific regions in brain up to 14 d after exposure	Seegal et al. 1986b
Aroclor 1254	Wistar rats, adult male	500-1000 mg/kg bw, oral exposure for 30 d, postnatal	Dopamine and its metabolites decreased, PCB concentrations in brain after 30 d were 75-82 μg/g, 6 di-*ortho* and 3 mono-*ortho* congeners dominated	Seegal et al. 1991a
Aroclor 1254	Wistar rats, adult female	5 and 25 mg/kg bw from day 10 to 16 of gestation, prenatal, oral	Alterations in seratonin metabolism in the brains of offspring after 21 and 90 d of birth, other biogenic amines (e.g., dopamine norepinephrine) in brain were unaffected, effect was significant at dose 25 mg/kg bw	Morse et al. 1996a
Aroclor 1254 and 3,3′,4,4′-(PCB 77)	Wistar rats, adult female	5 and 25 mg/kg bw from day 10 to 16 of gestation, prenatal, oral	Reduced plasma thyroid hormone, plasma concentrations of hydroxylated metabolite of PCB 153 was greater than the 153 in fetus, neonates and weanling rats, fetus brain thyroid residues affected, effect of OH-PCBs on brain is discussed	Morse et al. 1996b
Clophen A30	Wistar rats, adult female	5 and 30 mg/kg bw in diet or intake of 0.4 and 2.4 mg/kg/d, from 60 d prior to mating until 21 d after birth, oral	Behavioral effects, PCDF contamination in Clophen - 2.5 mg/kg, brain concentration = 60 ng/g after 420 d of exposure, PCBs 28, 52 and 101 were the prevalent ones	Lilienthal et al. 1990
Aroclor 1016	Pig-tailed macaque (*Macaca nemestrina*), male, 3-5 yr	0.8-3.2 mg/kg bw/d, for 20 wk, oral, postnatal	Persistent reduction in brain dopamine, brain PCB concentrations 1-5 μg/g, only PCBs 28, 47 and 52 accumulated in brain, lightly chlorinated PCB mixtures are more effective than heavily chlorinated ones	Seegal et al. 1990

TABLE G-1 *(Continued)*

PCB Congener/ Mixture	Species, Sex, Age	Dose and Exposure	Effects and Effective Doses	Reference
Aroclor 1260	Pig-tailed macaque (*Macaca nemestrina*), male, 3-5 yr	0.8-3.2 mg/kg bw/d, for 20 wk, oral, postnatal	Persistent reduction in brain dopamine, brain PCB concentrations 18-28 μg/g, di-*ortho* substituted hexa- and heptaCBs accumulated in brain, less effective to reduce dopamine as compared to Arolcor 1016 exposure	Seegal et al. 1990
Aroclor 1248	Rhesus monkeys, adult female	0.5-2.5 mg/kg in diet, exposed before and during gestation, oral, perinatal, cumulative PCB intake was 293 mg	Hyperactivity in offspring, behavioral deficits, PCB concentrations in body fat was 20 μg/g	Bowman et al. 1981
In vitro or ex vivo studies				
Aroclors 1254: 1260 (1:1)	Wistar rats, male, 65 d	10-100 μg/g in media, ex vivo brain tissue, 6 h exposure	Decrease in dopamine and its metabolites at 20 μg/g or above, brain total PCB concentration at the effective dose was >15 μg/g	Chishti et al. 1996
Aroclor 1254	PC-12 cells	1-100 μg/g, in vitro, 6 h exposure	Increase followed by a decrease in cellular catecholamine	Seegal et al. 1989
2,2′- (PCB 4)	Long-Evans hooded rats, adult male	50-200 μM, in vitro, cerebellar granule cells exposed	Altered Ca^{2+} homeostasis in cerebellar granule cells, IC_{50} = 6.17 μM, more effective than PCB 126	Kodavanti et al. 1993a,b
3,3′,4,4′,5- (PCB 126)	Long-Evans hooded rats, adult male	50-200 μM, in vitro, cerebellar granule cells exposed	Altered Ca^{2+} homeostasis in cerebellar granule cells, IC_{50} = 7.61 μM	Kodavanti et al. 1993a,b

2,2′- (PCB 4)	Long-Evans hooded, male, adult rats, 40-90 d	10-100 μM, in vitro, mitochondrial and synaptosomal preparations from brain exposed	Mg^{2+}-ATPase activity inhibited, but not Na^+/K^+-ATPase activity, ED_{50} is roughly 25 μM	Maier et al. 1994
3,3′,4,4′,5- (PCB 126)	Long-Evans hooded, male, adult rats, 40-90 d	10-100 μM, in vitro, mitochondrial and synaptosomal preparations from brain exposed	Mg^{2+}-ATPase activity was not inhibited up to the dose of 100 μM	Maier et al. 1994
2,2′,3,5′,6- (PCB 95)	Sprague-Dawley rats, male	1-200 μM, in vitro, microsomes of rat brain hippocampus	Alterations in neuronal Ca^{2+} signal and neuroplasticity, EC_{50} = 12 mM	Wong et al. 1997
2,3′,4,4′- (PCB 66)	Sprague-Dawley rats, male	1-200 μM, in vitro, microsomes of rat brain hippocampus	No effect was found on [^{3}H] ryanodine receptors suggesting no alterations in neuronal Ca^{2+} signal up to 200 μM	Wong et al. 1997

2,2′,3,5′,6-(PCB 95), a di-ortho substituted congener, altered Ca^{2+} transport in rat brain microsomes (Wong et al. 1997). Another di-ortho substituted congener, 2,2′-di-CB, interfered with oxidative phosphorylation by inhibiting mitochondrial Mg^{2+}-ATPase activity in mitochondrial and synaptosomal preparations of rat brain (Maier et al. 1994). Alterations in hormone levels involved in regulating neuronal growth and development, including thyroid hormones, could also contribute to PCB-induced neurotoxicity (Seegal 1996).

Coplanar[3] HAHs inhibit estradiol-induced cell clumping in the MCF-7 breast cancer cell line (Gierthy and Crane 1984) by affecting the metabolism of estradiol to 2- and 4-hydroxy estradiol (Lloyd and Weisz 1978; Foreman and Porter 1980). Similarly, the hydroxy metabolites of PCBs are antiestrogenic (Kramer et al. 1997).

A few experimental studies have examined the effects of ortho-substituted PCBs in fish (Fingerman and Russel 1980) and birds (Kreitzer and Heinz 1974). Dietary exposure of Japanese quail (*Coturnix coturnix japonica*) to Aroclor 1254 at 200 μg/g for 8 d showed a suppressed avoidance response (Kreitzer and Heinz 1974). Further studies on the behavioral effects of ortho-substituted PCB congeners in fish and other wildlife are needed for risk assessment.

Few studies have examined the effects of ortho-substituted PCBs on behavioral alterations or neurotoxic effects in wildlife. Dietary exposure of mink to 2,2′,4,4′,5,5′-HxCB (PCB 153) and 2,2′,3,3′,6,6′-HxCB (PCB 136) at 5 μg/g for over 3 months did not produce significant changes in concentrations of DA, norepinephrine, or seratonin in the brain (Aulerich et al. 1985). Exposure to the di-ortho congeners 2,2′,5,5′-(PCB 52), 2,2′,4,5,5′-(PCB 101), 2,2′,3,3′,4,4′-(PCB 128), 2,2′,3,4,4′,5′-(PCB 138), 2,2′,4,4′,5,5′-(PCB 153), and 2,2′,3,4,4′,5,5′-(PCB 180) did not affect survival, growth, or reproduction in the fathead minnow *Pimephales promelas*, despite accumulating up to 183 μg/g (wet weight (wt)) in tissues (Suedel et al. 1997). Behavioral effects, however, were not examined in that study. Studies have demonstrated the presence of ortho-substituted tetra- through hexa-CB PCB congeners in the brains of fish. 2,2′,4,4′,5,5′-(PCB 153) was the predominant di-ortho congener; lesser chlorinated ortho-substituted congeners, such as 2,4,4′- (PCB 28) and 2,2′,5,5′-(PCB 52), did not accumulate (Qi et al. 1997). Slightly chlorinated PCBs have been shown to be metabolized in fish (Willman et al. 1997).

Studies with wildlife have demonstrated a causal link between adverse health effects and PCB exposure (Kennedy et al. 1996a,b; Giesy et al. 1994a; Bowerman et al. 1995). The observed toxicity to birds and mammals, how-

---

[3]There is also evidence that noncoplanar PCBs can have antiestrogenic effects.

ever, correlates more strongly to TCDD equivalents (TEQs) than to total PCBs (Giesy et al. 1994a; Leonards et al. 1995).

Although neurotoxicological effects of PCBs have been seen following dietary exposure of rats, mice, or nonhuman primates with technical mixtures of PCBs, those mixtures might not represent the PCB mixtures found in environmental matrices, and the doses used in those studies are higher than those seen in the environment. Similarly, in vitro assays use high concentrations of PCBs and the EC_{50} values (effective concentration in 50% of the test population) for various endpoints were generally high (greater than 50 μM). The EC_{50} values for neurotoxicological effects in in vitro studies are presented in Table G-1.

Few studies have examined which PCB congeners are present in the brains of humans and wildlife. PCBs were not detected in brain tissues obtained from two men with Parkinson's disease (Corrigan et al. 1996). Concentrations of total PCBs in the brain of a Yu-Cheng victim was 80 ng/g, whereas those in fat tissues ranged up to 11 μg/g (Chen and Hsu 1986). PCB 153 (2,2′,4,4′,5,5′-HxCB 1.6 ng/g, wet wt) and PCB 138 (2,2′,3,4,4′,5′-HxCB 0.96 ng/g, wet wt) were the only two congeners detected in the brain of grey seals; the concentration was only 1% of that measured in the blubber (Jenssen et al. 1996). In harbor porpoises, the PCB profile in brain tissue resembled those in other tissues, with PCB 153 > PCB 138 > PCB 187 (Tilbury et al. 1997), suggesting that there was no preferential enrichment of lesser chlorinated ortho-substituted PCBs in wildlife. PCB concentrations in the brains were 1.5% of those found in the blubber. Similarly, concentrations of total PCBs in the brains of marine mammals from Greek waters were 1-2% of those found in the blubber (Georgakopoulou-Gregoriadou et al. 1995). Lesser chlorinated ortho-substituted PCB congeners are metabolized in humans (Tanabe et al. 1988) and dolphins (Kannan et al. 1994; Boon et al. 1997; Leonards et al. 1997). Therefore, the accumulation of the lesser chlorinated PCBs might be small following chronic exposure. Laboratory studies have shown the presence of PCBs at greater than 1 μg/g (wet wt) in brains of exposed rats and mice, but that amount could be from exposure at higher concentrations. Therefore, the neurotoxicological effects observed in laboratory studies might occur only at higher exposures, such as those seen in Yusho and Yu-Cheng or following occupational exposures.

TOXIC EQUIVALENCY FACTORS

In the environment, PCBs usually exist as mixtures, which complicates their risk assessment. One approach to congener-specific hazard assessment for complex mixtures is to develop relative potency factors for individual congeners on the basis of its mechanism of action. The complex nature of PCB mixtures found in environmental and biological samples make this a daunting, if not impossible, task. If each congener causes a different toxic response by an independent mechanism, then the relative toxicities of each congener must be determined separately. 2,3,7,8-Tetrachlorodibenzo-*p*-dioxin (TCDD) and structurally related halogenated aromatics invoke a number of common toxic responses that are mediated through the AhR (Poland et al. 1976, 1979; Poland and Glover 1977; Poland and Knutson 1982; Safe 1990; Whitlock 1987; Goldstein and Safe 1989). TCDD has the highest affinity for the AhR and is the most toxic HAH. Structurally similar polychlorinated dibenzo-*p*-dioxin (PCDD), PCDF, and PCB congeners have similar effects but are less potent.

The TCDD equivalency factor (TEF) approach has been developed on the basis of AhR-binding, structure-activity relationships, and cellular responses to express the potency of various PCBs, PCDDs, and PCDFs relative to TCDD. In that way, data on TCDD, for which there is information on several endpoints in many different species, can be used to derive a maximum allowable tissue concentration (MATC) for the related comments. If congeners have the same rank order across endpoints and species and relative potencies are available for PCB congeners for a few endpoints and species, then TEFs can be developed for each congener. TCDD, the most potent AhR agonist, is designated to have a TEF of 1, and other compounds are assigned a TEF that is some fraction of 1, depending upon its characteristic responses. TEFs, however, are endpoint- and species-dependent for all PCB congeners that have been tested (Safe 1990). Regulatory agencies have established consensus TEF values for individual congeners. The World Health Organization (WHO) has proposed tentative TEFs for mammals, birds, and fish that are updated as more data become available (see Table G-2) (van den Berg et al. 1998). The mechanistic considerations for the development of TEFs for the risk assessment of PCBs, including human, teleost, and avian risk assessments, are described elsewhere (Safe 1990, 1994). Considerations for the use of mammalian, teleost, and avian TGFs are discussed briefly below, as well as the limitations of this approach.

Although TEFs have limitations and uncertainties associated with them, when used appropriately they are currently considered acceptable for use in the risk assessment of planar HAHs. Some of the uncertainties resulting from interactive effects among planar HAHs can be quantified in vitro bioassay techniques.

Mammalian TEFs

The largest amount of data for the development of TEFs comes from research in mammals. Initially, all PCB congeners were regarded as toxic,

TABLE G-2 TCDD Toxicity Equivalency Factors (TEFs) for Several Dioxin-like PCB Congeners for Fish, Birds, and Mammals

PCB Congener (IUPAC No)	Fish TEF	Bird TEF	Mammal TEF
2,3,7,8-Tetra-CDD	1	1	1
3,3′,4,4′-Tetra-CB (77)	0.0001	0.05	0.0001
3,4,4′,5-Tetra-CB (81)	0.0005	0.1	0.0001
3,3′,4,4′,5-Penta-CB (126)	0.005	0.1	0.1
3,3′,4,4′,5,5′-Hexa-CB (169)	0.00005	0.001	0.01
2,3,3′,4,4′-Penta-CB (105)	<0.000005	0.0001	0.0001
2,3,4,4′,5-Penta-CB (114)	<0.000005	0.0001	0.0005
2,3′,4,4′,5-Penta-CB (118)	<0.000005	0.00001	0.0001
2′,3,4,4′,5-Penta-CB (123)	<0.000005	0.00001	0.0001
2,3,3′,4,4′5-Hexa-CB (156)	<0.000005	0.0001	0.0005
2,3,3′,4,4′,5′-Hexa-CB (157)	<0.000005	0.0001	0.0005
2,3,4,4′,5,5′-Hexa-CB (167)	<0.000005	0.00001	0.00001
2,3,3′,4,4′,5,5′-Hepta-CB (189)	<0.000005	0.00001	0.0001

Source: Data from van den Berg et al. (1998).

and early text books suggested that the toxicity of PCB congeners was proportional to the degree of chlorination. In vivo studies conducted with rodents in the 1970s and the 1980s, however, found that the toxicity of PCB congeners varied greatly, with a small group of congeners having great toxic potential (Safe et al. 1982; Silberhorn et al. 1990). The location of chlorine atoms was found to be more important to PCB toxicity than the number of chlorine atoms. Studies using mammalian models found a correlation between the structure and the AhR-binding affinity for certain PCB congeners that exhibit "dioxin-like" activities, such as induction of cytochrome P-450s (e.g., increased arylhydrocarbon hydroxylase (AHH) activity and ethoxyresorufin *O*-deethylase (EROD) activity), body-weight loss, hypothyroidism, decreased hepatic or plasma vitamin A levels, porphyria, thymic atrophy, immunotoxicity, and teratogenicity (Safe 1984; Poland and Glover 1977; Goldstein and Safe 1989; Parkinson et al. 1980; Bandiera et al. 1982; Parkinson et al. 1983; Yoshimura et al. 1985; Leece et al. 1985). The non-ortho-substituted coplanar PCBs 3,4,4′,5-tetra-CB (PCB 81), 3,3′,4,4′-tera-CB (PCB 77), 3,3′,4,4′,5-penta-CB (PCB 126), and 3,3′,4,4′,5,5′-hexa-CB (PCB 169), which are substituted in both para, at least two meta, and no ortho positions, are the most toxic PCB congeners. It is hypothesized that the lack of chlorine substitution at opposing ortho positions allows the two phenyl rings to rotate into the same plane; therefore, these congeners are commonly referred to as coplanar PCBs. The toxic potencies vary; for example, the potency ratio of PCB 126 to TCDD was 66 for body-weight loss in rats, 8.1 for thymic atrophy in rats, 10 for mouse fetal thymic lymphoid development, 125 for AHH induction in rats, and 3.3 for AHH induction in rat hepatoma H4IIE cells. Based on those data, TEFs in the range of 0.008-0.3 could be derived. A consensus mammalian TEF of 0.1 was assigned to this congener (van den Berg et al. 1998).

Similar to non-ortho coplanar PCBs, chlorobiphenyl congeners with chlorine substitution at only one ortho position (mono-ortho PCBs) can achieve partial coplanarity and exhibit AhR agonist activity. TEFs have been proposed for those PCBs based on their potency relative to TCDD in various assays (Safe 1994). Those potencies can vary by 2 to 3 orders of magnitude, depending on the species and the endpoint used to derive values. When the TEF values were calculated, in vivo studies were given more strength than biochemical changes. In mammalian studies, chronic in vivo exposures were given more weight than acute exposure studies. Current TEF values are considered tentative and subject to modification as new data become available. Recognizing the need for a more consistent approach for setting internationally accepted TEFs, the World Health Organization–European Centre for Environment and Health (WHO-ECEH) and the International Program of Chemical Safety (IPCS) initiated a project in the early 1990s to create a database containing information relevant to the setting of TEFs and to derive consensus TEFs for halogenated aromatics (Ahlborg et al. 1994). The first international TEFs for dioxin-like PCBs were proposed in 1994; those values have been revised and updated (van den Berg et al. 1998).

Teleost TEFs

Because the toxicity of coplanar PCBs vary among vertebrate taxa, recent studies have focused on determining TEFs for coplanar PCBs in fish and bird models (Bosveld and van den Berg 1994; Brunstrom et al. 1995; Newsted et al. 1995; Zabel et al. 1995; Kennedy et al. 1996a,b). The AhR is present in several fish species and cell lines (Lorenzen and Okey 1990; Swanson and Perdew 1991; Hahn et al. 1992). Therefore, the mechanistic basis for TEFs in aquatic species should be similar to that observed in mammalian systems, but at present few reports are available for the development of TEFs in fish. Fish-specific TEFs have been estimated using induction of AHH and EROD in salmonid species (Janz and Metcalf 1991; Williams and Giesy 1992; Newsted et al. 1995), and embryo mortalities in salmonids (Walker and Peterson 1991; Zabel et al. 1995; Zabel et al. 1996) and Japanese medaka (Harris et al. 1994). Generally, TEFs for coplanar PCBs derived from early-life-stage mortality studies with rainbow trout are less than mammalian TEFs (Zabel et al. 1995). Little information is available on the toxic potencies of mono-ortho-substituted congeners in fish to speculate on appropriate teleost TEFs (Metcalf and Haffner 1995). Mono-ortho-substituted congeners (IUPAC Nos. 105,118, and 156; 0.45 to 4.4 mg/kg) do not induce EROD activity in rainbow trout (Metcalf and Haffner 1995), and studies of cytochrome P-450 1A1 induction and mortality show that ortho-substituted PCBs lack biological activity in trout (Walker and Peterson 1991; Newsted et al. 1995; Zabel et al. 1995; Hornung et al. 1996). Congener 156, however, did induce cytochrome P-450 1A messenger RNA in a rainbow trout gonadal cell line (RTG-2), but its potency was weak relative to non-ortho-substituted congeners (Zabel et al. 1996). The use of TEFs for di-ortho PCBs derived from mammalian exposure studies, therefore, would overestimate the effects of these mixtures on fish. Fish-specific TEFs have been developed for non-ortho and mono-ortho PCBs (Walker and Peterson 1991; Walker et al. 1996) and have been tentatively adopted for use by EPA (EPA 1993). It should be noted, however, that TEF values for fish are based primarily on acute exposure data. Long-term toxicity studies with rainbow trout exposed to lower concentrations of TCDD indicate that many toxic responses can develop after a few weeks (Van der Weiden et al. 1992). Additional research is needed to refine TEFs in fish and establish consensus teleost TEFs.

Avian TEFs

Endpoints used to estimate TEFs for birds include in vitro and in ovo EROD induction (Yao et al. 1990; Bosveld et al. 1992; Kennedy et al. 1996a,b) and embryo mortalities (Brunström 1989); most are based on EROD induction potencies. Unlike rodent bioassay results (Safe 1994), PCB 169 was less potent than PCB 77 in chicken embryo hepatocytes (Brunström 1990; Kennedy et al. 1996a,b) and in embryo lethality assays in birds (Brunström 1989). TCDF was also a more potent inducer of EROD activity in avian assays than TCDD (Bosveld et al. 1997). Mono-ortho PCBs were less potent inducers of EROD than non-ortho congeners in bird models, but mono-ortho congeners were more potent in birds than in teleosts and rodents. In most environmental PCB mixtures, PCB 77 (3,3′,4,4′-tetra-CB) contributes the greatest proportion of TEQs based on avian TEFs. Therefore, more information on the TEF and environmental fate of this congener, particularly on its pharmacokinetics in birds, is necessary for an accurate avian risk assessment.

Furthermore, avian TEFs are difficult to estimate, because there is considerable variation in the toxicity of PCB congeners among birds (Bosveld and van den Berg 1994). Although TEFs in birds are based on EROD induction with eggs and cell cultures from chickens, the preferred endpoint is embryo lethality following in ovo exposure. Domestic chickens and their embryos are considerably more sensitive to AhR-mediated responses than other avian species (Kennedy et al. 1996a,b; Lorenzen et al. 1997) and, in general, fish-eating bird species are at lest an order of magnitude less sensitive than the domestic chicken. In contrast to what was seen in mammalian and teleost cell lines, TCDF was 1.2- to 3.4-fold more potent than TCDD in the white leghorn chicken (Bosveld et al. 1997). Based on in vitro EROD induction potency of coplanar PCB congeners in several bird species, the order of sensitivity is the following: domestic chicken > ring-necked pheasant > turkey ≈ double-crested cormorant ≈ great blue heron ≈ ring-billed gull ≈ duck ≈ herring gull ≈ common tern > Foster's tern (Sanderson et al. 1998). Therefore, toxicological information for the chicken is less appropriate than other species for risk assessment of avian wildlife species, and use of RfD values based on chicken would be over protective of most species. If chicken TEFs and RfDs are used as a surrogate for wild birds, no uncertainty factors (Ufs) should be applied.

Application of the TEF Approach

TEFs have been used to assess the risk of a mixture of PCB congeners measured in biota and or environmental matrices. The concentration of each non-ortho or mono-ortho congener detected in the biota is multiplied by its corresponding TEF to yield a TCDD-equivalent concentration, or TEQ (Table G-3) (Tanabe et al. 1989). A total TEQ for a sample can be calculated by summing the TEQs for each congener present. In that way, the toxic potential of a mixture of individual congeners can be expressed as one integrated parameter relative to the potency of TCDD.

The utility of the TEF approach to environmental risk assessment is shown by the correlation between total TEQs and adverse effects in populations of birds and fish (Tillitt et al. 1992; Giesy et al. 1994a,b). Negative correlations were reported between measured TEQs and the incidence of deformities in cormorant populations from the Great Lakes (Yamashita et al. 1993), egg volume in populations of common terns in The Netherlands (Bosveld and van den Berg 1994), survival of early life stages in populations of lake trout from the Great Lakes (Mac et al. 1993), and hatching success in Forster's tern (Kubiak et al. 1989).

TABLE G-3 An Example for Deriving 2,3,7,8-TCDD Toxicity Equivalents (TEQs) by the TEF Approach

Congener	TEF[a]	Concentration (pg/g, wet wt)[b]	TEQ (pg/g, wet wt)
Dioxins			
2,3,7,8-Tetra-CDD	1	3.7	3.7
1,2,3,7,8-Penta-CDD	1	6.4	6.4
1,2,3,4,7,8-Hexa-CDD	0.1	3.9	0.39
1,2,3,6,7,8-Hexa-CDD	0.1	34	3.4
1,2,3,4,6,7,8-Hepta-CDD	0.01	33	0.33
OCDD	0.0001	510	0.051
Furans			
2,3,7,8-Tetra-CDF	0.1	3.1	0.31
1,2,3,7,8-Penta-CDF	0.05	0.5	0.025
2,3,4,7,8-Penta-CDF	0.5	11	5.5
1,2,3,4,7,8-Hexa-CDF	0.1	5.6	0.56
1,2,3,4,6,7,8-Hepta-CDF	0.01	2.9	0.029
Non-*ortho* PCBs			
3,3′,4,4′-Tetra-CB	0.0001	350	0.035
3,3′,4,4′,5-Penta-CB	0.1	330	33
3,3′,4,4′,5,5′-Hexa-CB	0.01	90	0.9
Total TEQs			54.69

[a]From World Health Organization (1997).
[b]From Tanabe et al. (1989) for human adipose tissue.

Limitations of TEF Approach

Interactive Effects of PCBs

Despite the ability of the TEF approach to predict the potency of some mixtures of planar HAHs, it assumes that toxic responses to planar HAHs are additive and that other classes of contaminants do not modify or add to the toxicity. Those assumptions are equivocal (Giesy et al. 1994a; Safe 1994). Some rodent data indicate that responses to mixtures of planar HAHs are additive (Sawyer and Safe 1985; Pluess et al. 1988), but other rodent data show antagonistic (Haake et al. 1987; Biegel et al. 1989) or synergistic responses (Birnbaum et al. 1985; Bannister and Safe 1987). Recent studies have also reported additive effects (Walker et al. 1996), but other data indicated interactive effects (Pohjanvirta et al. 1995; Harper et al. 1995; Van Birgelen et al. 1995; Li and Hansen 1996, 1997; Vamvakas et al. 1996). TEQs estimated on the basis of instrumental analysis do not account for those interactions. In contrast, bioassay-derived TEQs integrate potential interactions among AhR agonists and other compounds by measuring a receptor-mediated response (Tillitt et al. 1991; Sanderson et al. 1998). Comparison of bioassay-derived TEQs with those of TEQs estimated for the same sample by instrumental analysis also suggest that additive and nonadditive interactions exist in biota (Tillitt et al. 1992; Williams and Giesy 1992; Mac et al. 1993). Often those interactions are ignored in the application of TEFs to risk assessment. Ignoring nonadditive interactions in the TEF approach has been justified because (1) the antagonistic or synergistic effects are observed at only very high doses, and the magnitude of those interactions are smaller than the uncertainties already present in the TEF values; (2) the observed nonadditive effects are highly species-, response- and dose-dependent, and their relevance might be of minimal importance; and (3) the mechanisms responsible for these nonadditive effects are unknown (Ahlborg et al. 1992). In general, complex mixtures of PCBs are slightly less than additive; therefore, calculating a TEQ based on an additive model is considered to be a conservative approach (i.e., protective).

Specificity of TEFs

The TEF approach assumes that the rank order of relative potencies of congeners are the same among species. There are, however, considerable variations in the potency of mono-ortho and non-ortho PCBs among mammalian, teleost, and avian models. Therefore, the application of TEFs from rodent bioassays for the assessment of risks in aquatic mammals (e.g., dolphins and whales) might not be appropriate. In addition, PCB congeners have different potencies for various endpoints, resulting in a range of potency values from which the congener-specific TEF is derived. Therefore, the predictive ability of the TEF approach is species- and endpoint-dependent (Safe 1994; Metcalf and Haffner 1995; Seed et al. 1995), which can add uncertainties of a few orders of magnitude to the risk assessment. There is also evidence of age- and sex-specific differences in sensitivity to PCB toxicity (Bosveld et al. 1997) that could add uncertainty in the derivation of TEFs.

Many TEFs are based, in part, on a congener's ability to induce P-450 enzymes. The induction of P-450 enzymes could be an adaptive mechanism and might not necessarily indicate a toxic effect. That induction is also sometimes nonspecific. Therefore, the use of EROD cytochrome P-450 induction (e.g., measuring EROD activity) to measure TEF and, eventually in risk assessment, requires careful interpretation.

Presence of Non-AhR-Mediated Effects

The TEF approach has been validated for estimating the risks of non-ortho-substituted planar PCB congeners that exhibit dioxin-like activities based on their ability to interact with and activate the Ah receptor. The TEF approach does not consider potential adverse effects of ortho-substituted nonplanar PCB congeners that do not interact with the AhR, but elicit nondioxin-like effects (Seegal 1996). Therefore, the use of TEQs for assessing the potential toxicity of PCBs might not address all the issues of potential adverse effects by PCBs, and the risk to humans or wildlife following exposure to complex mixtures of PCBs might be biased or underestimated (Safe 1994; Birnbaum and DeVito 1995; Seegal 1996). That possibility is especially important because only a small portion of the total mass of PCB mixtures are coplanar non-ortho congeners that elicit dioxin-like activities (Birnbaum and DeVito 1995; Safe 1990; Neubert et al. 1992; Neumann 1996). If the dioxin-like PCBs are the critical contaminant, then variation among mixtures can be reduced by the TEQ approach. For example, application of the TEF concept to examine the tumor promotion potential of Aroclor 1260 underpredicted the observed carcinogenic potential, because ortho-substituted PCB congeners, such as 2,2′,4,4′,5,5′-hexachlorobiphenyl (PCB 153), which are major constituents of Aroclor 1260, are potent tumor promoters (Smith 1997). However, if the critical mechanism of action is caused by non-TCDD-like compounds, the use of the TEQ approach would not be accurate.

The effects of non-TCDD-like PCBs are probably not the critical toxic effects of PCBs. First, relatively high concentrations of lesser chlorinated di-ortho-substituted congeners need to accumulate in the brain to cause the observed effects. Second, field studies suggest that the active congeners do not tend to accumulate in the brains of animals exposed to complex mixtures of PCBs in the environment. Finally, some neurotoxic effects are mediated by the less-chlorinated PCBs, which are more easily degraded in the environment, bioconcentrate less, and are more readily metabolized and excreted.

Toxicokinetics

The TEF values for dioxin-like PCBs have been derived mainly from short-term tests and in vitro assays (Safe 1994). Such studies might not reflect the pharmacokinetics, metabolism, and excretion that affect the concentration of a chemical at the target organ (De Vito et al. 1995; Lawrence and Gobas 1997). When extrapolating across species, the toxicokinetics must be identical, or differences have to be taken into account. Some factors that affect the interspecies differences have been reviewed (Barrett 1995). In addition, species- and tissue-specific differences in the binding properties, specificity, and physical-chemical properties of the AhR, and the contribution of other P-450 genes to HAH-induced activities challenge the generalities of assumptions of the TEF approach.

Exposure of experimental animals to weathered PCBs might provide more realistic estimates for the risk assessment of ortho-substituted PCB congeners. A few studies have examined behavioral alterations in rats following exposure to contaminated fish from the Great Lakes (Hertzler 1990; Daly 1993). Rats fed different amounts of Great Lakes fish (8%, 15%, and 30% of the diet) for 20 days exhibited behavioral alterations, including reduced exploratory activity, and decreased rearing and nose-poke behavior, which were not exhibited by controls. PCB concentrations in the fish were in the range of 4 to 19 μg/g (wet wt), and total PCB concentrations in the brains of the rat after the exposure period was 50-78 ng/g (wet wt). In contrast, in another study, no significant behavioral effects were seen in rats following a 90-day exposure to PCB-contaminated Great Lakes fish, even though accumulation of ortho-substituted congeners, such as 2,2′,4,4′- (PCB 47), 2,2′,5,5′- (PCB 48), 2,2′,4,4′,5,5′- (PCB 153), 2,2′,5,5′- (PCB 52), 2,4,4′,5- (PCB 74), and 2,2′,4,5′- (PCB 49), in the brain was in the range of 2.5 to 18 ng/g (wet wt) (Beattie et al. 1996). However, the presence of several other contaminants in the diet, such as methylmercury, could confound those studies.

Dose-Response Relationships

When developing relative potency values, a number of assumptions are made about the dose responses for the various compounds. One assumption is that the maximum achievable response for the endpoint of interest is identical for the chemicals evaluated and TCDD (i.e., the congener of interest, although less potent, must have the same efficacy as TCDD). It is also assumed that the dose-response curves are parallel and that they have the same origin. Based on both theoretical analyses and empirical examples from some studies that developed TEFs, it has been demonstrated that these assumptions for dose-response relationships are seldom met (Putzrath 1997). For example, the slopes of the dose-response curves for many endpoints are different (De Vito et al. 1994). It has been suggested that the relative potency among chemicals would be more accurately represented by a function rather than a point estimate, such as the EC_{50} or LD_{50} (lethal dose to 50% of the test population), which are generally used to estimate relative potency (Neubert et al. 1992; Putzrath 1997). That could be accomplished by the use of probability functions.

Extrapolation of Dose Ranges and Routes of Exposure

Most of the information used for establishing TEFs has come from in vitro studies of the induction of monooxygenases and, more recently, from subchronic toxicity studies. Many of the in vivo studies examined acute high-dose effects, such as lethality. Chronic, low-level effects are more relevant to real-world scenarios; therefore, TEFs derived from high-dose exposures might be questionable. The use of different dose regimens in toxicity studies adds further uncertainty to the derived TEF value and eventually to the risk-assessment process.

TOXICITY REFERENCE VALUES

A toxicity reference value (TRV) is a concentration of a chemical in water, food, or a tissue[4] that is not expected to cause toxicological effects in the organism of concern. Ideally, TRVs are derived from chronic toxicity studies in which an endpoint relevant to ecological systems or humans was assessed in the species of concern or a closely related species. TRVs are usually derived from the no-observed-adverse-effect level (NOAEL) or the lowest-observed-adverse-effect level (LOAEL). Alternatively, TRVs can be expressed as the geometric mean of the NOAEL and LOAEL to provide a conservative estimate of a threshold of effect (Tillitt et al. 1996).

TRVs that are based on data in the species of interest are not available for the majority of wildlife. Therefore, it is often necessary to use experimental data from a surrogate species to derive a TRV. Uncertainties are associated with the extrapolation of laboratory toxicity data to species exposed in the environment, and the magnitude of those uncertainties should be accounted

---

[4]That can be derived as part of a risk assessment.

for in the TRV. First, there is a wide range of sensitivities to AhR-active chemicals among even closely related species (Gasiewicz 1997). Second, most laboratory studies of toxicity are based on exposure to a parent Aroclor mixture. That mixture might be substantially different from the mixture to which animals in the environment are exposed. Third, if a TEF-TEQ approach was used in the derivation of the TRV, uncertainty could result from the appropriateness of the TEFs. For example, Elliott et al. (1996) derived TEQ-based TRVs for bald eagles using a mammal-based set of TEFs, because bird-specific TEFs were not available. If new data are not available with which to calculate the appropriate TEFs, the resulting uncertainty should be accounted for.

Because of those uncertainties, it is essential to evaluate the applicability of the toxicological data on a site-specific basis for the different exposure pathways and organisms of concern. Appropriate UFs should be applied for each scenario. Uncertainty concerning interpretation of the toxicity test information among different species, different laboratory endpoints, and differences in experimental design (e.g., age of test animals and duration of test) are addressed by applying UFs to the toxicology data to derive the final TRV (TetraTech 1998).

Methods for applying UFs have been published (Opresko et al. 1994; EPA 1995). For example, the method published by EPA Region 8 for the Rocky Mountain Arsenal (RMA) (EPA 1997) uses three UFs: intertaxon variability extrapolation, where values range from 1 to 5; exposure duration extrapolation, where values range from 0.75 to 15; and toxicological endpoint extrapolation, where values range from 1 to 15. Modifying factors, which incorporate other sources of uncertainty, are also used, including threatened, or listed, and endangered species, where values range from 0 to 2; relevance of endpoint to ecological health, where values range from 0 to 2; extrapolation from laboratory to field, where values range from -1 to 2; study conducted with relevant co-contaminants, where values range from -1 to 2; endpoint is mechanistically unclear (versus clear), where values range from 0 to 2; study species is either highly sensitive or highly resistant, where values range from -1 to 2; ratios used to estimate whole-body burden from tissue or egg, where values range from 0 to 2; intraspecific variability, where values range from 0 to 2; and other applicable modifiers, where values range from -1 to 2. The TRV is calculated by dividing the NOAEL or LOAEL from the critical experimental study by the product of the UFs and by the sum of the modifying factors.

In addition to dietary and media-specific TRVs for PCBs, tissue-residue-based TRVs are being used increasingly to evaluate the potential for adverse effects due to PCBs. Tissue-residue-based TRVs have been compiled for fish, birds, and mammals. To derive tissue-residue-based TRVs, site-specific parameters for exposure to upper trophic level species, including concentrations of contaminants in prey, are used to estimate relevant tissue concentrations in the species of concern. It is important to note that the accuracy of this approach depends on the availability of sufficient data to properly verify food-chain exposure models. Tissue-residue-based effect level data are gaining increasing regulatory acceptance as evidenced in the "Canadian Tissue Residue Guidelines (TRG) for Polychlorinated Biphenyls for the Protection of Wildlife Consumers of Aquatic Biota" (CCME 1998).

Fish Effect-Based Data

Adult fish are exposed to PCBs and related compounds via water, sediment, and food. Eggs and embryos can accumulate these highly lipophilic chemicals from the female during vitellogenesis. Bioaccumulation of PCBs by fish is dependent on the physical and chemical characteristics of individual congeners and on the biotransformation and elimination rates of congeners by fish. The log octanol/water partition coefficient (K_{ow}) of PBCs increases with molecular weight from approximately 4.5 to 8.2 (Mackay et al. 1992), indicating that all but the most highly chlorinated congeners are efficiently bioaccumulated. As a result of these factors, fish preferentially bioaccumulate highly chlorinated penta-, hexa-, and hepta-chlorinated biphenyls, but not deca-chlorinated biphenyls (Walker and Peterson 1994).

Early life stages generally represent the most sensitive developmental stage of fish to chemical contaminants. Thus, accumulation of these persistent chemicals during early life stages in fish is critical to the characterization of risks posed by exposure to PCBs. Signs of toxicity and histopathological lesions produced by PCBs and related compounds in juvenile fish are similar to those seen in higher vertebrates and include decreased food intake, wasting syndrome, delayed mortality, and lesions in epithelial and lymphomyeloid tissues. Toxicity and histopathological lesions produced by PCBs during early development in fish are characterized primarily by cardiovascular and circulatory changes, edema, hemorrhages, and mortality (Walker and Peterson 1991). As with higher vertebrates, there is a great variability in species sensitivities of fish to PCBs. Freshwater salmonid species, particularly lake trout and rainbow trout, are the most sensitive of fish species.

The adverse effects of PCBs on fish have been studied by two primary experimental methods: (1) laboratory exposure of fish to single congeners or technical mixtures via water, diet, or intraperitoneal or in ovo injection; or (2) correlation of concentrations of PCBs and related compounds in the environment with abnormalities in fish populations (e.g., mortality during early development or thyroid hyperplasia in adult fish (Walker and Peterson 1991). Although field research can integrate the impact of multiple environmental contaminants on fish, laboratory exposures can identify the specific responses associated with exposure to a single toxicant and determine the dose-response relationships for those responses. Thus, laboratory research can elucidate whether the body burden of a particular contaminant in fish in the environment is capable of producing the abnormalities observed in feral fish populations. Both field research and laboratory research are vital to understanding the toxicity of PCBs to fish and in predicting the risk that these compounds pose to fish in the environment. The available in vivo, in ovo, and in vitro toxic responses of fish characterized in laboratory and field studies were evaluated. However, for the purposes of summarizing effect levels in fish, only data from selected studies using the most relevant aquatic organisms were evaluated.

The majority of the information available on the toxicity of PCBs to fish is from laboratory water-exposure experiments; however, the major route of exposure is likely to be via the food chain. Thus, it is difficult to make accurate estimates of risk when the exposure pathways for actual exposure and laboratory studies are different. As an alternative approach, available studies reporting correlations between concentrations of PCBs in tissue and observed effects were evaluated. Recently, Jarvinen and Ankley (1999) compiled such data for a variety of chemicals, including PCBs. With the use of that data set, effect levels were determined for marine fish species in which critical life stages were evaluated for effects, and tissue concentrations were measured (Table G-4). Studies were evaluated for comparability to the potential species of interest, strength of the cause-effect linkage, exposures to critical life stages (embryos, fry, and juveniles), and sensitive developmental toxicity endpoints, including decreased survival and decreased growth. These tissue-residue effect levels incorporate all of the possible fish exposure pathways, including water, diet, and sediment ingestion. Because the data are so limited in number, a NOEL (no-observed-adverse-effect level) and LOEL (lowest-observed-adverse-effect level) could not be readily determined. Thus, a geometric mean of the available data was calculated to estimate a toxicity threshold in tissue (whole-body or fillet). Niimi et al.(1996) presented a similar table with greater tissue-residue-based effect threshold values (Table G-5). To illustrate the relevance of a tissue-based effect threshold and independence of exposure route, Walker and Peterson (1994) measured concentrations of TCDD in the eggs of rainbow trout after exposure by maternal transfer, water uptake, and injection. The results from this study show that the tissue-based effect level is consistent regardless of exposure route (Table G-6).

TABLE G-4 Tissue-based Effect Concentrations of Total PCBs in Early Developmental Stages of Marine Fish

Effect Level/Species	Life Stage	Exposure Route	Duration (Days)	Whole body[a]	Fillet[a]	Effect	Reference
NOAEL							
Sheepshead minnow, *Cyprinodon variegatus*	Embryo	Adult fish; 49 µg/g	5	27	9	Survival - no effect	Hansen et al. 1973
Sheepshead minnow, *Cyprinodon variegatus*	Embryo-larvae	Adult fish; 1.9-2.5 µg/g	28	0.88	0.3	Survival - no effect	Hansen et al. 1973
Pinfish, *Lagodon rhomboides*	Juvenile	water; 100/L	2	17	6	Survival - no effect	Duke et al. 1970
Spot, *Leiostomus xanthrus*	Juvenile	water; 1µg/L	33-56	27	9	Survival - no effect	Hansen et al. 1971
LOAEL							
Sheepshead minnow, *Cyprinodon variegatus*	Embryo	Adult fish	5	170	59	Reduced survival	Hansen et al. 1973
Sheepshead minnow, *Cyprinodon variegatus*	Embryo-larvae	Adult fish; 9.3-9.7 µg/g	28	5.1	2	Reduced survival	Hansen et al. 1973
Pinfish, *Lagodon rhomboides*	Juvenile	Water; 5 µg/L	14	14	5	Reduced survival	Hansen et al. 1971
Spot, *Leiostomus xanthrus*	Juvenile	Water; 5 µg/L	20-26	46	16	Reduced survival	Hansen et al. 1971
GEOMEAN OF ALL VALUES for PCBs			9.2	3.2			

[a]A relationship between whole body and skin-off fillet was utilized to derive fillet concentrations. The units are mg/kg, wet weight.

TABLE G-5 Summary of PCB Concentrations in Algae, Zooplankton, and Macroinvertebrates at Which Adverse and Chronic Effects, Cytological Changes, and Changes in Biochemical Activity Levels Occur Based on Short- and Long-term Laboratory Studies[a]

Response	Algae	Zooplankton	Macroinvertebrates
Lethality	>0.5-1 μg/L	>0.5-1 μg/L	>25 mg/kg
Growth	>0.5-1 μg/L	>0.5-1 μg/L	>25 mg/kg
Reproduction	>0.5-1 μg/L	>0.5-1 μg/L	>25 mg/kg
Behavior			>100 μg/L
Cellular changes			Low mg/kg

[a]Estimates for algae include those for phytoplankton. Zooplankton estimates were primarily based on *Daphnia*. Some threshold concentrations are expressed on a waterborne exposure basis, where estimates of tissue concentrations were poorly defined (Niimi et al.1996).

Concentrations of dissolved PCBs that are toxic to fish also are very low, particularly following chronic exposure. The 96-hr LC_{50} values for fathead minnows are 8-15 μg/L for different PCB mixtures (Nebecker et al. 1974). The threshold levels for effects of TCDD in 960-hr exposures to fathead minnows vary from 0.0001 μg/L (for retarded growth and development) to 0.01 μg/L for 100% mortality (Helder 1980, 1981). A no-observed-effective concentration for TCDD has been estimated to be less than 38 pg/L in fingerling rainbow trout (one of the most sensitive fish species to TCDD) exposed for 28 days (Mehrle et al. 1988). However, such relatively short-term exposures do not generally address the chronic effects of dioxin-like chemicals, particularly because delayed mortality, usually after several weeks, occurs at lesser concentrations than acute effects.

Aroclor-based Data—Several water exposure and a few dietary toxicity studies have been conducted with Aroclors. Some of the limitations of the Aroclor-based data are that (1) very few include marine species, and (2) as discussed previously, it might be inappropriate to compare laboratory exposures to Aroclors with field exposures to weathered PCB mixtures. Thus, consideration of these limitations should be made to understand the uncertainty associated with estimation of risk based on TRVs from laboratory exposure to technical PCB mixtures.

Total PCB Data—Considerable field data and limited laboratory data are available in which the concentrations of total PCBs in fish have been

TABLE G-6 Effect of Exposure Route on the Lethal Potency of TCDD to Rainbow Trout Eggs

Exposure Route	NOAEL μg/kg egg	LOAEL μg/kg egg	LD_{50} μg/kg egg	LD_{100} μg/kg egg
Maternal	0.023	0.05	0.058	0.145
Water uptake	0.034	0.04	0.069	0.119
Egg injection	0.044	0.055	0.080	0.154

Source: Data from Walker and Peterson (1994).

measured and related to particular adverse effects. The main advantage of these studies is that PCBs were quantified in the tissue of several feral species. One of the main limitations of these data is that there are potential co-contaminants that might confound the interpretation of effect levels from field studies for fish.

PCB Congener and TCDD-Equivalent Data—Considerable field and laboratory data (water exposure, dietary, intraperitoneal injection, and egg injection) are available in which concentrations of PCB congeners, TCDD, or TEQs in eggs or other fish tissue have been measured and related to a particular adverse effect. The main advantage of these studies is that there are several laboratory studies that have been conducted under controlled conditions. Some of the limitations of this data are that (1) very few of these studies have been conducted on marine species, and (2) for the field studies, there are potentially co-contaminants that might confound the interpretation of effect levels. (In many cases, however, co-contaminant data are available from these same studies.)

Avian Effect-Based Data

There is a great deal of information available on the toxicity of PCBs to birds compared with other biota. The information includes both dietary and tissue-residue-based effect levels of PCBs. Some toxic effects of PCBs and related compounds are listed in Table G-7. Although several earlier studies with juvenile and adult birds have shown lethal and biochemical effects of PCBs and related compounds, few studies were designed in such a way that TRVs could be determined reliably. Results of earlier studies, conducted before 1996, have been critically examined in Hoffman et al. (1996). Toxic effects of PCBs and related compounds in birds have been studied by in vivo exposure, in ovo exposure by

TABLE G-7 Toxic Effects of PCBs and TCDD-Equivalents Observed in Birds

Embryo lethality
Decreased productivity
Liver mfo induction
Unabsorbed yolk sacs
Vitamin A depletion
Porphyria
Teratogenesis:
• Gastroschisis
• Crossed bills
• Clubfoot
• Dwarfed appendages
• Edema/ascites
• Hemorrhaging
• Abnormal feathering
• Abnormal eyes
• Hydrocephaly
• Anencephaly

egg injection, and in vitro exposure with cultured avian hepatocytes. Due to better control of exposure dose and timing, egg injection studies are often of more use in deriving TRVs than in vivo or in vitro exposure studies. In most cases, embryotoxic and teratogenic effects of PCBs seem to be the most sensitive and ecologically relevant endpoints in birds (Hoffman et al. 1998). Thus, results from in ovo studies are particularly relevant for developing tissue-residue-based toxicity thresholds.

Recently, in ovo studies of dioxin-like compounds have been described (Powell et al. 1996a; Hoffman et al. 1998). Several of these studies were well-conducted and feature an adequate number of replications; several sensitive parameters were monitored. Values for LOAEL, NOAEL, and EC_{50} that are useful for derivation of TRVs have been estimated for certain PCB congeners from in ovo studies. Additionally, for many bird species, the most sensitive dose metric or effects predictor for PCBs and other dioxin-like chemicals is PCB concentrations in eggs rather than adult tissue (Giesy et al. 1994a,b). That result is in part due to the sensitivity of developing embryos of many species (birds, fish, and mammals) and to the relative tolerance of adults to the effects of dioxin-like chemicals. For example, the LD_{50} for chicken eggs (Henshel et al. 1993) is 200-fold less than the LD_{50} for an adult chicken on a wet-weight basis (Greig et al. 1973). Furthermore, LOAEL values for developmental toxicity occur at doses that are approximately 10-fold lower than LD_{50} endpoints. Thus, when trying to characterize risk to avian species for dioxin-like compounds, the most sensitive endpoint is developmental toxicity.

Available toxicological studies that correlated effects with PCB concentrations in eggs were evaluated (Tables G-8 through G-12). Similar tissue-residue effects thresholds have been summarized elsewhere (Hoffman et al. 1996) (Table G-13). Some of the dietary avian toxicological studies from the 1970s are still the most useful for deriving PCB reference doses. Alternatively, a tissue-residue-based approach can be used in which observed effects are compared with a known dose to the egg or tissue residues of total PCBs (from a congener-specific analysis) or TEQs. The results of egg-injection studies for predicting potential embryotoxicity of PCBs and TCDD compare favorably with those of feeding studies. In studies in which the same chemicals have been administered by both methods, the egg concentrations required to elicit effects are quite similar for both methods. The biological effects of Aroclor mixtures, individual PCB congeners, TCDD, and TCDD equivalents have been assessed with egg-injection experiments. For risk-assessment purposes, it is possible to model concentrations of PCBs in bird eggs by using published biomagnification factors (Table G-14). The collection of site-specific foraging information and determination of PCB congener concentrations in prey and receptor tissues can improve the accuracy of the predicted egg concentrations. The predicted or measured concentrations can then be compared with TRVs to estimate the magnitude of possible risks.

Birds demonstrate considerable differences in species sensitivities to PCBs and related dioxin-like chemicals among species. In particular, chickens, which are the most frequently used species for PCB exposures are among the most sensitive of the avian species to the effects of PCBs and dioxin-like chemicals. Thus, in most cases, it would be inappropriate to use the chicken as a surrogate species when the application of UFs would result in unrealistically low TRVs for the risk assessment of less-sensitive species, such as bald eagles, gulls, and terns. It is recommended that, wherever possible, family-specific, if not species-specific, toxicity data (whether based on dietary or tissue-residue TRVs) be chosen to most closely match the receptor of concern.

A number of studies with hepatocytes prepared from domestic and wild birds have examined the toxicity of PCBs and related compounds. EROD (a mixed function oxygenase enzyme) activity has been the most commonly measured endpoint in these studies. These studies suggest that, in general, EROD induction is not a toxic response per se but an adaptive biochemical response that is associated with exposure and some of the toxic effects of these chemicals. This reaction is catalyzed predominately by CYP1A1 with some contribution from CYP1A2 and CYP1B1. Excessive induction of mixed function oxygenase activity contributes to TCDD toxicity. Studies with cultured chicken embryo hepatocytes indicate that

TABLE G-8 Avian PCB Toxicity Summary for Dietary Exposures to Aroclors

Species	Adverse Effects Evaluated	Congener or Mixture	NOAEL (μg/kg/day)	LOAEL (μg/kg/day)	References
Chicken	Hatching success	A 1242	980	9,800	Lillie et al. 1974
Chicken	Chick growth	A 1242	980		Lillie et al. 1974
Chicken	Hatching success	A 1242	2,440		Britton and Huston 1973
Chicken	Hatching success	A 1242	2,440	4,880	Lillie et al. 1975
Chicken	Hatching success	A 1254	9,760		Cecil et al. 1974
Chicken	Hatching success	A 1254	980	9,800	Lillie et al. 1974
Chicken	Chick growth	A 1254		980	Lillie et al. 1974
Chicken	Hatching success	A 1254	2,440		Lillie et al. 1975
Chicken	Reproductive success	A 1254	244	2,440	Platonow and Reinhart 1973
Pheasant	Hatching success	A 1254	180	1,800	Dahlgren et al. 1972
Mallard	Reproductive success	A 1254	1,450		Custer and Heinz 1980
Chicken	Hatching success	A 1248	2,440	4,880	Lillie et al. 1975
Chicken	Hatching success	A 1248	980	9,800	Lillie et al. 1974
Chicken	Chick growth	A 1248		980	Lillie et al. 1974

Note: All studies are laboratory studies.
Abbreviations: NOAEL, no-observed-adverse-effect-level; LOAEL, lowest-observed-adverse-effect level.

TABLE G-9 Avian PCB Toxicity Summary for Tissue Residue Effect Levels for Aroclors

Species (Study Type)	Adverse Effects Evaluated	Congener or Mixture	Metric (Unit)	LD_{64}[a]	NOAEL	LOAEL	References
Chicken (laboratory)	Embryo mortality	A 1242	Egg injection (μg/kg egg)	10,000			Blazak and Marcum 1975
Chicken (laboratory)	Chick growth	A 1242	Tissue (μg/kg egg)		670	6,700	Gould et al. 1997
Mallard (field)		A 1242	Tissue (μg/kg egg)			105,000	Haseltine and Prouty 1980
Chicken (laboratory)	Chick growth	A 1254	Tissue (μg/kg egg)		670	6,700	Gould et al. 1997
Ringed Turtle Dove (laboratory)	Hatching success	A 1254	Tissue (μg/kg egg)			16,000	Peakall and Peakall 1973
Chicken (laboratory)	Hatching success	A 1248	Dose (μg/kg day)		490	4,900	Scott 1977
Screech Owls (laboratory)	Hatching Success	A 1248	Dose (μg/kg day)		410		McLane and Hughes 1980
Chicken (laboratory)		A 1232	Dose (μg/kg day)		980		Lillie et al. 1974
Chicken (laboratory)	Hatching success	A 1232	Dose (μg/kg day)		980	9,800	Lillie et al. 1974
Chicken (laboratory)	Hatching success	A 1232	Dose (μg/kg day)		2,440	4,880	Lillie et al. 1975
Chicken (laboratory)	Hatching success	A 1016	Dose (μg/kg day)		2,440		Lillie et al. 1975

[a]LD_{64}: The dose that is lethal to 64% of a test population.

non-ortho PCB congeners 126, 81, 77, and 169 are typically the most potent compounds, and the mono-ortho PCBs 66, 70, 105, 118, 122, 156, 157, and 167 and di-ortho PCBs 128, 138, 170, and 180 can also induce EROD activity (Kennedy et al. 1996b). Although in vitro studies can provide sound indications of the relative potency of different congeners, these experiments are of limited value for establishing defensible whole-organism TRVs for individual congeners in birds.

TABLE G-10 Avian PCB Toxicity Summary for Tissue Residue Effect Levels for Total PCBs

Species (Study Type)	Adverse Effects Evaluated	NOAEL[a] (Concentration in Egg, μg/kg)	LOAEL (Concentration in Egg, μg/kg)	References
Chicken (laboratory)	Hatching success	360	2,500	Scott 1977
Chicken (laboratory)	Hatching success	950	1,500	Britton and Huston 1973
Chicken (laboratory)	Hatching success deformities		4,000	Tumasonis et al. 1973
Tree Swallow (field)	Reproductive behavior		5,000-7,000	McCarty and Secord 1999
Bald Eagle (field)	Reproductive success		4,000	Ludwig et al. 1993
Bald Eagle (field)	Reproductive success	1300	7,200	Wiemeyer et al. 1984
Bald Eagle (field)	Reproductive success		13,000	Bosveld and van den Berg 1994
Bald Eagle (field)	Hatching success	400	4,000	180, source document
Double-crested Cormorant (field)	Hatching success	350	3,500	23,24, source document
Common tern (field)	Reproductive success	7000	8,000	Bosveld and van den Berg 1994
Common tern (laboratory)	Hatching success, deformities	4800	10,000	Hoffman et al. 1993
Common tern (field)	Hatching success	5,200-5,600	7,000	Becker et al. 1993
Forster's tern (both)	Hatching success	4500	22,000[a]	Kubiak et al. 1989
Forster's tern (field)	Reproductive success	7000	19,000	Bosveld and van den Berg 1994
Caspian terns (field)	Hatching success, deformities	420	4,200	Yamashita et al. 1993; Giesy et al. 1994a
Herring gulls (field)	Hatching success deformities	500		Weseloh et al. 1994; Giesy et al. 1994a

[a]NOAEL = 2.2 μg/kg TEQ in egg.

TABLE G-11 Avian PCB Toxicity Summary for Tissue Residue Effect Levels for PCB Congeners

Species	Adverse Effects Evaluated	Congener or mixture	Metric	LD_{50}[a] (µg/kg/egg)	NOAEL (µg/kg/egg)	LOAEL (µg/kg/egg)	TEQ LD_{50}	References
Chicken	Hatching success	PCB 77	Tissue	8.6			0.43	Brunstrom and Andersson 1988
Chicken		PCB 77	Egg injection			30		Nikolaidis et al. 1988
Chicken		PCB 77	Egg injection	2.6	0.12	1.2		Hoffman et al. 1998
Chicken	Decreased hatch weight	PCB 77	Egg injection	8.8	1	3		Powell et al. 1996a
Chicken	Embryo mortality	PCB 77	Egg injection	40				van den Berg et al 1998
Turkey	Hatching success	PCB 77	Tissue	~800				Brunstrom and Lund 1988
Ring-necked pheasant	Hatching success	PCB 77	Tissue		100			Brunstrom and Reutergardh 1986
American kestrel	Embryo mortality	PCB 77	Egg Injection	316		100		Hoffman et al. 1998
Goldeneye	Hatching success	PCB 77	Tissue	>1,000			>50	Brunstrom and Reutergardh 1986
Domestic	Hatching success	PCB 77	Tissue	>1,000			>50	Brunstrom 1988
Mallard	Hatching success	PCB 77	Tissue	>5,000			>250	Brunstrom 1988
Blackheaded gull	Hatching success	PCB 77	Tissue	<1,000			<50	Brunstrom 1988
Herring gull	Hatching success	PCB 77	Tissue	>1,000			36892	Brunstrom 1988

TABLE G-11 *(Continued)*

Species	Adverse Effects Evaluated	Congener or mixture	Metric	LD_{50}[a] (µg/kg/egg)	NOAEL (µg/kg/egg)	LOAEL (µg/kg/egg)	TEQ LD_{50}	References
Chicken	Hatching success	PCB 126	Tissue	3.2			0.32	Brunstrom and Andersson 1988
Chicken	Hatching success	PCB 126	Tissue	2.3			0.23	Powell et al. 1996b
Chicken	Hatching success	PCB 126	Tissue	0.4			0.04	Hoffman et al. 1998
Chicken	Reproductive behavior	PCB 126	Egg injection		0.5	1		Zhao et al. 1997
Chicken	Deformities; decreased hatch weight	PCB 126	Egg injection	0.4		0.3		Hoffman et al. 1998
Bobwhite	Deformities	PCB 126	Tissue	24			2.4	Hoffman et al. 1998
American kestrel	Deformities	PCB 126	Egg injection	65	2.3	23	6.5	Hoffman et al. 1998
Dbl-crested cormorant	Hatching success	PCB 126	Tissue	158			16	Powell et al. 1997b
Dbl-crested cormorant	Embryo mortality	PCB 126	Egg injection		200	400		Powell et al. 1997a
Common tern	Hatching success; Embryo mortality	PCB 126	Egg injection	104		44	10.4	Hoffman et al. 1998
Common tern	Embryo mortality	PCB 126	Egg injection	45				Hoffman et al. 1998
Chicken	Embryo mortality	PCB 157	Egg injection	2,000				van den Berg et al 1998

Chicken	Embryo mortality	PCB 157	Egg injection	1,500				Brunstrom 1990
Chicken	Hatching success	PCB 105	Tissue	2,200			0.22	Brunstrom 1990
Chicken	Decreased hatch weight	PCB 105	Egg injection		100	300		Powell et al. 1996a
Chicken	Hatching success	PCB 118	Tissue	8,000			0.08	Brunstrom 1989
Chicken	Hatching success	PCB 156	Tissue	1,500			0.15	Brunstrom 1990
Chicken	Hatching success	PCB 167	Tissue	>4,000			>0.04	Brunstrom 1990
Chicken	Hatching success	PCB 169	Tissue	170			0.17	Brunstrom and Andersson 1988

Note: All studies are laboratory studies.

[a]LD_{50}: The dose that is lethal to 50% of a test population.

TABLE G-12 Avian PCB Toxicity Summary for Dietary Exposures and Tissue Residue Effect Levels for TCDD and TCDD Equivalents

Species (Study Type)	Adverse Effects Evaluated	Congener or Mixture	Metric (μg/kg/egg)	LD_{50}[a]	NOAEL	LOAEL	References
Ring-neck pheasants (laboratory)	Embryo mortality	TCDD	Dose		0.014	0.14	Nosek et al. 1992a,b, 1993
Chicken (laboratory)	Embryo mortality	TCDD	Egg injection	0.115			Henshel et al. 1993
Chicken (laboratory)	Embryo mortality	TCDD	Egg injection	0.18			Henshel et al. 1993
Chicken (laboratory)	Embryo mortality	TCDD	Egg injection	0.24			Allred and Strange 1977
Chicken (laboratory)	Embryo mortality	TCDD	Egg injection	LD_{100} 1.0			Higginbotham et al. 1968
Chicken (laboratory)	Embryo mortality	TCDD	Egg injection	0.15	0.08	0.16	Powell et al. 1996b
Chicken (laboratory)	Decreased hatch weight	TCDD	Egg injection		0.06	0.01	Henshel et al. 1997a
Chicken (laboratory)	Decreased hatch weight	TCDD	Egg injection		0.1	0.3	Henshel et al. 1997a
Chicken (laboratory)	Deformities	TCDD	Egg injection			0.32	Walker et al. 1997
Chicken (laboratory)	Deformities	TCDD	Egg injection		0.001	0.01	Henshel et al. 1997b
Chicken (laboratory)		TEQ	Tissue	LD_{100} 1.0			Giesy et al. 1994a
Chicken (laboratory)		TEQ	Tissue	0.14			Cheung et al. 1981; Giesy et al. 1994a
Chicken (laboratory)	Deformities	TEQ	Tissue	0.65			Giesy et al. 1994a
Chicken (laboratory)	Deformities	TEQ	Tissue			0.0064	Giesy et al. 1994a
Chicken (laboratory)	Embryo mortality	TEQ	Tissue	LD_{100} 1.0			Higgenbotham et al. 1968; Giesy et al. 1994a
Pheasant (laboratory)	Embryo mortality	TEQ	Egg injection	1.4-2.2			Nosek et al. 1993
Wood duck (field)	Reproductive success	TEQ	Tissue			0.02	White and Hoffman 1995; Giesy et al. 1994a
Wood duck (field)	Hatching success	TEQ	Tissue		£5	>20-50	White and Hoffman 1995

TABLE G-12 *(Continued)*

Species (Study Type)	Adverse Effects Evaluated	Congener or Mixture	Metric (μg/kg/egg)	LD_{50}[a]	NOAEL	LOAEL	References
Dbl-crested cormorant (laboratory)	Embryo mortality	TCDD	Egg injection		1	4	Powell et al. 1997a
Dbl-crested cormorant (field)	Hatching success	TEQ	Tissue	~0.55			Tillet et al. 1992
Dbl-crested cormorant (field)	Hatching success	TEQ	Tissue	LD_{100} 1.03			Giesy et al. 1994a
Dbl-crested cormorant (field)	Hatching success	TEQ	Tissue	LD_{37} 0.344			Giesy et al. 1994a
Dbl-crested cormorant (field)	Hatching success	TEQ	Tissue	LD_{27} 0.217			Giesy et al. 1994a
Dbl-crested cormorant (field)	Hatching success	TEQ	Tissue	LD_{8} 0.35			Giesy et al. 1994a
Dbl-crested cormorant (field)	Embryo mortality	TEQ	Tissue	0.46			Tillitt 1989
Dbl-crested cormorant (field)	Hatching success	TEQ	Tissue	0.46			Tillitt et al. 1992; Giesy et al. 1994a
Caspian tern (field)	Hatching success	TEQ	Tissue	0.75			Giesy et al. 1994a
Common tern (laboratory)	Hatching success	TEQ	Tissue		<1		Bosveld and van den Berg 1994
Herring gull (field)	Hatching success	TEQ	Tissue		36892		Ludwig et al. 1993
Herring gull (field)	Hatching success	TEQ	Tissue	LD_{19} 0.557			Giesy et al. 1994a
Osprey (field)	Reproductive success	TEQ	Tissue	0.14			Woodford et al. 1998
Bald eagle (field)	Reproductive success	TEQ		0.2			Elliott et al. 1996
Great blue heron (field)	Deformities; Chick growth	TEQ			0.02	0.245	Hart et al. 1991

[a]LD_{50}: The dose that is lethal to 50% of a test population.

TABLE G-13 Summary of PCB and TCDD Threshold Effect Levels in Birds

Concentration	Effect
20 to 50 ppt of TCDD in eggs	Embryo mortality and teratogenesis in chickens, decreased productivity and teratogenesis for wood ducks
150 to 250 ppt of TCDD in eggs	Decreased embryonic growth, edema in herons
618 to 7,366 ppt of TCDD equivalents (congener chemistry)	Embryotoxicity in Forester's tern
1,000 ppt of TCDD in eggs	Embryo mortality in pheasants
1 to 5 ppm of total PCBs in eggs	Decreased hatching success for chickens
8 to 25 ppm of total PCBs in eggs	Decreased hatching success for terns, cormorants, doves, eagles

Source: Data from Hoffman et al. (1996).

TABLE G-14 Biomagnification Factors from Alewife to Herring Gull Egg for Dioxin-like Compounds

Compound	Biomagnification Factor
2, 3, 7, 8-TCDD	21
Total PCBs	32
2, 3, 3′, 4, 4′-Penta-CB (105)	20
2, 3, 4, 4′, 5-Penta-CB (114)	31
2, 2′, 3, 4, 4′, 5-Hexa-CB (138)	42
3, 3′, 4-Tri-CB (35)	0.8
3, 3′, 4, 4′-Tetra-CB (77)	1.8
3, 3′, 4, 4′, 5-Penta-CB (126)	29
3, 3′, 4, 4′, 5, 5′-Hexa-CB (169)	46

Source: Data from Hoffman et al. (1996).

In addition, in vitro studies do not account for pharmacokinetic and pharmacodynamic parameters, which could alter the toxic potential of a congener.

For a tissue-residue-based TRV, several studies for wildlife have been evaluated. A specific recommendation cannot be made at this point, because the TRVs are usually species-specific and thus should be selected based on similarity to receptors of concern. It is, therefore, essential to perform a critical evaluation of the applicability of the toxicological data to the site-specific receptors of concern and exposure pathways. TRVs derived in the same species are not available for the majority of wildlife receptors and, therefore, it is necessary to derive TRVs using toxicological data for surrogate species in combination with UFs.

Tissue-residue-based TRVs are being used increasingly to evaluate the potential for adverse effects due to PCBs. For the purposes of this report, the term "tissue-residue-based TRV" is synonymous with "maximum allowable tissue concentration (MATC)," a term that is sometimes used by agencies and reported in the literature. To derive tissue-residue-based TRVs, site-specific parameters for exposure to upper trophic level receptors, including concentrations of contaminants in prey, are used to estimate relevant tissue concentrations in the receptors of concern. It is important to note that the accuracy of this approach depends on the availability of sufficient data to properly verify food-chain exposure models. Tissue-residue-based effect level data are gaining increasing regulatory acceptance as evidenced in the "Canadian Tissue Residue Guidelines (TRG) for Polychlorinated Biphenyls for the Protection of Wildlife Consumers of Aquatic Biota" (CCME 1998).

When deriving TRVs for a particular receptor species, there are a number of uncertainties that need to be taken into account. One way to do that is to assign UFs to assure that the derived TRV is protective of the receptor of interest. Thus, it is important to remember that the derivation and application of TRVs is part of a conservative process, meant to be protective rather than predictive. For this reason, a tiered risk-assessment approach that allows for more and more refined estimates of TRVs needs to be derived. Each level of assessment requires more sophisticated estimates of both exposure and response. The TRV provides information only on the response parameters. Refining TRV estimates might require additional toxicity testing either for a particular mechanism of action, species, or vector of exposure.

For bird species, the most sensitive dose metric for dioxin-like chemicals are egg concentrations rather than adult tissue concentrations (Giesy et al. 1994a,b). That is in part due to stage-specific sensitivity of many species (birds, fish, and mammals) during development and relative tolerance of adults to the effects of dioxin-like chemicals. In other words, effects can be seen at lesser doses for young, developing animals than for adults. For example, the LD_{50} for chicken eggs (Henshel et al. 1993) is 200-fold less than the LD_{50} for an adult chicken (on a wet-weight basis) (Greig et al. 1973). Furthermore, LOAEL values for developmental toxicity occur at doses that are approximately 10-fold lower than LD_{50} endpoints. Thus, when trying to characterize risk to avian species for dioxin-like compounds, the most sensitive endpoint is developmental toxicity.

The biological effects of Aroclor mixtures, individual PCB congeners, TCDF, and TCDD equivalents have been assessed with egg-injection experiments. For risk-assessment purposes, it is possible to model concentrations of PCBs in bird eggs using published biomagnification factors. The collection of site-specific foraging information and determination of PCB congener concentrations in prey and receptor tissues can improve the accuracy of the predicted egg concentrations. The predicted or measured concentration can then be compared with TRVs to estimate the magnitude of possible risks.

Aroclor-Based Data—Some of the most frequently cited bird studies for controlled, laboratory, dietary exposures are Aroclor-based exposure studies (Table G-8). In particular, the pheasant study of Dahlgren et al. (1972) and the chicken study of Platonow and Reinhart (1973) are often selected for development of TRVs. The Great Lakes Water Quality Initiative report (EPA 1995a) used the pheasant study of Dahlgren et al., (1972) because it is a wildlife species and because the study evaluated a critical life stage. However, to compare estimated exposures for receptors in a risk assessment with this type of benchmark, an estimate of total PCBs from a PCB congener-specific analysis, rather than an Aroclor-based method, is recommended because of environmental weathering processes. Furthermore, although the predominant route of exposure for toxicity studies for Aroclors is dietary, few of the available dietary studies have evaluated the concentration of PCBs in eggs during and after the exposure (Table G-9). Some of the limitations of this Aroclor-based data are that (1) very few of these studies have been conducted on wildlife species, and (2) as discussed previously, it might not be appropriate to compare laboratory exposures to technical Aroclors (with potential contamination by other more potent dioxin-like chemicals) with field exposures to weathered PCBs. These factors need to be considered when estimating TRVs and the uncertainty associated with them.

Total PCB Data—There are many field and laboratory studies in which concentrations of total PCBs in eggs have been measured and related to adverse effects (Table G-10). The main advantages of these studies, especially those conducted in the last 10 years, are that the total weathered PCBs (often by PCB congener analysis as either total PCBs or TEQs) are measured in the tissue of wildlife species and that these concentrations were related to ecologically relevant endpoints. Some of the limitations of this total PCB data are that (1) in a few cases, the individual PCB congeners were not quantified (studies from the 1970s), and (2) there are potential co-contaminants that might confound the interpretation of effect levels from field studies. (Data on some co-contaminants are available from these studies.)

PCB Congener and TCDD-Equivalent Data—Considerable field and laboratory data are available in which the concentrations of PCB congeners, TCDD, or TEQs in eggs have been measured and related to adverse effects (Table G-11). The main advantages of these studies are that there are several laboratory studies that have been conducted under controlled conditions, and many of the available studies have been conducted with wildlife species. Some of the limitations of these data are that (1) there are no dietary exposures, and (2) for the field studies, there are potential co-contaminants that might confound the interpretation of effect levels. (In many cases, co-contaminant data are available from these studies.)

Aquatic Mammal Effect-Based Data

Aroclor-based Data—No data are available for Aroclor-based toxicological studies.

Total PCB Data—The evaluation of the effects of environmental contamination on the health of aquatic mammals represents a considerable challenge. Logistical considerations make the sampling of large numbers of wild animals difficult, and ethical concerns discourage in vivo studies on captive animals. A few semi-field studies involving the exposure of seals to contaminated fish were used to derive TRVs for PCBs (de Swart et al. 1994; Ross et al. 1995). Immunotoxic endpoints, such as natural killer cell activity in the blood, vitamin A levels, and reproductive effects were measured in these studies. The TRVs derived for exposure studies with seals were compared with those reported for otter (Smit et al. 1996) and mink (Leonards et al. 1995; Tillitt et al. 1996). Because a majority of reports of PCBs in aquatic mammals have been from blubber samples, a blubber-based TRV can be derived by lipid normalizing the blood or liver-based PCB concentrations. Although the results of semi-field studies have been confounded by the occurrence of co-contaminants in the diet and by the lack of dose-response relationships, these studies have suggested that PCBs were the main cause for the observed toxic effects. Furthermore, the exposures mimicked observed field situations and, therefore, these values were considered suitable for deriving a toxicity benchmark to assess risks of PCBs to aquatic mammals. The lipid normalized NOAEL value in seal blood of 11 μg/g was comparable to the otter liver NOAEL of 9 μg/g (Kannan et al. 2000; Table G-15). After reviewing all of the available literature, a tissue-residue NOAEL for PCBs in aquatic mammals of 10 μg/g, lipid wt, has been suggested (Kannan et al. 2000) (Table G-15).

PCB Congener and TCDD-Equivalent Data—Based on immunotoxicological studies in seals, a LOAEL for blubber TEQ of 210 pg/g, lipid wt, has been suggested (Ross et al. 1995). However, the lipid normalized liver TEQs for otters and mink have been estimated to be in the range of 400-2,000 pg/g. That estimate

TABLE G-15 Summary of NOAEL and LOAEL for Total PCBs and TEQs in Semi-Field Investigations with Seals, Dolphins and European Otter

Exposure	PCBs	TEQs
Seals and Dolphins		
Daily dose NOAEL	5.2 μg/kg bw/d	0.58 ng/kg bw/d
Daily dose LOAEL	28.9 μg/kg bw/d	5.8 ng/kg bw/d
Dietary NOAEL	100 ng/g, wet wt	NA
Dietary LOAEL	200 ng/g, wet wt	NA
Seal blood NOAEL (lipid 0.05-0.32%)	5.2 μg/kg, lipid wt	NA
Seal blood LOAEL	25 μg/kg, lipid wt	NA
Seal blubber NOAEL	NA	90 pg/g, lipid wt
Seal blubber LOAEL	NA	286 pg/g, lipid wt
Dolphin blood (in vitro)	26 ng/g, wet wt	NA
European Otter		
Dietary NOAEL (lipid 6.2%)	12 ng/g, wet wt (or) 200 ng/g, lipid wt	1 pg/g, wet wt or 16 pg/g, lipid wt
Dietary LOAEL	33 ng/g, wet wt (or) 530 ng/g, lipid wt	2 pg/g, wet wt 33 pg/g, lipid wt
Otter liver NOAEL	170 ng/g, wet wt (or) 4 μg/g, lipid wt	42 pg/g, wet wt 1 ng/g, lipid wt
Otter liver LOAEL	460 ng/g, wet wt (or) 11 μg/g, lipid wt	84 pg/g, wet wt or 2 ng/g, lipid wt

implies that sensitivity to TEQs might be greater in seals than in mink or otter. Thus, TRVs for TEQs might be dependent on the aquatic mammal species in question.

Uncertainties

This discussion of uncertainty is designed to assist in the understanding of the relative degree of confidence in the toxicity benchmarks and available data. An uncertainty analysis is required for ecological risk assessments under EPA guidance and should be performed for the quantitative and qualitative parameters that are included in risk assessments. When appropriate, sensitivity analyses should be performed to illustrate how the results would change if different toxicity benchmarks and other assumptions were incorporated into a particular analyses. Some common factors that contribute to uncertainty include assumptions relating to exposure models and toxicity thresholds. Additional parameters that contribute to uncertainty and their associated concerns include the following:

- Spatial distribution of contaminants. (What is the exposure from hot spots compared to more diffuse concentrations of contaminants?)
- Temporal (including seasonal) distribution of contaminants. (Is there temporal variability for exposure pathways? Is there any evidence of natural attenuation and/or a reduction in source inputs?)
- Bioavailability issues. (Are there factors that cause variability in the bioavailability of PCBs?)
- Co-contaminants and unknown interactions. (Are there other contaminants besides PCBs or a noncontaminant stressor that is predicted to cause unacceptable risk to the receptors of concern?)
- Species-specific differences in TEFs.

CONCLUSIONS AND RECOMMENDATIONS

In this appendix, available toxicological data for PCBs derived from dietary or media-specific exposures and tissue residues were evaluated and discussed. A database of studies is available; however, for each taxonomic group, evaluations must be made about the appropriateness and usefulness of data for risk assessments, especially for ecological risk assessments for which the database is more limited. Recommendations for each taxonomic group have been provided separately. In general, a weight-of-evidence approach can be utilized in which multiple measurement endpoint approaches (dietary TRVs, tissue-residue-based TRVs, and field studies) provide separate lines of evidence.

General recommendations are the following:

- An estimate of total PCBs (from a PCB congener-specific analysis) is sufficient to characterize potential risk to invertebrates. Individual PCB congener data or total PCB data can then be used, if necessary, as input to dietary food-chain models for biota that consume invertebrates.
- An estimate of tissue residues of total PCBs (from a congener-specific analysis) or TEQs is sufficient to characterize risk to fish populations. Water-based concentration effect levels are not appropriate to evaluate the potential risk

of PCBs to fish, because the predominant exposure pathway is not from the water but from dietary sources. Individual PCB congener data or total PCB data can then be used, if necessary, as input to dietary food-chain models for biota that consume fish.

- An estimate of total PCBs (from a congener-specific analysis) or TEQs is sufficient to characterize risk to birds. The dietary exposure TRV, tissue-residue TRV, or both can be used to characterize risk.
- An estimate of tissue residues of total PCBs (from a congener-specific analysis) or TEQs is sufficient to characterize risk to marine mammals.

Recommendations for Selection of Toxicity Reference Values

TRVs are ideally derived from chronic toxicity studies in which an ecologically relevant endpoint was assessed in the species of concern, or a closely related species. In this section, available toxicological data for PCBs derived from dietary or media-specific exposures and tissue residues were evaluated and discussed. A database of thousands of studies is available. However, for each group of biota, the appropriateness and usefulness of data must be considered for ecological risk assessments. Thus, for each group of biota, recommendations have been provided separately. In general, a weight-of-evidence approach should be utilized in which multiple measurement endpoint approaches (dietary TRVs, tissue-residue-based TRVs, and field studies) provide separate lines of evidence.

Benthic Invertebrate and Other Lower Trophic Level Biota

Generally, PCBs and other dioxin-like chemicals are not particularly toxic to lower trophic level biota, including algae, zooplankton, and invertebrates as these species lack the cellular Ah receptor that is required to mediate toxicity. It is recommended that an estimate of total PCBs (from a PCB congener-specific analysis) is sufficient to characterize potential risk to invertebrates. Individual PCB congener data or total PCB data can then be used, if necessary, as input to dietary food chain models for biota that consume invertebrates.

Fish

It is recommended that an estimate of tissue residues of total PCBs (from a congener-specific analysis) or TEQs is sufficient to characterize risk to fish. Water-based concentration effect levels are not appropriate to evaluate the potential risk from PCBs to fish because the predominant exposure pathway is not from the water but from dietary sources. Individual PCB congener data or total PCB data can then be used, if necessary, as input to dietary food chain models for biota that consume fish. To illustrate the relevance of a tissue-based effect threshold and independence of exposure route, Walker and Peterson, (1994) measured concentrations of TCDD in the eggs of rainbow trout after exposure by maternal transfer, water uptake, and injection. The results from this study show that the tissue-based effect level is consistent regardless of exposure route. A geometric mean of the available NOAEC and LOAEC data for marine species of fish was calculated (9.2 mg/kg) to estimate a toxicity threshold in tissue (whole body) based on survival of juveniles (a sensitive life stage).

Birds

For birds, it is recommended that a combination of dietary exposure modeling and tissue-residue-based effect levels be used. The potential for exposure is greatest for top-level predators; thus, several lines of evidence should be evaluated (possibly in a phased approach).

Considerable field and laboratory data are available in which the concentrations of total PCBs, PCB congeners, TCDD, or TEQs in eggs have been measured and related to adverse effects. The main advantages of these studies, especially those conducted in the last 10 years, are that the total weathered PCBs (often by PCB congener analysis as either total PCBs or TEQs) in the tissue of wildlife species was measured and that these concentrations were related to ecologically relevant endpoints.

Very few studies have been conducted from which a dietary TRV for PCB exposure to birds can be derived. Attempts to derive TRVs for birds at Navy sites in the San Francisco Bay have focused on a dietary study by Platonow and Reinhart (1973) (Tetra Tech 1998). However, there appears to be confusion regarding the calculation of the daily intake rate from this study. In different parts of the same report (Tetra Tech 1998) and in an EPA (1995) document, four different NOAEL and LOAEL values have been reported (Table G-16). Because the 1995 EPA document states clearly how its value was calculated, it appears to have the most support. However, for a dietary exposure-based TRV, it is recommended that the study by Dahlgren et al. (1972) be used, because it was conducted on a wildlife species and it evaluated a sensitive life stage. This study was also selected by EPA (1995) to provide a basis for water-quality values protective of wildlife. The NOEL and LOEL values from this study are 0.18 and 1.8 mg/kg/d, respectively.

TABLE G-16 Avian TRVs for PCBs

Original Study	NOAEL (mg/kg/d)	LOAEL (mg/kg/d)	TRV Citation
Platonow and Reinhart 1973	0.09	0.88	TetraTech 1998, pp. 5-68
Platonow and Reinhart 1973	0.034	0.34	TetraTech 1998, p. D-67
Platonow and Reinhart 1973	0.244	2.44	EPA (1995)
Dahlgren et al. 1972	0.18	1.8	EPA (1995

For a tissue-residue-based TRV, several studies for wildlife have been evaluated (refer to Chapter 6). A specific recommendation cannot be made at this point, because the TRVs are usually species-specific and thus should be selected on the basis of similarity to the receptors of concern. For example, to be protective of the embryo mortality for double-crested cormorants, a tissue-residue-based TRV would be between 350 (NOAEL) and 3,500 μg/kg (LOAEL) for total PCBs, respectively, and between 1 (NOAEL) and 4 μg/kg (LOAEL) for TEQs, respectively (Giesy et al. 1994a,b).

Aquatic Mammals

It is recommended that a tissue-residue threshold effect concentration for PCBs in marine mammals be 11 μg/g, lipid wt (Kannan et al. 2000) to be protective of immune function.

PCB Congener and TCDD-Equivalent Data—Based on immunotoxicological studies in seals, a threshold effect concentration for blubber TEQ of 520 pg/g, lipid wt, has been suggested (Ross et al. 1995).

REFERENCES

Agarwal, A.K., H.A. Tilson, H.A., and S.C. Bondy. 1981. 3,3′,4,4′-tetrachlorobiphenyl given to mice prenatally produces long term decreases in striatal dopamine and receptor binding sites in the caudate nucleus. Toxicol. Lett. 7(6):417-424.

Ahlborg, U.G., A. Brouwer, M.A. Fingerhut, J.L. Jacobson, S.W. Jacobson, S.W. Kennedy, A.A. Kettrup, J.H. Koeman, H. Poiger, C. Rappe, S.H. Safe, R.F. Seegal, J. Tuomisto, and M. van den Berg. 1992. Impact of polychlorinated dibenzo-*p*-dioxins, dibenzofurans, and biphenyls on human and environmental health, with special emphasis on application of the toxic equivalency factor concept. Eur. J. Pharmacol. 228(4):179-199.

Ahlborg, U.G., G.C. Becking, L.S. Birnbaum, A. Brouwer, H.J.G.M. Derks, M. Feeley, G. Golor, A. Hanberg, J.C. Larsen, A.K.D. Liem, S.H. Safe, C. Schlatter, F. Wærn, M. Younes, and E. Yrjänheikki. 1994. Toxic equivalency factors for dioxin-like PCBs. Chemosphere 28(6):1049-1067.

Ahlborg, U.G., L. Lipworth, L. Titus-Ernstoff, C.C. Hsieh, A. Hanberg, J. Baron, D. Trichopoulos, and H.O. Adami. 1995. Organochlorine compounds in relation to breast cancer, endometrial cancer, and endometriosis: an assessment of the biological and epidemiological evidence. Crit. Rev. Toxicol. 25(6): 463-531.

Allred, P.M., and J.R. Strange. 1977. The effects of 2,4,5-trichlorophenoxyacetic acid and 2,3,7,8-tetrachloro-dibenzo-*p*-dioxin on developing chick embryos. Arch. Environ. Contam. Toxicol. 6(4):483-489.

Angus, W.G., and M.L. Contreras. 1996. Effects of polychlorinated biphenyls on dopamine release from PC12 cells. Toxicol. Lett. 89(3):191-199.

Aulerich, R.J., S.J. Bursian, W.J. Breslin, B.A. Olson, and R.K. Ringer. 1985. Toxicological manifestations of 2,4,5,2′,4′,5′-, 2,3,6,2′,3′,6′- and 3,4,5,3′,4′,5′-hexachlorobiphenyls and Aroclor 1254 in mink. J. Toxicol. Environ. Health. 15(1):63-79.

Bandiera, S., S. Safe, and A.B. Okey. 1982. Binding of polychlorinated biphenyls classified as either phenobarbitone-, 3-methylcholanthrene- or mixed-type inducers to cytosolic Ah receptor. Chem. Biol. Interact. 39(3):259-277.

Bannister, R., and S. Safe. 1987. Synergistic interactions of 2,3,7,8-TCDD and 2,2′,4,4′,5,5′- hexachlorobiphenyl in C57BL/6J and DBA/2J mice: role of the Ah receptor. Toxicology 44(2):159-169.

Barrett, J.C. 1995. Mechanisms for species differences in receptor-mediated carcinogenesis. Mutat. Res. 333(1-2):189-202.

Beattie, M.K., S. Gerstenberger, R. Hoffman, and J.A. Dellinger. 1996. Rodent neurotoxicity bioassays for screening contaminated Great Lakes fish. Environ. Toxicol. Chem. 15(3):313-318.

Becker, P.H., S. Schuhmann, and C. Koepff. 1993. Hatching failure in common terns (Sterna hirundo) in relation to environmental chemicals. Environ. Pollut. 79(2):207-213.

Bernhoft, A., I. Nafstad, P. Engen, and J.U. Skaare. 1994. Effects of pre- and postnatal exposure to 3,3′,4,4′,5-pentachlorobiphenyl on physical development, neurobehavior and xenobiotic metabolizing enzymes in rat. Environ. Toxicol. Chem. 13(10):1589-1597.

Biegel, L., M. Harris, D. Davis, R. Rosengren, L. Safe, and S. Safe. 1989. 2,2′,4,4′,5,-5′- hexachlorobiphenyl as a 2,3,7,8-tetrachlorodibenzo-*p*-dioxin antagonist in C57BL/6J mice. Toxicol. Appl. Pharmacol. 97(3):561-571.

Birnbaum, L.S., and M.J. DeVito. 1995. Use of toxic equivalency factors for risk assessment for dioxins and related compounds. Toxicology 105(2/3):391-402.

Birnbaum, L.S., H. Weber, M.W. Harris, J.C. Lamb, and J.D. McKinney. 1985. Toxic interaction of specific polychlorinated biphenyls and 2,3,7,8-tetrachlorodibenzo-*p*-dioxin: increased incidence of cleft palate in mice. Toxicol. Appl. Pharmacol. 77(2):292-302.

Blankenship, A.L., K. Kannan, S.A. Villalobos, J. Falandysz, and J.P. Giesy. 2000. Relative potencies of individual polychlorinated naphthalenes and halowax mixtures to induce Ah receptor-mediated responses. Environ. Sci. Tehnol. 34(15): 3153-3158.

Blazak, W.F., and J.B. Marcum. 1975. Attempts to induce chromosomal breakage in chicken embryos with Aroclor 1242. Poult. Sci. 54(1):310-312.

Boon, J.P., J. van der Meer, C.R. Allchin, R.J. Law, J. Klungsoyr, P.E.G. Leonards, H. Spliid, E. Storr-Hansen, C. Mckenzie, and D.E. Wells. 1997. Concentration, dependent changes of PCB patterns in fish-eating mammals: Structure evidence for induction of cytochrome P450. Arch. Environ. Contam. Toxicol. 33(3):298-311.

Bosveld, A.T.C., and M. van den Berg. 1994. Effects of polychlorinated biphenyls, dibenzo-*p*-dioxins and dibenzofurans on fish-eating birds. Environ. Rev. 2(2):147-166.

Bosveld, A.T.C., S.W. Kennedy, W. Seinen, and M. van den Berg. 1997. Ethoxyresorufin-*O*-deethylase (EROD) inducing potencies of planar chlorinated aromatic hydrocarbons in primary cultures of hepatocytes from different developmental stages of the chicken. Arch. Toxicol. 71(12): 746-750.

Bosveld, B.A., M. van den Berg, and R.M. Theelen. 1992. Assessment of the EROD inducting potency of eleven 2,3,7,8-substituted PCDD/Fs and three coplanar PCBs in the chick embryo. Chemosphere 25(7-10):911-916.

Bowerman, W.W., J.P. Giesy, D.A. Best, and V.J. Kramer. 1995. A review of factors affecting productivity of bald eagles in the Great Lakes region: implications for recovery. Environ. Health Perspect. 103(Suppl. 4):51-59.

Bowman, R.E., M.P. Heironimus, and J.R. Allen. 1978. Correlation of PCB body burden with behavioral toxicology in monkeys. Pharmacol. Biochem. Behav. 9(1):49-56.

Bowman, R.E., M.P. Heironimus, and D.A. Barsotti. 1981. Locomotor hyperactivity in PCB-exposed rhesus monkeys. Neurotoxicology 2(2):251-268.

Britton, W.M., and T.M. Huston. 1973. Influence of polychlorinated biphenyls in the laying hen. Poult. Sci. 52(4):1620-1624.

Brouwer, A. 1991. Role of biotransformation in PCB-induced alterations in vitamin A and thyroid hormone metabolism in laboratory and wildlife species. Biochem. Soc. Trans. 19(3):731-737.

Brouwer, A., U.G. Ahlborg, M. van den Berg, L.S. Birnbaum, E.R. Boersma, B. Bosveld, M.S. Denison, L.E. Gray, L. Hagmar, E. Holene, et al. 1995. Functional aspects of developmental toxicity of polyhalogenated aromatic hydrocarbons in experimental animals and human infants. Eur. J. Pharmacol. 293(1):1-40.

Brunström, B. 1988. Sensitivity of embryos from duck, goose, herring gull, and various chicken breeds to 3,3′,4,4′-tetrachlorobiphenyl. Poult. Sci. 67(1):52-57.

Brunström, B. 1989. Toxicity of coplanar polychlorinated biphenyls to avian embryos. Chemosphere 19(1-6):765-768.

Brunström, B. 1990. Mono-ortho-chlorinated chlorobiphenyls: toxicity and induction of 7-ethoxyresorufin-*O*-deethylase.(EROD) activity in chick embryos. Arch. Toxicol. 64(3):188-192.

Brunström, B., and L Andersson. 1988. Toxicity and 7-ethoxyresorufin *O*-deethylase-inducing potency of coplanar polychlorinated biphenyls (PCBs) in chick embryos. Arch. Toxicol. 62(4):263-266.

Brunstöom, B., and J. Lund. 1988. Differences between chick and turkey embryos in sensitivity to 3,3′,4,4′-tetrachlorobiphenyl and in concentration/affinity of the hepatic receptor for 2,3,7,8-tetrachlorodibenzo-*p*-dioxin. Com. Biochem. Physiol. C 91(2): 507-512.

Brunström, B., and L Reutergardh. 1986. Differences in sensitivity of some avian species to the embryotoxicity of a polychlorinated biphenyl 3,3′,4,4′-tetrachlorobiphenyl injected into eggs. Environ. Pollut. Ser. A Ecol. Biol. 42 (1):37-46.

Brunström, B., M. Engwall, K. Hjelm, L. Lindqvist, and Y. Zebühr. 1995. EROD induction in cultured chick embryo liver: a sensitive bioassay for dioxin-like environmental pollutants. Environ. Toxicol. Chem. 14(5):837-842.

CCME (Canadian Council of Ministers of the Environment). 1998. Canadian Tissue Residue Guidelines for Polychlorinated Biphenyls for the Protection of Wildlife Consumers of Biota. Prepared by the Guidelines and Standards Division, Science Policy and Environmental Quality Branch, Environment Canada, Hull, Quebec.

Cecil, H.C., J. Bitman, R.J. Lillie, G.F. Fries, and J. Verrett. 1974. Embryotoxic and teratogenic effects in unhatched fertile eggs from hens fed polychlorinated biphenyls (PCBs). Bull. Environ. Contam. Toxicol. 11(6):489-495.

Chen, P.H., and S.T. Hsu. 1986. PCB poisoning from toxic rice-bran oil in Taiwan. Pp. 27-38 in PCBs and the Environment, Vol. III, J.S. Waid, J.S., ed.. Boca Raton, FL: CRC Press.

Chen, Y.J., and C.C. Hsu. 1994. Effects of prenatal exposure to PCBs on the neurological function of children: a neuropsychological and neurophysiological study. Dev. Med. Child Neurol. 36(4): 312-320.

Chen, Y.C., M.L. Yu, W.J. Rogan, B.C. Gladen, and C.C. Hsu. 1994. A 6-year follow up of bahavior and activity disorders in the Taiwan Yu-cheng children. Am. J. Public Health 84(3):415- 421.

Cheung, M.O., E.F. Gilbert, and R.E. Peterson. 1981. Cardiovascular teratogenicity of 2,3,7,8 tatrachlorodibenze-*p*-dioxin in the chick embryo. Toxicol. Appl. Pharmacol. 61(2):197-204.

Chishti, M.A., and R.F. Seegal. 1992. Intrastriatal injection of PCBs decresases striatal dopamine concentrations in rats. Toxicologist 12:320.

Chishti, M.A., J.P. Fisher, and R.F. Seegal. 1996. Aroclors 1254 and 1260 reduce dopamine concentrations in rat striatal slices. Neurotoxicology 17(3/4):653-660.

Chu, I., D.C. Villeneuve, A. Yagminas, P. Lecavalier, R. Poon, M. Feeley, S.W. Kennedy, R.F. Seegal, H. Håkansson, U.G. Ahlborg, and V.E. Valli. 1994. Subchronic toxicity of 3,3′,4,4′5-pentachlorobiphenyl in the rat: I. Clinical, biochemical and hematological, and histopathological changes. Fundam. Appl. Toxicol. 22(3):457-468.

Chu, I., D.C. Villeneuve, A. Yagminas, P. Lecavalier, H. Håkansson, U.G. Ahlborg, V.E. Valli, S.W. Kennedy, Å. Bergman, R.F. Seegal, and M. Feeley. 1995. Toxicity of PCB 77 (3,3′,4,4′-tetrachlorobiphenyl) and PCB 118 (2,3′,4,4′,5-pentachlorobiphenyl) in the rat following subchronic dietary exposure. Fundam. Appl. Toxicol. 26(2):282-292.

Chu, I., D.C. Villeneuve, A. Yagminas, P. Lecavalier, R. Poon, M. Feeley, S.W. Kennedy, R.F. Seegal, H. Håkansson, U.G. Ahlborg, V.E. Valli, and Å Bergman. 1996. Toxicity of 2,2′,4,4′,5,5′-hexachlorobiphenyl in rats: effects following 90-day oral exposure. J. Appl. Toxicol. 16(2):121-128.

Corrigan, F.M., M. French, and L. Murray. 1996. Organochlorine compounds in human brain. Hum. Exp. Toxicol. 15(3):262-264.

Custer, T.W., and G.H. Heinz. 1980. Reproductive success and nest attentiveness of mallard ducks fed Aroclor 1254. Environ. Pollut. 21:313-318.

Dahlgren, R.B., R.L. Linder, and C.W. Carlson. 1972. Polychlorinated biphenyls: their effects on penned pheasants. Environ. Heath Perspect. 1(1):89-101.

Daly, H.B. 1993. Laboratory rat experiments show consumption of Lake Ontario salmon causes behavioral changes: support for wildlife and human research results. J. Great Lakes Res. 19(4): 784-788.

de Swart, R.L., P.S. Ross, L.J. Vedder, H.H. Timmerman, S. Heisterkamp, H. van Loveren, J.V. Vos, P.J.H. Reijnders, and A.D.M.E. Osterhaus. 1994. Impairment of immune function in harbor seals (Phoca vitulina) feeding on fish from polluted waters. Ambio 23(2):155-159.

DeVito, M.J., L.S. Birnbaum, W.H. Farland, and T.A. Gasiewicz. 1995. Comparisons of estimated body burdens of dioxinlike chemicals and TCDD body burdens in experimentally exposed animals. Environ. Health Perspect. 103(9):820-831.

DeVito, M.J., X. Ma, J.G. Babish, M. Menache, and L.S. Birnbaum. 1994. Dose-response relationships in mice following subchronic exposure to 2,3,7,8-tetrachlorodibenzo-*p*-dioxin: CYP1A1, CYP1A2, estrogen receptor, and protein tyrosine phosphorylation. Toxicol. Appl. Pharmacol. 124(1):82-90.

Duke, T.W., J.I. Lowe, and A.J. Wilson, Jr. 1970. A polychlorinated biphenyl (Aroclor 1254) in the water, sediment, and biota of Escambia Bay, Florida. Bull Environ. Contam. Toxicol. 5(2):171-180.

Elliot, J.E., R.J. Norstrom, A. Lorenzen, L.E. Hart, H. Philibert, S.W. Kennedy, J.J. Stegeman, G.D. Bellward, and K.M. Cheng. 1996. Biological effects of polychlorinated dibenzo-*p*-dioxins, dibenzofurans, and biphenyls, in bald eagle chicks (Haliaeetus leucocephalus) chicks. Environ. Toxicol. Chem. 15(5):782-793.

EPA (U.S. Environmental Protection Agency). 1993. Interim Report on Data and Methods for Assessment of 2,3,7,8-tetrachlorodibenzo-*p*-dioxin Risks to Aquatic Life and Associated Wildlife. EPA/600/R-93/005. Office of Research and Development, U.S. Environmental Protection Agency, Washington, DC. March.

EPA (U.S. Environmental Protection Agency). 1995. Final Water Quality Guidance for the Great Lakes System. Fed. Regist. 60(56):15365-15425.

EPA (U.S. Environmental Protection Agency). 1995a. Great Lakes Water Quality Initiative Technical Support Document for Wildlife Criteria. EPA-820-B-95-009. Office of Water, U.S. Environmental Protection Agency, Washington, DC. NTIS PB95-187332.

EPA (U.S. Environmental Protection Agency). 1997. Uncertainty Factor Protocol for Ecological Risk Assessment. Toxicological Extrapolations to Wildlife Receptors. Ecosystems Protection and Remediation Division, U.S. Environmental Protection Agency, Region VIII, Denver, CO. [Online]. Available: http://www.epa.gov/Region8/superfund/risksf/trv-ucfs.pdf . February 1997.

EPA (U.S. Environmental Protection Agency). 2000. Draft Exposure and Human Health Reassessment of 2,3,7,8-Tetrachlorodibenzo-*p*-Dioxin (TCDD) and Related Compounds. Office of Research and Development, U.S. Environmental Protection Agency. [Online]. Available: http://www.epa.gov/ncea/pdfs/dioxin/dioxreass.htm. [Nov. 14, 2000].

Eriksson, P., U. Lundkvist, and A. Fredriksson. 1991. Neonatal exposure to 3,3′,4,4′-tetrachlorobiphenyl: changes in spontaneous behavior and cholinergic muscarinic receptors in the adult mouse. Toxicology 69(1):27-34.

Eriksson, P., and A. Fredriksson. 1996a. Neonatal exposure to 2,2′,5,5′-tetrachlorobiphenyl causes increased susceptibility in the cholinergic transmitter system at adult age. Environ. Toxicol. Pharmacol. 1(3):217-220.

Eriksson, P., and A. Fredriksson. 1996b. Developmental neurotoxicity of four ortho-substituted polychlorinated biphenyls in the neonatal mouse. Environ. Toxicol. Pharmacol. 1(3):155-165.

Eriksson, P., and A. Fredriksson. 1998. Neurotoxic effects in adult mice neonatally exposed to 3,3′,4,4′,5-pentachlorobiphenyl or 2,3,3′,4,4′-pentachlorobiphenyl. Changes in brain nicotinic receptors and behaviour. Environ. Toxicol. Pharmacol. 5(1):17-27.

Fingerman, S.W., and L.C. Russel. 1980. Effects of the polychlorinated biphenyl Aroclor 1242 on locomotor activity and on the neurotransmitters dopamine and norepinephrine in the brain of the Gulf killifish, Fundulus grandis. Bull. Environ. Contam. Toxicol. 25(4):682-687.

Foreman, M.M., and J.C. Porter. 1980. Effects of catechol estrogens and catecholamines on hypothalamic and corpus striatal tyrosine hydroxylase activity. J. Neurochem. 34(5):1175-1183.

Gasiewicz, T.A. 1997. Dioxins and the Ah receptor: probes to uncover processes in neuroendocrine development. Neurotoxicology 18(2):393-413.

Georgakopoulou-Gregoriadou, E., R. Psyllidou-Giouranovits, F. Voutsinou-Taliadouri, and U.A. Catsiki. 1995. Organochlorine residues in marine mammals from the Greek waters. Fresenius Environ. Bull. 4(6): 375-380.

Gierthy, J.F., and D. Crane. 1984. Reverse inhibition of in vitro epithelial cell proliferation by 2,3,7,8-tetrachlorodibenzo-*p*-dioxin. Toxicol. Appl. Pharmacol. 74(1):91-98.

Giesy, J.P., J.P. Ludwig, and D.E. Tillitt. 1994a. Dioxins, dibenzofurans, PCBs, and colonial fish-eating water birds. Pp. 249-307 in Dioxins and Health, A. Schecter, ed.. New York: Plenum Press.

Giesy, J.P., J.P. Ludwig and D.E. Tillitt. 1994b. Deformities in birds of the Great Lakes region: assigning causality. Environ. Sci. Technol. 28(3):128A-135A.

Giesy, J.P., D.A. Verbrugge, R.A. Othout, W.W. Bowerman, M.A. Mora, P.D. Jones, J.L. Newsted, C. Vandervoort, S.N. Heaton, R.J. Aulerich, et al. 1994c. Contaminants in fishes from Great Lakes-influenced sections and above dams of three Michigan rivers. 2. Implications for health of mink. Arch. Environ. Contam. Toxicol. 27(2):213-223.

Goldstein, J.A., and S. Safe. 1989. Mechanism of action and structure-activity relationships for the chlorinated dibenzo-*p*-dioxins and related compounds. Pp. 239-293 in Halogenated Biphenyls, Terphenyls, Naphthalenes, Dibenzodioxins and Related

Products, 2nd Ed., R.D. Kimbrough, and A.A. Jensen, eds. Amsterdam: Elsevier.
Gould, J.C., K.R. Cooper, and C.G. Scanes. 1997. Effects of polychlorinated biphenyl mixtures and three specific congeners on growth and circulating growth-related hormones. Gen. Comp. Endocrinol. 106(2):221-230.
Greene, L.A., and G. Rein. 1977. Release, storage and uptake of catecholamines by a clonal cell line of nerve growth factor (NGF) responsive pheochromocytoma cells. Brain Res. 129(2):247-263.
Grieg, J.B., G. Jones, W.H. Butler, and J.M. Barnes. 1973. Toxic effects of 2,3,7,8-tetra chlorodibenzo-*p*-dioxin. Fd Cosmet. Toxicol. 11:585-595.
Haake, J.M., S. Safe, K. Mayura, and T.D. Phillips. 1987. Aroclor 1254 as an antagonist of the teratogenicity of 2,3,7,8-tetrachlorodibenzo-*p*-dioxin. Toxicol. Lett. 38(3):299-306.
Hahn, M.E., A. Poland, E. Glover, and J.J. Stegeman. 1992. The Ah receptor in marine animals: phylogenetic distribution and relationship to P4501A inducibility. Mar. Environ. Res. 34(1/4):87-92.
Hansen, D.J., S.C. Schimmel, and J. Forester. 1973. Aroclor 1254 in eggs of sheepshead minnows; effect on fertilization success and survival of embryos and fry. Pp. 420-426 in Proceedings of 27th Annual Conference, Southeastern Association of Game and Fish Commissioners: October 14-17, 1973, Hot Springs, Arkansas, A.L. Mitchell, ed. Hot Springs, Ark: Southeastern Association of Game and Fish Commissioners.
Hansen, D.J., P.R. Parrish, J.L. Lowe, A.J. Wilson, Jr., and P.D. Wilson. 1971. Chronic toxicity, uptake, and retention of Aroclor 1254 in two estuarine fishes. Bull. Environ. Contam. Toxicol. 6(2):113-119.
Harper, N., K. Connor, M. Steinberg, and S. Safe. 1995. Immunosuppressive activity of polychlorinated biphenyl mixtures and congeners: nonadditive (antagonistic) interactions. Fundam. Appl. Toxicol. 27(1):131-139.
Harris, G.E., T.L. Metcalfe, C.D. Metcalfe, and S.Y. Huestis. 1994. Embryotoxicity of extracts from Lake Ontario rainbow trout (Oncorhynchus mykiss) to Japanese medaka (Oryzias latipes). Environ. Toxicol. Chem. 13(9):1393-1404.
Hart, L.E., K.M. Cheng, P.E. Whitehead, R.M. Shah, R.J. Lewis, S.R. Ruschkowski, R.W. Blair, D.C. Bennett, S.M. Bandiera, R.J. Norstrom, et al. 1991. Dioxin contamination and growth and development in great blue heron embryos. J. Toxicol.. Environ. Health 32(3):331-344..
Haseltine, S.D., and R.M. Prouty. 1980. Aroclor 1242 and reproductive success of adult mallards (Anas platyrhynchos). Environ. Res. 23(1):29-34.
Helder, T. 1980. Effects of 2, 3, 7, 8-tetrachlorodibenzo-p-dioxin (TCDD) on early life stages of the pike (Esox lucius L.). Sci. Total Environ. 14(3):255-264.
Helder T. 1981. Effects of 2,3,7,8-tetrachlorodibenzo-{Ip}-dioxin (TCDD) on early life stages of rainbow trout (Salmo gairdneri, Richardson). Toxicology 19(2):101-112.
Henshel, D.S., B.M. Hehn, M.T. Vo, and J.D. Steeves. 1993. A short term test for dioxin teratogenicity using chicken embryos. Pp. 159-174 in Environmental Toxicology and Risk Assessment, Vol.2., J.W. Gorsuch, F.J. Dwyer, C.G. Ingesoll, and T.W. LaPoint, eds. Philadelphia, PA: ASTM.
Henshel, D.S., B. Hehn, R. Wagey, M. Vo, and J.D. Steeves. 1997a. The relative sensi-

tivity of chicken embryos to yolk- or aircell-injected 2,3,7,8-tetrachlorodibenzo-*p*-dioxin. Environ. Toxicol Chem. 16(4):725-732.

Henshel, D.S., J.W. Martin, and J.C. DeWitt. 1997b. Brain asymmetry as a potential biomarker for developmental TCDD intoxication: a dose-response study. Environ. Health Perspect. 105(7):718-725.

Hertzler, D.R. 1990. Neurotoxic behavioral effects of Lake Ontario salmon diets in rats. Neurotoxicol. Teratol. 12(2):139-143.

Higginbotham, G.R., A. Huang, D. Firestone, J. Verrett, J. Ress, and A.D. Campbell. 1968. Chemical and toxicological evaluations of isolated and synthetic chloroderivatives of dibenzo-*p*-dioxin. Nature 220(168):702-703.

Hoffman, D.J., G.J. Smith, and B.A. Rattner. 1993. Biomarkers of contaminant exposure in common terns and black-crowned night herons in the Great Lakes. Environ. Toxicol. Chem. 12(6):1095-1103.

Hoffman, D.J., M.J. Melancon, P.N. Klein, C.P. Rice, J.D. Eisemann, R.K. Hines, J.W. Spann, and G.W. Pendleton. 1996. Developmental toxicity of PCB 126 (3,3′,4,4′,5-pentachlorobiphenyl) in nesting American kestrels (Falco sparverius). Fundam. Appl. Toxicol. 34(2)188-200.

Hoffman, D.J., M.J. Melancon, P.N. Klein, J.D. Eisemann, and J.W. Spann. 1998. Comparative developmental toxicity of planar polychlorinated biphenyl congeners in chicken, American kestrels, and common terms. Environ. Toxicol. Chem. 17(4):747-757.

Holene, E., I. Nafstad, J.U. Skaare, A. Bernhoft, P. Engen, and T. Sagvolden. 1995. Behavioral effects of pre- and postnatal exposure to individual polychlorinated biphenyl congeners in rat. Environ. Toxicol. Chem. 14(6):967-976.

Hornung, M.W., E.W. Zabel, and R.E. Peterson. 1996. Toxic equivalency factors of polybrominated dibenzo-*p*-dioxin, dibenzofuran, biphenyl, and polyhalogenated diphenyl ether congeners based on rainbow trout early life stage mortality. Toxicol. Appl. Pharmacol. 140(2):227-234.

Hsu, C.-C., T.C. Chen, W.T. Soong, C.C. Tsung, S.J. Sue, and C.Y. Liu. 1988. A six-year follow up study of intellectual and behavioral development of Yu-Cheng children: cross-sectional findings of the first field work. [in Chinese]. Clin. Psychiatry 2:27-40.

Huisman, M., C. Koopman-Esseboom, V. Fidler, M. Hadders-Algra, C.G. van der Paauw, L.G. Tuinstra, N. Weisglas-Kuperus, P.J. Sauer, B.C. Touwen, and E.R. Boersma. 1995. Perinatal exposure to polychlorinated biphenyls and dioxins and its effect on neonatal neurological development. Early Hum. Dev. 41(2):111-127.

IOM (Institute of Medicine). 1994. Veterans and Agent Orange: Health Effects of Herbicides Used in Vietnam. Washington, DC: National Academy Press.

IOM (Institute of Medicine). 1996. Veterans and Agent Orange: Update 1996. Washington, DC: National Academy Press.

IOM (Institute of Medicine). 1999. Veterans and Agent Orange: Update 1998. Washington, DC: National Academy Press.

IOM (Institute of Medicine). 2000. Veterans and Agent Orange: Update 1999. Washington, DC: National Academy Press.

IOM (Institute of Medicine). 2001. Veterans and Agent Orange: Update 2000. Washington, DC: National Academy Press. In press.

Jacobson, J.L., and S.W. Jacobson. 1997. Evidence for PCBs as neurodevelopmental toxicants in humans. Neurotoxicology 18(2):415-424.

Jacobson, J.L., S.W. Jacobson, and H.E. Humphrey. 1990. Effects of exposure to PCBs and related compounds on growth and activity in children. Neurotoxicol. Teratol. 12(4):319-326.

Janz, D.M., and C.D. Metcalfe. 1991. Relative induction of aryl hydrocarbon hydroxylase by 2,3,7,8-TCDD and two coplanar PCBs in rainbow trout (Oncorhynchus mykiss). Environ. Toxicol. Chem. 10(7):917-924.

Jarvinen, A.W., and G.T. Ankley. 1999. Linkage of Effects to Tissue Residues: Development of a Comprehensive Database for Aquatic Organisms Exposed to Inorganic and Organic Chemicals. Pensacola, FL: Society of Environmental Toxicology and Chemistry (SETAC). 364 pp.

Jenssen, B.M., J.U. Skaare, M. Ekker, D. Vongraven, and S.H. Lorentsen. 1996. Organochlorine compounds in blubber, liver and brain in neonatal grey seal pups. Chemosphere 32(11):2115-2125.

Kannan, K., S. Tanabe, R. Tatsukawa, and R.K. Sinha. 1994. Biodegradation capacity and residue pattern of organochlorines in Ganges river dolphins from India. Toxicol. Environ. Chem. 42(3/4): 249-261.

Kannan, K., A.L. Blankenship, P.D. Jones, and J.P. Giesy. 2000. Toxicity reference values for the toxic effects of polychlorinated biphenyls to aquatic mammals. Hum. Ecol. Risk Assess. 6(1):181-201.

Kennedy, S.W., A. Lorenzen, S.P. Jones, M.E. Hahn, and J.J. Stegeman. 1996a. Cytochrome P4501A induction in avian hepatocyte cultures: a promising approach for predicting the sensitivity of avian species to toxic effects of halogenated aromatic hydrocarbons. Toxicol. Appl. Pharmacol. 141(1):214-230.

Kennedy, S.W., A. Lorenzen, and R.J. Norstrom. 1996b. Chick embryo hepatocyte bioassay for measuring cytochrome P4501A-based 2,3,7,8-tetrachlorodibenzo-*p*-dioxin equivalent concentrations in environmental samples. Environ. Sci. Technol. 30(2):706-715.

Kittner, B., M. Brautigam, and H. Herken. 1987. PC12 cells: a model system for studying drug effects on dopamine synthesis and release. Arch. Int. Pharamacodyn. Ther. 286(2):181-194.

Kodavanti, P.R., and H.A. Tilson. 1997. Structure-activity relationships of potentially neurotoxic PCB congeners in the rat. Neurotoxicology 18(2):425-441.

Kodavanti, P.R., W.R. Mundy, H.A. Tilson, and G.J. Harry. 1993a. Effects of selected neuroactive chemicals on calcium transporting systems in rat cerebellum and on survival of cerebellar granule cells. Fundam. Appl. Toxicol. 21(3):308-316.

Kodavanti, P.R., D.S. Shin, H.A. Tilson, and G.J. Harry. 1993b. Comparative effects of two polychlorinated biphenyl congeners on calcium homeostasis in rat cerebellar granule cells. Toxicol. Appl. Pharmacol. 123(1):97-106.

Kodavanti, P.R., T.J. Shafer, T.R. Ward, W.R. Mundy, T. Freudenrich, G.J. Harry, and H.A. Tilson. 1994. Differential effects of PCB congeners on phosphoinositide hydrolysis and protein kinase C translocation in rat cerebellar granule cells. Brain Res. 662(1-2):75-82.

Kodavanti, P.R., T.R. Ward, J.D. McKinney, and H.A. Tilson. 1995. Increased [^{3}H]phorbol ester binding in rat cerebellar granule cells by polychlorinated mixtures and congeners: structure-activity relationships. Toxicol. Appl. Pharmacol. 130(1): 140-148.

Kodavanti, P.R., T.R. Ward, J.D. McKinney, and H.A. Tilson. 1996. Inhibition of microsomal and mitochondrial Ca^{2+} sequestration in rat cerebellum by polychlorinated biphenyl mixtures and congeners: structure-activity relationships. Arch. Toxicol. 70(3-4):150-157.

Kramer, V.J., W.G. Helferich, Å Bergman, E. Klasson-Wehler, and J.P. Giesy. 1997. Hydroxylated polychlorinated biphenyl metabolites are anti-estrogenic in a stably transfected human breast adenocarcinoma (MCF7) cell line. Toxicol. Appl. Pharmacol. 144(2):363-376.

Kreitzer, J.F., and G.H. Heinz. 1974. The effect of sublethal dosages of five pesticides and a polychlorinated biphenyl on the avoidance response of coturnix quail chicks. Environ. Pollut. 6(1):21-29.

Kubiak, T.J., H.J. Harris, L.M. Smith, T.R. Schwartz, D.L. Stalling, J.A. Trick, L. Sileo, D.E. Docherty, and T.C. Erdman. 1989. Microcontaminants and reproductive impairment of the Forster's tern on Green Bay, Lake Michigan--1983. Arch. Environ. Contam. Toxicol 18(5):706-727.

Kuratsune, M., T. Yoshimura, J. Matsuzaka, and A. Yamaguchi. 1972. Epidemiological study on Yusho, a poisoning caused by ingestion of rice oil contaminated with a commercial brand of polychlorinated biphenyls. Environ. Health Perspect. 1:119-128.

Lawrence, G.S., and F.A. Gobas. 1997. A pharmacokinetic analysis of interspecies extrapolation in dioxin risk assessment. Chemosphere 35(3):427-452.

Lecavalier, P., I. Chu, A. Yagminas, D.C. Villeneuve, R. Poon, M. Feeley, H. Hakansson, U.G. Ahlborg, V.E. Valli, A. Bergman, R.F. Seegal, and S.W. Kennedy. 1997. Subchronic toxicity of 2,2′,3,3′,4,4′-hexachlorobiphenyl in rats. J. Toxicol. Environ. Health 51(3):265-277.

Leece, B., M.A. Denomme, R. Towner, S.M. Li, and S. Safe. 1985. Polychlorinated biphenyls: correlation between in vivo and in vitro quantitative structure-activity relationships (QSARs). J. Toxicol. Environ. Health 16(3-4):379-388.

Leonards, P.E.G., T.H. de Vries, W. Minnaard, S. Stuijfzand, P. de Voogt, W.P. Cofino, N.M. van Straalen, and B. van Hattum. 1995. Assessment of experimental data on PCB-induced reproduction inhibition in mink, based on an isomer- and congener-specific approach using 2,3,7,8-tetrachlorodibenzo-*p*-dioxin toxic equivalency. Environ. Toxicol. Chem. 14(4):639-652.

Leonards, P.E., Y. Zierikzee, U.A.T. Brinkman, W.P. Cofino, N.M. van Straalen, and B. van Hattum. 1997. The selective dietary accumulation of planar polychlorinated biphenyls in the otter (Lutra lutra). Environ. Toxicol. Chem. 16(9):1807-1815.

Li, M.H., and L.G. Hansen. 1996. Enzyme induction and acute endocrine effects in prepubertal female rats receiving environmental PCB/PCDF/PCDD mixtures. Environ. Health Perspect. 104(7):712-22

Li, M.H., and L.G. Hansen. 1997. Considerations of enzyme and endocrine interactions in the risk assessment of PCBs. Reviews in Toxicology, Series D. In Vitro Toxicology and Risk Assessment. 1:71-156.

Lilienthal, H., M. Neuf, C. Munoz, and G. Winneke. 1990. Behavioral effects of pre-

and postnatal exposure to a mixture of low chlorinated PCBs in rats. Fundam. Appl. Toxicol. 15(3):457-467.

Lillie, R.J., H.C. Cecil, J. Bitman, and G.F. Fries. 1974. Differences in response of caged White Leghorn layers to various polychlorinated biphenyls (PCBs) in the diet. Poult. Sci. 53(2):726-732.

Lillie, R.J., H.C. Cecil, J. Bitman, G.F. Fries, and J. Verrett. 1975. Toxicity of certain polychlorinated and polybrominated biphenyls on reproductive efficiency of caged chickens. Poult. Sci. 54(5):1550-1555.

Lloyd, T., and J. Weisz. 1978. Direct inhibition of tyrosine hydroxylase activity by catechol estrogens. J. Biol. Chem. 253(14): 4841-4843.

Lorenzen, A., and A.B. Okey. 1990. Detection and characterization of [^{3}H] 2,3,7,8-tetrachlorodibenzo-*p*-dioxin binding to Ah receptor in a rainbow trout hepatoma cell line. Toxicol. Appl. Pharmacol. 106(1):53-62.

Lorenzen, A., J.L. Shutt, and S.W. Kennedy. 1997. Sensitivity of common tern (Sterna hirundo) embryo Hepatocyte Cultures to CYP1A induction and porphyrin accumulation by halogenated aromatic hydrocarbons and common tern egg extracts. Arch. Environ. Contam. Toxicol. 32(2):126-134.

Ludwig, J.P., J.P. Giesy, C.L. Summer, W. Bowerman, R. Aulerich, S. Bursian, H.J. Auman, P.D. Jones, L.L. Williams, et al. 1993. A comparison of water quality criteria in the Great Lakes basin based on human and wildlife health. J. Great Lakes Res. 19(4):789-807.

Mac, M.J., T.R. Schwartz, C.C. Edsall, and A.M. Frank. 1993. Polychlorinated biphenyls in Great Lakes lake trout and their eggs: relation to survival and congener composition, 1979-1988. J. Great Lakes Res. 19(4):752-765.

Mackay, D., W.Y. Shiu, and K.C. Ma. 1992. Illustrated Handbook of Physic-Chemical Properties and Environmental Fate for Organic Chemicals: Monoaromatic Hydrocarbons, Chlorobenzenes, and PC. Chelsea, MI: Lewis.

Maier, W.H., PRAISE Kodavanti, G.J. Harry, and H.A. Tilson. 1994. Sensitivity of adenosine triphosphatases in different brain regions to polychlorinated biphenyl congeners. J. Appl. Toxicol. 14(3):225-229.

Masuda, Y. 1985. Health status of Japanese and Taiwanese after exposure to contaminated rice oil. Environ Health Perspect. 60:321-325.

McCarty, J.P., and A.L. Secord. 1999. Nest-buildling behavior in PCB-contaminated tree swallows. AUK 116(1):55-63.

McLane, M.A., and D.L. Hughes. 1980. Reproductive success of screech owls fed Aroclor 1248. Arch. Environ. Contam. Toxicol. 9(6):661-665.

Mehrle, P.M., D.R. Buckler, E.W. Little, L.M. Smith, J.D. Petty, P.H. Peterman, D.L. Stalling, G.M. De Graeve, J.J. Coyle, and W.J. Adams. 1988. Toxicity and bioconcentration of 2,3,7,8-tetrachlorodibenzodioxin and 2,3,7,8-tetrachlorodibenzofuran in rainbow trout. Environ. Toxicol. Chem. 7(1):47-62.

Metcalfe, C.D., and G.D. Haffner. 1995. The ecotoxicology of coplanar polychlorinated biphenyls. Environ. Rev. 3(2):171-190.

Morse, D.C., R.F. Seegal, K.O. Borsch, and A. Brouwer. 1996a. Long-term alterations in regional brain serotonin metabolism following maternal polychlorinated biphenyl exposure in the rat. Neurotoxicology 17(3-4):631-638.

Morse, D.C., E.K. Wehler, W. Wesseling, J.H. Koeman, and A. Brouwer. 1996b.

Alterations in rat brain thyroid hormone status following pre- and postnatal exposure to polychlorinated biphenyls (Aroclor 1254). Toxicol. Appl. Pharmacol. 136(2):269-279.

Nebecker, A.V., and F.A. Puglisi, and D.L. Defoe. 1974. Effect of polychlorinated biphenyl compounds on survival and reproduction of the fathead minnow and flagfish. Trans. Am. Fish. Soc. 103(3):562-568.

Neubert, D., G. Color, and R. Neubert. 1992. TCDD-toxicity equivalencies for PCDD/PCDF congeners: prerequisites and limitations. Chemosphere 25(1-2):65-70.

Neumann, H.G. 1996. Toxic equivalence factors, problems and limitations. Food Chem. Toxicol. 34(11-12):1045-1051.

Newsted, J.L., J.P. Giesy, G.T. Ankley, D.E. Tillitt, R.A. Crawford, J.W. Gooch, P.D. Jones, and M.S. Denison. 1995. Development of toxic equivalency factors for PCB congeners and the assessment of TCDD and PCB mixtures in rainbow trout. Environ. Toxicol. Chem. 14(5):861-871.

Niimi, A.J., H.B. Lee and D.C.G. Muir. 1996. Environmental assessment and ecotoxicological implications of the co-elution of PCB congeners 132 and 153. Chemosphere 32(4):627-638.

Nikolaidis, E., B. Brunstrom, and L. Dencker. 1988. Effects of TCDD and its congeners 3,3′,4,4′-tetrachloroazoxybenzene and 3,3′,4,4′-tetrachlorobiphenyl on lymphoid development in the thymus of avian embryos. Pharmacol. Toxicol. 63(5):333-336.

Nosek, J.A., S.R. Craven, J.R. Sullivan, S.S. Hurley, and R.E. Peterson. 1992a. Toxicity and reproductive effects of 2,3,7,8-tetrachlorodibenzo-p-dioxin in ring-necked pheasant hens. J. Toxicol. Environ. Health 35(3):187-198.

Nosek, J.A., S.R. Craven, J.R. Sullivan, J.R. Olson, and R.E. Peterson. 1992b. Metabolism and disposition of 2,3,7,8-tetrachlorodibenzo-p-dioxin in ring-necked pheasant hens, chicks, and eggs. J. Toxicol. Environ. Health 35(3):153-164.

Nosek, J.A., J.R. Sullivan, and A. Fitzpatrick-Gendron. 1993. Embryotoxicity of 2,3,7,8-tetrachlorodibenzo-*p*-dioxin in the ring-necked pheasant. Environ. Toxicol. Chem 12(7):1215-1222.

Opresko, D.M., B.E. Sample, and G.W. Suter. 1994. Toxicological Benchmarks for Wildlife: 1994 Revision. Oak Ridge National Laboratory, Health Sciences Research Division, Oak Ridge, TN. NTIS/DE95002293.

Overmann, S.R., J. Kostas, L.R. Wilson, W. Shain, and B. Bush. 1987. Neurobehavioral and somatic effects of perinatal PCB exposure to rats. Environ. Res. 44(1):56-70.

Pantaleoni, G., D. Fannini, A.M. Sponta, G. Palumbo, R. Giorgi, and P.M. Adams. 1988. Effects of maternal exposure to polychlorinated biphenyls (PCBs) on F1 generation behavior in the rat. Fundam. Appl. Toxicol. 113:440-449.

Parkinson, A., L. Robertson, L. Safe, and S. Safe. 1980. Polychlorinated biphenyls as inducers of hepatic microsomal enzymes: structure-activity rules. Chem. Biol. Interact. 30(3):271-285.

Parkinson, A., S.H. Safe, L.W. Robertson, P.E. Thomas, D.E. Ryan, L.M. Reik, and W. Levin. 1983. Immunochemical quantitation of cytochrome P-450 isozymes and epoxide hydrolase in liver microsomes from polychlorinated and polybrominated biphenyls: a study of structure-activity relationships. J. Biol. Chem. 258(9):5967-5976.

Peakall, D.B., and M.L. Peakall. 1973. Effects of a polychlorinated biphenyl on the reproduction of artificially and naturally incubated dove eggs. J. Appl. Ecol. 10(3):863-868.

Platonow, N.S., and B.S. Reinhart. 1973. The effects of polychlorinated biphenyls (Aroclor 1254) on chicken egg production, fertility and hatchability. Can. J. Comp. Med. Vet. Sci. 37(4):341-346.

Pluess, N., H. Poiger, C. Hohbach, and C. Schlatter. 1988. Subchronic toxicity of some chlorinated dibenzofurans (PCDFs) and a mixture of PCDFs and chlorinated dibenzodioxins (PCDDs) in rats. Chemosphere 17(5):973-984.

Pohjanvirta, R., M. Unkila, J. Linden, J.T. Tuomisto, and J. Tuomisto. 1995. Toxic equivalency factors do not predict the acute toxicities of dioxins in rats. Eur. J. Pharmacol. Environ. Toxicol. Pharmacol. Sec. 293(4-5):341-353.

Poland, A., and E. Glover. 1977. Chlorinated biphenyl induction of aryl hydrocarbon hydroxylase activity: a study of the structure-activity relationship. Mol. Pharmacol. 13(5):924-938.

Poland, A., and J.C. Knutson. 1982. 2,3,7,8-tetrachlorodibenzo-*p*-dioxin and related halogenated aromatic hydrocarbons: examination of the mechanism of toxicity. Ann. Rev. Pharmacol. Toxicol. 22:517-554.

Poland, A., E. Glover, and A.S. Kende. 1976. Stereospecific, high affinity binding of 2,3,7,8-tetrachlorodibenzo-*p*-dioxin by hepatic cytosol: evidence that the binding species is receptor for induction of aryl hydrocarbon hydroxylase. J. Biol. Chem. 251(16):4936-4946.

Poland, A., W.F. Greenlee, and A.S. Kende. 1979. Studies on the mechanism of action of chlorinated dibenzo-*p*-dioxins and related compounds. Ann. N.Y. Acad. Sci. 320:214-230.

Powell, D.C., R.J. Aulerich, J.C. Meadows, D.E. Tillitt, J.P. Giesy, K.L. Stromberg, and S.J. Bursian. 1996a. Effects of 3,3′,4,4′,5-pentachlorobiphenyl (PCB 126) and 2,3,7,8-tetrachlorodibenzo-*p*-dioxin (TCDD) injected into the yolks of chicken (Gallus domesticus) eggs prior to incubation. Arch. Environ. Contam. Toxicol 31(3):404-409.

Powell, D.C., R.J. Aulerich, K.L. Stromborg, and S.J. Bursian. 1996b. Effects of 3,3′,4,4′-tetrachlorobiphenyl, 2,3,3′,4,4′- pentachlorobiphenyl, and 3,3′,4,4′,5-pentachlorobiphenyl on the developing chicken embryo when injected prior to incubation. J. Toxicol. Environ. Health 49(3):319-338.

Powell, D.C., R.J. Aulerich, J.C. Meadows, D.E. Tillitt, J.F. Powell, J.C. Restum, K.L. Stromborg, J.P. Giesy, and S.J. Bursian. 1997a. Effects of 3,3′,4,4′,5-Pentachlorobiphenyl (PCB 126), 2,3,7,8-tetrachlorodibenzo-*p*-dioxin (TCDD), or an extract derived from field-collected cormorant eggs injected into double-crested cormorant (Phalacrocorax auritus) eggs. Environ. Toxicol. Chem. 16(7):1450-1455.

Powell, D.C., R.J. Aulerich, J.C. Meadows, D.E. Tillitt, K.L. Stromborg, T.J. Kubiak, J.P. Giesy, and S.J. Bursian. 1997b. Organochlorine contaminants in double-crested cormorants from Green Bay, Wisconsin: 2. Effects of an extract derived from cormorant eggs on the chicken embryo. Arch. Environ. Contam. Toxicol 32(3):316-322.

Putzrath, R.M. 1997. Estimating relative potency for receptor-mediated toxicity: reevaluating the toxicity equivalence factor (TEF) model. Regul. Toxicol. Pharmacol. 25(1):68-78.

Qi, M., M. Anderson, S. Meyer, and J. Carson. 1997. Contamination of PCB congeners in bear lake fish tissues, livers, and brains. Toxicol. Environ. Chem. 61(1-4):147-161.

Rogan, W.J., and B.C. Gladen. 1992. Neurotoxicology of PCBs and related compounds. Neurotoxicology 13(1):27-35.

Ross, P.S., R.L. de Swart, P.J.H. Reijnders, H. van Loveren, J.G. Vos, and A.D.M.E. Osterhaus. 1995. Contaminant-related suppression of delayed-type hypersensitivity and antibody responses in harbor seals fed herring from the Baltic Sea. Environ. Health Perspect. 103(2):162-167.

Safe, S. 1984. Polychlorinated biphenyls (PCBs) and polybrominated biphenyls (PBBs): biochemistry, toxicology and mechanism of action. Crit. Rev. Toxicol. 13(4):319-395.

Safe, S. 1990. Polychlorinated biphenyls(PCBs), dibenzo-*p*-dioxins(PCDDs), dibenzofurans (PCDFs), and related compounds: environmental and mechanistic considerations which support the development of toxic equivalency factors (TEFs). Crit. Rev. Toxicol. 21(1):51-88.

Safe, S.H. 1994. Polychlorinated Biphenyls (PCBs): environmental impact, biochemical and toxic responses, and implications for risk assessment. Crit. Rev. Toxicol. 24(2):87-149.

Safe, S., L.W. Robertson, L. Safe, A. Parkinson, S. Bandiera, T. Sawyer, and M.A. Campbell. 1982. Halogenated biphenyls: molecular toxicology. Can. J. Physiol. Pharmacol. 60(7):1057-1064.

Sanderson, J.T., S.W. Kennedy, and J.P. Giesy. 1998. In vitro induction of ethoxyresorufin-O-deethylase and porphyrins by halogenated aromatic hydrocarbons in avian primary hepatocytes. Environ. Toxicol. Chem. 17(10):2006-2018.

Sawyer, T.W., and S. Safe. 1985. In vitro AHH induction by polychlorinated biphenyl and dibenzofuran mixtures: additive effects. Chemosphere 14(1):79-84.

Schantz, S.L., E.D. Levin, and R.E. Bowman. 1991. Long-term neurobehavioral effects of perinatal polychlorinated biphenyl (PCB) exposure in monkeys. Environ. Toxicol. Chem. 10(6): 747-756.

Schantz, S.L., J. Moshtaghian, and D.K. Ness. 1992. Long-term perinatal exposure to PCB congeners and mixtures on locomotor activity of rats. Teratology 45(5):524-525.

Scott, M.L. 1977. Effects of PCBs, DDT, and mercury compounds in chickens and Japanese quail. Fed. Proc. 36(6):1888-1893.

Seed, J., R.P. Brown, S.S. Olin, and J.A. Foran. 1995. Chemical mixtures: current risk assessment methodologies and future directions. Regul.. Toxicol. Pharmacol. 22(1):76-94.

Seegal, R.F. 1996. Epidemiological and laboratory evidence of PCB-induced neurotoxicity. Crit. Rev. Toxicol. 26(6):709-737.

Seegal, R.F., and S.L. Schantz. 1994. Neurochemical and behavioral sequelae of exposure to dioxins and PCBs. Pp. 409-447 in Dioxins and Health, A. Schecter, ed. New York: Plenum Press.

Seegal, R.F., B. Bush, and K.O. Brosch. 1985. Polychlorinated biphenyls induce regional changes in brain norepinephrine concentrations in adult rats. Neurotoxicology 6(3):13-24.

Seegal, R.F., K.O. Brosch, and B. Bush. 1986a. Polychlorinated biphenyls produce regional alterations of dopamine metabolism in rat brain. Toxicol. Lett. 30(2):197-202.

Seegal, R.F., K.O. Brosch, and B. Bush. 1986b. Regional alterations in serotonin metabolism induced by oral exposure of rats to polychlorinated biphenyls. Neurotoxicology 7(1):155-166.

Seegal, R.F., K. Brosch, B. Bush, M. Ritz, and W. Shain. 1989. Effects of Aroclor 1254 on dopamine and norepinephrine concentrations in pheochromocytoma (PC-12) cells. Neurotoxicology 10(4):757-764.

Seegal, R.F., B. Bush, and W. Shain. 1990. Lightly chlorinated ortho-substituted PCB congener decrease dopamine in nonhuman primate brain and in tissue culture. Toxicol. Appl. Pharmacol. 106(1):136-144.

Seegal, R.F., B. Bush, and W. Shain. 1991a. Neurotoxicology of ortho-substituted polychlorinated biphenyls. Chemosphere, 23(11-12):1941-1950.

Seegal, R.F., B. Bush, and K.O. Brosch, 1991b. Sub-chronic exposure of the adult rat to Aroclor 1254 yields regionally-specific changes in central dopaminergic function. Neurotoxicology 12(1): 55-66..

Seegal, R.F., B. Bush, and K.O. Brosch, 1991c. Comparison of effects of Aroclors 1016 and 1260 on non-human primate catecholamine function. Toxicology 66(2):145-164.

Seegal, R.F., B. Bush, and K.O. Brosch. 1994. Decreases in dopamine concentrations in adult, non-human primate brain persist following removal from polychlorinated biphenyls. Toxicology 86(1-2):71-87.

Shain, W., B. Bush, and R. Seegal. 1991. Neurotoxicity of polychlorinated biphenyls: structure-activity relationship of individual congeners. Toxicol. Appl. Pharmacol. 111(1):33-42.

Silberhorn, E.M., H.P. Glauert, and L.W. Robertson. 1990. Carcinogenicity of polyhalogenated biphenyls: PCBs and PBBs. Crit. Rev. Toxicol. 20(6):439-496.

Smit, M.D., P.E.G. Leonards, A.J. Murk, A.W.J.J. de Jongh, and B. van Hattum. 1996. Development of Otter-Based Quality Objectives for PCBs. Institute for Environmental Studies. V.U. Boekhandel/Uitgeverij, Amsterdam.

Smith, M.A. 1997. Reassessment of the carcinogenicity of polychlorinated biphenyls. J. Toxicol. Environ. Health 50(6):567-579.

Strang, C., S.P. Levine, B.P. Orlan, T.A. Gouda, and W.A. Saner. 1984. High resolution of gas chromatographic analysis of cytochrome P-448 inducing PCB congeners in hazardous waste. J. Chromatogr. 314:482-487.

Suedel, B.C., T.M. Dillon, and W.H. Benson. 1997. Subchronic effects of five di-ortho PCB congeners on survival, growth and reproduction in the fathead minnow Pimephales promelas. Environ. Toxicol. Chem. 16(7):1526-1532.

Swanson, H.I., and G.H. Perdew. 1991. Detection of the Ah receptor in rainbow trout: use of 2-azido-3[^{125}I]iodo-7,8-dibromodibenzo-*p*-dioxin in cell culture. Toxicol. Lett. 58(1):85-95.

Tanabe, S., S. Watanabe, H. Kan, and R. Tatsukawa. 1988. Capacity and mode of PCB metabolism in small cetaceans. Mar. Mamm. Sci. 4(2):103-124.

Tanabe, S., N. Kannan, T. Wakimoto, R. Tatsukawa, T. Okamoto, and Y. Masuda. 1989. Isomer-specific determination and toxic evaluation of potentially hazardous coplanar PCBs, dibenzofurans and dioxins in the tissues of "Yusho" PCB poisoning victim and in the causal oil. Toxicol. Environ. Chem. 24:215-231.

Tetra Tech. 1998. Development of Toxicity Reference Values for Conducting Ecological Risk Assessments at Naval Facilities in California. Interim Final Technical Memorandum. Prepared by Tetra Tech., Chicago, IL, for the Department of the Navy, Engineering Field Activity West, Naval Facilities Engineering Command, San Bruno, CA.

Tilbury, K.L., J.E. Stein, J.P. Meador, C.A. Krone, and S.L. Chan. 1997. Chemical contaminants in harbor porpoise (Phocoena phocoena) from the North Atlantic coast: tissue concentrations and intra- and inter- organ distribution. Chemosphere 34(9-10):2159-2181.

Tillitt, D.E. 1989. Characterization Studies of the H4IIE Bioassay for Assessment of Planar Halogenated Hydrocarbons in Fish and Wildlife. Ph.D. dissertation. Michigan State University, Ann Arbor, MI.

Tillitt, D.E., J.P. Giesy, and G.T. Ankley. 1991. Characterization of the H4IIE rat hepatoma cell bioassay as a tool for assessing toxic potency of planar halogenated hydrocarbons in environmental samples. Environ. Sci. Technol. 25(1):87-92.

Tillitt, D.E., G.T. Ankley, J.P. Giesy, J.P. Ludwig, H. Kurita-Matsuba, D.V. Weseloh, P.S. Ross, C.A. Bishop, L. Sileo, K.L. Stromborg, J. Larson, and T.J. Kubiak. 1992. Polychlorinated biphenyl residues and egg mortality in double crested cormorants from the Great Lakes. Environ. Toxicol. Chem. 11(9):1281-1288.

Tillitt, D.E., R.W. Gale, J.C. Meadows, J.L. Zajicek, P.H. Peterman, S.N. Heaton, P.D. Jones, S.J. Bursian, T.J. Kubiak, J.P. Giesy, and R.J. Aulerich. 1996. Dietary exposure of mink to carp from Saginaw Bay. 3. Characterization of dietary exposure to planar halogenated hydrocarbons, dioxin equivalents, and biomagnification, Environ. Sci. Technol. 30(1):283-291.

Tilson, H.A., G.J. Davis, J.A. McLachlan, and G.W. Lucier. 1979. The effects of PCBs given prenatally on the neurobehavioral development of mice. Environ. Res. 18(2):466-474.

Tumasonis, C.F., B. Bush, and F.D. Baker. 1973. PCB levels in egg yolks associated with embryonic mortality and deformity of hatched chicks. Arch. Environ. Contam. Toxicol. 1(4):312-324.

Vamvakas, A., J. Keller, and M. Dufresne. 1996. In vitro induction of CYP1A1-associated activities in human and rodent cell lines by commercial and tissue-extracted halogenated aromatic hydrocarbons. Environ. Toxicol. Chem. 15(6):814-823.

Van Birgelen, A.P.J.M., E.A. Smit, I.M. Kampen, C.N. Groeneveld, K.M. Fase, J. Van der Kolk, H. Poiger, M. Van den Berg, J.H. Koeman, and A. Brouwer. 1995. Subchronic effects of 2,3,7,8-TCDD or PCBs on thyroid hormone metabolism: use in risk assessment. Eur. J. Pharmacol. Environ. Toxicol. Pharmacol. Sec. 5(1):77-85.

Van den Berg, M., L. Birnbaum, A.T.C. Bosveld, B. Brunström, P. Cook, M. Feeley,

J.P. Giesy, A. Hanberg, R. Hasegawa, S.W. Kennedy, S.W., T. Kubiak, J.C. Larsen, F.X. van Leeuwen, A.K. Liem, C. Nolt, R.E. Peterson, L. Poellinger, S. Safe, D. Schrenk, D. Tillitt, M. Tysklind, M. Younes, F. Waern, and T. Zacharewski. 1998. Toxic equivalency factors (TEFs) for PCBs, PCDDs, PCDFs for humans and wildlife. Environ. Health Perspect. 106(12):775-792.

Van der Weiden, M.E.J., J. van der Kolk, R. Bleumink, W. Seinen, and M. Van den Berg. 1992. Concurrence of P4501A1 induction and toxic effects after administration of a low dose of 2,3,7,8-tetrachlorodibenzo-*p*-dioxin (TCDD) in the rainbow trout (Oncorhynchus mykiss). Aquat. Toxicol. 24(½):123-142.

Walker, M.K., and R.E. Peterson. 1991. Potencies of polychlorinated dibenzo-*p*-dioxin, dibenzofuran and biphenyl congeners, relative to 2,3,7,8-tetrachlorodibenzo-*p*-dioxin for producing early life stage mortality in rainbow trout (Oncorhynchus mykiss), Aquatic Toxicol. 21(3/4):219-238.

Walker, M.K., and R.E. Peterson. 1994. Aquatic toxicity of dioxins and related chemicals. Pp.347-387 in Dioxins and Health, A. Schecter, ed. New York: Plenum.

Walker, M.K. P.M. Cook, B.C. Butterworth, E.W. Zabel, and R.E. Peterson. 1996. Potency of a complex mixture of polychlorinated dibenzo-*p*-dioxin, dibenzofuran, and biphenyl congeners compared to 2,3,7,8,-tetrachlorodibenzo-*p*-dioxin in causing early life stage mortality. Fundam. Appl. Toxicol. 30(2):178-186.

Walker, M.K., R.S. Pollenz, and S.M. Smith. 1997. Expression of the aryl hydrocarbon receptor (AhR) and AhR nuclear translocator during chick cardiogenesis is consistent with 2,3,7,8- tetrachlorodibenzo-*p*-dioxin-induced heart defects. Toxicol. Appl. Pharmacol. 143(2):407-419.

Weinand-Härer, A., H. Lilienthal, K.A. Bucholski, and G. Winneke. 1997. Behavioral effects of maternal exposure to an ortho-substituted or a coplanar PCB congener in rats. Environ. Toxicol. Pharmacol. 3(2):97-103.

Weseloh, D.V.C., P.J. Ewins, J. Struger, P. Mineau, and R.J. Norstrom. 1994. Geographical distribution of organochlorine contaminants and reproductive parameters in herring gulls on Lake Superior in 1983. Environ. Monit. Assess. 29(3):229-251.

White, D.H., and D.J. Hoffman. 1995. Effects of polychlorinated dibenzo-*p*-dioxins and dibenzofurans on nesting wood ducks (Aix sponsa) at Bayou Meto, Arkansas. Environ. Health Perspect. 103(Suppl.4):37-39.

Whitlock, Jr., J.P. 1987. The regulation of gene expression by 2,3,7,8 tetrachlorodibenzo-*p*-dioxin. Pharmacol. Rev. 39(2):147-161.

Wiemeyer, S.N., T.G. Lamont, C.M. Bunck, C.R. Sindelar, F.J. Gramlich, J.D. Fraser, and M.A. Byrd. 1984. Organochlorine pesticide, polychlorobiphenyl, and mercury residues in bald eagle eggs--1969-79--and their relationships to shell thinning and reproduction. Arch. Environ. Contam. Toxicol. 13(5):529-549.

Williams, L.L., and J.P. Giesy. 1992. Relationship among concentrations of individual polychlorinated biphenyl (PCB) congeners, 2,3,7,8-tetrachlorodibenzo-*p*-dioxin equivalents (TCDD-EQ) and rearing mortality of chinook salmon (Oncorhynchus tshawytscha) eggs from Lake Michigan. J. Great Lakes Res. 18(1):108-124.

Willman, E.J., I.B. Manchester-Neesvig, and D.E. Armstrong. 1997. Influence of ortho-substitution on patterns of PCB accumulation in sediment, plankton and fish in a freshwater estuary. Environ. Sci. Technol. 31(12):3712-3718.

Wong, P.W., R.M. Joy, T.E. Albertson, S.L. Schantz, and I.N. Pessah. 1997. Ortho-substituted 2,2′,3,5′,6-pentachlorobiphenyl (PCB 95) alters rat hippocampal ryanodine receptors and neuroplasticity in vitro: evidence for altered hippocampal function. Neurotoxicology 18(2):443-456.

Woodford, J.E., W.H. Karasov, M.W. Meyer, and L. Chambers. 1998. Impact of 2,3,7,8,-TCDD exposure on survival, growth, and behavior of Ospreys breeding in Wisconsin, USA. Environ. Toxicol. Chem. 17(7):1323-1331.

Yamashita, N., S. Tanabe, J.P. Ludwig, H. Kurita, M.E. Ludwig, and R. Tatsukawa. 1993. Embryonic abnormalities and organochlorine contamination in double-crested cormorants (Phalacrocorax auritus) and Caspian terns (Hydropogne caspia) from the upper Great Lakes in 1988. Environ. Pollut. 79(2):163-173.

Yao, C., B. Panigrahy, and S. Safe. 1990. Utilization of cultured chick embryo hepatocytes as in vitro bioassays for polychlorinated biphenyls (PCBs): quantitative-structure-induction relationships. Chemosphere 21(8):1007-1016.

Yoshimura, H., S. Yoshihara, N. Koga, K. Nagata, I. Wada, J. Kuroki, and Y. Hokama. 1985. Inductive effect on hepatic enzymes and toxicity of congeners of PCBs and PCDFs. Environ. Health Perspect. 59:113-119.

Yu, M.L., C.C. Hsu, B.C. Gladen, and W.J. Rogan. 1991. In utero PCB/PCDF exposure: relation to developmental delay to dysmorphology and dose. Neurotoxicol. Teratol. 13(2):195-202.

Zabel, E.W., P.M. Cook, and R.E. Peterson. 1995. Toxic equivalency factors of polychlorinated dibenzo-*p*-dioxin, dibenzofuran and biphenyl congeners based on early life stage mortality in rainbow trout (Oncorhynchus mykiss). Aquat. Toxicol. 31:315-328.

Zabel, E.W., R. Pollenz, and R.E. Peterson. 1996. Relative potencies of individual polychlorinated dibenzo-*p*-dioxin, dibenzofuran, and biphenyl congener mixtures based on induction of cytochrome P4501A mRNA in a rainbow trout gonadal cell line (RTG-2). Environ. Toxicol. Chem. 15(12):2310-2318.

Zhao, F., K. Mayura, N. Kocurek, J.F. Edwards, L.F. Kubena, S.H. Safe, and T.D. Phillips. 1997. Inhibition of 3,3′,4,4′,5-pentachlorobiphenyl-induced chicken embryotoxicity by 2,2′,4,4′,5,5′-hexachlorobiphenyl. Fundam. Appl. Toxicol. 35(1):1-8.

Appendix H

Nomenclature of PCBs

TABLE H-1 Nomenclature of PCB Congeners

BZ No.[a]	Compound	CAS No.	BZ No.[a]	Compound	CAS No.
	Biphenyl	92-52-4	15	4,4′	2050-68-2
	Mono-CB	27323-18-8		Tri-CB	25323-68-6
1	2	2051-60-7	16	2,2′,3	38444-78-9
2	3	2051-61-8	17	2,2′,4	37680-66-3
3	4	2051-62-9	18	2,2′,5	37680-65-2
	Di-CB	25512-42-9	19	2,2′,6	38444-73-4
4	2,2′	13029-08-8	20	2,3,3′	38444-84-7
5	2,3	16605-91-7	21	2,3,4	55702-46-0
6	2,3′	25569-80-6	22	2,3,4′	38444-85-8
7	2,4	33284-50-3	23	2,3,5	55720-44-0
8	2,4′	34883-43-7	24	2,3,6	58702-45-9
9	2,5	34883-39-1	25	2,3′,4	55712-37-3
10	2,6	33146-45-1	26	2,3′,5	38444-81-4
11	3,3′	2050-67-1	27	2,3′,6	38444-76-7
12	3,4	2974-92-7	28	2,4,4′	7012-37-5
13	3,4′	2974-90-5	29	2,4,5	15862-07-4
14	3,5	34883-41-5	30	2,4,6	35693-92-6

TABLE H-1 *(Continued)*

BZ No.[a]	Compound	CAS No.	BZ No.[a]	Compound	CAS No.
31	2,4′,5	16606-02-3	58	2,3,3′,5′	41464-49-7
32	2,4′,6	38444-77-8	59	2,3,3′,6	74472-33-6
33[b]	2′,3,4	38444-86-9	60	2,3,4,4′	33025-41-1
34[b]	2′,3,5	37680-68-5	61	2,3,4,5	33284-53-6
35	3,3′,4	37680-69-6	62	2,3,4,6	54230-22-7
36	3,3′,5	38444-87-0	63	2,3,4′,5	74472-34-7
37	3,4,4′	38444-90-5	64	2,3,4′,6	52663-58-8
38	3,4,5	53555-66-1	65	2,3,5,6	33284-54-7
39	3,4′,5	38444-88-1	66	2,3′,4,4′	32598-10-0
	Tetra-CB	26914-33-0	67	2,3′,4,5	73575-53-8
40	2,2′,3,3′	38444-93-8	68	2,3′,4,5′	73575-52-7
41	2,2′,3,4	52663-59-9	69	2,3′,4,6	60233-24-1
42	2,2′,3,4′	36559-22-5	70	2,3′,4′,5	32598-11-1
43	2,2′,3,5	70362-46-8	71	2,3′,4′,6	41464-46-4
44	2,2′,3,5′	41464-39-5	72	2,3′,5,5′	41464-42-0
45	2,2′3,6	70362-45-7	73	2,3′,5′,6	74338-23-1
46	2,2′,3,6′	41464-47-5	74	2,4,4′,5	32690-93-0
47	2,2′,4,4′	2437-79-8	75	2,4,4′,6	32598-12-2
48	2,2′4,5	70362-47-9	76[b]	2′,3,4,5	70362-48-0
49	2,2′4,5′	41464-40-8	77	3,3′,4,4′	32598-13-3
50	2,2′4,6	62796-65-0	78	3,3′4,5	70362-49-1
51	2,2′4,6′	65194-04-7	79	3,3′4,5′	41464-48-6
52	2,2′5,5′	35693-99-3	80	3,3′5,5′	33284-52-5
53	2,2′5,6′	41464-41-9	81	3,4,4′,5	70362-50-4
54	2,2′,6,6′	15968-05-5		Penta-CB	25429-29-2
55	2,3,3′,4	74338-24-2	82	2,2′,3,3′,4	52663-62-4
56	2,3,3′,4′	41464-43-1	83	2,2′,3,3′,5	60145-20-2
57	2,3,3′,5	70424-67-8	84	2,2′,3,3′,6	52663-60-2

TABLE H-1 *(Continued)*

BZ No.[a]	Compound	CAS No.	BZ No.[a]	Compound	CAS No.
85	2,2′,3,4,4′	65510-45-4	114	2,3,4,4′,5	74472-37-0
86	2,2′,3,4,5	55312-69-1	115	2,3,4,4′,6	74472-38-1
87	2,2′,3,4,5′	38380-02-8	116	2,3,4,5,6	18259-05-7
88	2,2′,3,4,6	55215-17-3	117	2,3,4′,5,6	68194-11-6
89	2,2′,3,4,6′	73575-57-2	118	2,3′,4,4′,5	31508-00-6
90	2,2′,3,4′,5	68194-07-0	119	2,3′,4,4′,6	56558-17-9
91	2,2′,3,4′,6	58194-05-8	120	2,3′,4,5,5′	68194-12-7
92	2,2′,3,5,5′	52663-61-3	121	2,3′,4,5′,6	56558-18-0
93	2,2′,3,5,6	73575-56-1	122[b]	2′,3,3′,4,5	76842-07-4
94	2,2′,3,5,6′	73575-55-0	123[b]	2′,3,4,4′,5	65510-44-3
95	2,2′,3,5′,6	38379-99-6	124[b]	2′,3,4,5,5′	70424-70-3
96	2,2′,3,6,6′	73575-54-9	125[b]	2′,3,4,5,6′	74472-39-2
97[b]	2,2′,3′,4,5	41464-51-1	126	3,3′,4,4′,5	57465-28-8
98[b]	2,2′,3′,4,6	60233-25-2	127	3,3′,4,5,5′	39635-33-1
99	2,2′,4,4′,5	38380-01-7		Hexa-CB	26601-64-9
100	2,2′,4,4′,6	39485-83-1	128	2,2′,3,3′,4,4′	38380-07-3
101	2,2′,4,5,5′	37680-73-2	129	2,2′,3,3′,4,5	55215-18-4
102	2,2′,4,5,6′	68194-06-9	130	2,2′,3,3′,4,5′	52663-66-8
103	2,2′,4,5′,6	60145-21-3	131	2,2′,3,3′,4,6	61798-70-7
104	2,2′,4,6,6′	56558-16-8	132	2,2′,3,3′,4,6′	38380-05-1
105	2,3,3′,4,4′	32598-14-4	133	2,2′,3,3′,5,5′	35694-04-3
106	2,3,3′,4,5	70424-69-0	134	2,2′,3,3′,5,6	52704-70-8
107	2,3,3′,4′,5	70424-68-9	135	2,2′,3,3′,5,6′	52744-13-5
108	2,3,3′,4,5′	70362-41-3	136	2,2′,3,3′,6,6′	38411-22-2
109	2,3,3′,4,6	74472-35-8	137	2,2′,3,4,4′,5	35694-06-5
110	2,3,3′,4′,6	38380-03-9	138	2,2′,3,4,4′,5′	35065-28-2
111	2,3,3′,5,5′	39635-32-0	139	2,2′,3,4,4′,6	56030-56-9
112	2,3,3′,5,6	74472-36-9	140	2,2′,3,4,4′,6′	59291-64-4
113	2,3,3′,5′,6	68194-10-5	141	2,2′,3,4,5,5′	52712-04-6

TABLE H-1 *(Continued)*

BZ No.[a]	Compound	CAS No.	BZ No.[a]	Compound	CAS No.
142	2,2′,3,4,5,6	41411-61-4	170	2,2′,3,3′,4,4′,5	35065-30-6
143	2,2′,3,4,5,6′	68194-15-0	171	2,2′,3,3′,4,4′,6	52663-71-5
144	2,2′,3,4,5′,6	68194-14-9	172	2,2′,3,3′,4,5,5′	52663-74-8
145	2,2′,3,4,6,6′	74472-40-5	173	2,2′,3,3′,4,5,6	68194-16-1
146	2,2′,3,4′,5,5′	51908-16-8	174	2,2′,3,3′,4,5,6′	38411-25-5
147	2,2′,3,4′,5,6	68194-13-8	175	2,2′,3,3′,4,5′,6	40186-70-7
148	2,2′,3,4′,5,6′	74472-41-6	176	2,2′,3,3′,4,6,6′	52663-65-7
149	2,2′,3,4′,5′,6	38380-04-0	177[b]	2,2′,3,3′,4′,5,6	52663-70-4
150	2,2′,3,4′,6,6′	68194-08-1	178	2,2′,3,3′,5,5′6	52663-67-9
151	2,2′,3,5,5′,6	52663-63-5	179	2,2′,3,3′,5,6,6′	52663-64-6
152	2,2′,3,5,6,6′	68194-09-2	180	2,2′,3,4,4′,5,5′	35065-29-3
153	2,2′,4,4′,5,5′	35065-27-1	181	2,2′,3,4,4′,5,6	74472-47-2
154	2,2′,4,4′,5,6′	60145-22-4	182	2,2′,3,4,4′,5,6′	60145-23-5
155	2,2′,4,4′,6,6′	33979-03-2	183	2,2′,3,4,4′,5′,6	52663-69-1
156	2,3,3′,4,4′,5	38380-08-4	184	2,2′,3,4,4′,6,6′	74472-48-3
157	2,3,3′,4,4′,5′	69782-90-7	185	2,2′,3,4,5,5′,6	52712-05-7
158	2,3,3′,4,4′,6	74472-42-7	186	2,2′,3,4,5,6,6′	74472-49-4
159	2,3,3′,4,5,5′	39635-35-3	187	2,2′,3,4′,5,5′,6	52663-68-0
160	2,3,3′,4,5,6	41411-62-5	188	2,2′,3,4′,5,6,6′	74487-85-7
161	2,3,3′,4,5′,6	74472-43-8	189	2,3,3′,4,4′,5,5′	39635-31-9
162	2,3,3′,4′,5,5′	39635-34-2	190	2,3,3′,4,4′,5,6	41411-64-7
163	2,3,3′,4′,5,6	74472-44-9	191	2,3,3′,4,4′,5′,6	74472-50-7
164	2,3,3′4′,5′,6	74472-45-0	192	2,3,3′,4,5,5′,6	74472-51-8
165	2,3,3′5,5′,6	74472-46-1	193	2,3,3′,4′,5,5′,6	69782-91-8
166	2,3,4,4′,5,6	41411-63-6		Octa-CB	31472-83-0[b]
167	2,3′,4,4′,5,5′	52663-72-6	194	2,2′,3,3′,4,4′,5,5′	35694-08-7
168	2,3′,4,4′,5′,6	59291-65-5	195	2,2′,3,3′,4,4′5,6	52663-78-2
169	3,3′,4,4′,5,5′	32774-16-6	196	2,2′,3,3′,4,4′,5,6′	42740-50-1
	Hepta-CB	28655-71-2	197	2,2′,3,3′,4,4′,6,6′	33091-17-7

TABLE H-1 *(Continued)*

BZ No.[a]	Compound	CAS No.	BZ No.[a]	Compound	CAS No.
198	2,2′,3,3′,4,5,5′,6	68194-17-2	205	2,3,3′,4,4′,5,5′,6	74472-53-0
199	2,2′,3,3′,4,5,5′,6′	52663-75-9		Nona-CB	53742-07-7
200	2,2′,3,3′,4,5,6,6′	52663-73-7	206	2,2′,3,3′,4,4′,5,5′,6	40186-72-9
201	2,2′,3,3′,4,5′,6,6′	40186-71-8	207	2,2′,3,3′,4,4′,5,6,6′	52663-79-3
202	2,2′,3,3′,5,5′,6,6′	2136-99-4	208	2,2′,3,3′,4,5,5′,6,6′	52663-77-1
203	2,2′,3,4,4′,5,5′,6	52663-76-0		Deca-CB	2051-24-3
204	2,2′,3,4,4′,5,6,6′	74472-52-9	209	2,2′,3,3′,4,4′,5,5′,6,6′	2051-24-3

[a]BZ = Ballschmiter and Zell (1980)
[b]IUPAC (International Union of Pure and Applied Chemistry) names differ from BZ names for the following PCBs: BZ 33 = IUPAC 2,3′,4′; BZ 34 = IUPAC 2,3′,5′; BZ 76 = IUPAC 2,3′,4′,5′; BZ 97 = IUPAC 2,2′,3,4′,5′; BZ 98 = IUPAC 2,2′,3,4′,6′; BZ 122 = IUPAC 2,3,3′,4′,5′; BZ 123 = IUPAC 2,3′,4,4′,5′; BZ 124 = IUPAC 2,3′4′,5,5′; BZ 125 = IUPAC 2,3′4′,5′,6; and BZ 177 = IUPAC 2,2′,3,3′,4,5′,6′.